The Struggle for the Long-Term in Transnational Science and Politics

This book reconsiders the power of the idea of the future. Bringing together perspectives from cultural history, environmental history, political history and the history of science, it investigates how the future became a specific field of action in liberal democratic, state socialist and post-colonial regimes after the Second World War. It highlights the emergence of new forms of predictive scientific expertise in this period, and shows how such forms of expertise interacted with political systems of the Cold War world order, as the future became the prism for dealing with post-industrialisation, technoscientific progress, changing social values, Cold War tensions and an emerging Third World. A forgotten problem of cultural history, the future re-emerges in this volume as a fundamentally contested field in which forms of control and central forms of resistance met, as different actors set out to colonise and control and others to liberate. The individual studies of this book show how the West European, African, Romanian and Czechoslovak "long term" was constructed through forms of expertise, computer simulations and models, and they reveal how such constructions both opened up new realities but also imposed limits on possible futures.

Jenny Andersson is CNRS research professor at the Center for European Studies of Sciences Po, Paris.

Eglė Rindzevičiūtė is a researcher at the Center for European Studies of Sciences Po, Paris, and Associate Professor in Culture Studies at Linköping University, Sweden.

Routledge Approaches to History

1 **Imprisoned by History**
Aspects of Historicized Life
Martin Davies

2 **Narrative Projections of a Black British History**
Eva Ulrike Pirker

3 **Integrity in Historical Research**
Edited by Tony Gibbons and Emily Sutherland

4 **History, Memory, and State-Sponsored Violence**
Time and Justice
Berber Bevernage

5 **Frank Ankersmit's Lost Historical Cause**
A Journey from Language to Experience
Peter P. Icke

6 **Popularizing National Pasts**
1800 to the Present
Edited by Stefan Berger, Chris Lorenz and Billie Melman

7 **The Fiction of History**
Alexander Lyon Macfie

8 **The Rise and Propagation of Historical Professionalism**
Rolf Torstendahl

9 **The Material of World History**
Edited by Tina Mai Chen and David S. Churchill

10 **Modernity, Metatheory and the Temporal-Spatial Divide**
From *Mythos* to *Techne*
Michael Kimaid

11 **The Struggle for the Long-Term in Transnational Science and Politics**
Forging the Future
Edited by Jenny Andersson and Eglė Rindzevičiūtė

The Struggle for the Long-Term in Transnational Science and Politics

Forging the Future

Edited by Jenny Andersson and Eglė Rindzevičiūtė

Routledge
Taylor & Francis Group
LONDON AND NEW YORK

First published 2015 by Routledge

2 Park Square, Milton Park, Abingdon, Oxfordshire OX14 4RN
52 Vanderbilt Avenue, New York, NY 10017

Routledge is an imprint of the Taylor & Francis Group, an informa business

First issued in paperback 2019

Library of Congress Cataloging-in-Publication Data

The struggle for the long-term in transnational science and politics : forging the future / edited by Jenny Andersson and Eglė Rindzevičiūtė.
pages cm. — (Routledge approaches to history ; 11)
1. Science—Forecasting—History—20th century. 2. World politics—Forecasting—History—20th century. 3. Transnationalism—Political aspects—History—20th century. 4. Social prediction—Political aspects—History—20th century. 5. Forecasting—Political aspects—History—20th century. 6. Historiography—Political aspects—History—20th century. 7. World politics—1945–1989. 8. Cold War. 9. Social control—History—20th century. 10. Government, Resistance to—History—20th century. I. Andersson, Jenny, 1974– II. Rindzevičiūtė, Eglė.
Q172.S77 2015
338.9'26—dc23 2014045743

ISBN: 978-1-138-85853-4 (hbk)
ISBN: 978-0-367-26382-9 (pbk)

Typeset in Sabon LT
by Apex CoVantage, LLC

Contents

Figures

Abbreviations

ACNSAS	Arhiva Consiliului Naţional pentru Studierea Arhivelor Securităţii
ADELA	Atlantic Development Group for Latin America
A.N.I.C.-C.C. P.C.R.	Arhivele Naţionale ale României, Fond C.C. al P.C.R.
ARAN	Archives of the Russian Academy of Sciences
BESM	Large electronic calculating machine (bol'shaia elektronnaia shchetnaia mashina)
CSNRD	Consortium for the Study of Nigerian Rural Development
CDU	Christian Democrats
CEPECA	Centrul de Perfecţionare a Cadrelor de Conducere din Întreprinderi
CIA	Central Intelligence Agency
CLASS	Computer Library for Agricultural Systems Simulation
CSAS	Czechoslovak Academy of Sciences
DICE	Dynamic Integrated Climate-Economy Model
EWI	Energiewirtschaftliches Institut
GCM	Global Circulation Model
GDR	German Democratic Republic
Glavlit	Main Directorate for the Protection of State Secrets under the Council of Ministers of the Soviet Union
GKNT	State Committee of Science and Technology of the Soviet Union
Gosplan	State Planning Committee of the Soviet Union
GPID	Goals, Processs and Indicators of Development
ICSU	International Council of Science Unions
IIASA	International Institute for Applied Systems Analysis

INION	Institute of Scientific Information on the Social Sciences of the Academy of Sciences of the Soviet Union
KFA	Kernforschungszentrum
MIT	Massachusetts Institute of Technology
MSU	Michigan State University
MUA	Archive of Masaryk Institute of Czech Academy of Sciences Prague
NASS	Nigerian Agricultural Sector Simulation
OECD	Organization for Economic Co-operation and Development
OECEI	Oficina de Estudios para la Colaboración Económica Internacional
PPBS	Program-Planning-Budgeting System
RAND	RAND Corporation (Research and Development)
RCP	Romanian Communist Party
RGAE	Russian State Economic Archives
RWE	Rheinisch-Westfälisches Elektrizitätswerk
RWI	Rheinisch-Westfälisches Institut für Wirtschaftsforschung
RWP	Romanian Workers' Party
SCOPE	Scientific Committee on Problems of the Environment
SPD	Social Democrats
SPRU	Science Policy Research Unity at the University of Sussex
STR	Scientific-technical revolution
STS	Science and Technology Studies
TsEMI	Central Institute for Economic Mathematics of the Academy of Sciences of the Soviet Union
UN	United Nations
USAID	United States Agency for International Development
WFSF	World Future Studies Federation
WWIC	Woodrow Wilson International Center
VNIISI	All-Union Scientific Institute for Systems Research of the Academy of Sciences of the Soviet Union
YMCA	Young Men's Christian Association

Foreword

Michael D. Gordin

The central, overwhelming contention of this volume is that the future has a history. I do not mean by this the basic observation that, in order to transition into the present, there must have been a future in the past. Of course the future has a past. What is more significant, and less often recognized, is that it has a *history*: that is, an account that is built in the present from the shards and traces that have descended to us from the past. History is written by a historian in his or her present to answer pressing contemporary questions using the past. Questions about the future, asked either in the present or the past, partake of the same techniques of history construction, and are just as powerful. Precisely in the same manner that at every moment in the past individuals constructed histories out of the times that had preceded them, so too did they build futures for themselves.

Some of the futures they imagined were reasonable, some dull, some fantastical, some delusional, some obscure, and some revelatory. As a point of historical methodology it matters less what the contents of these various past-futures were—and how sensible or dreamy their creators appeared to themselves or others—but rather that we see those futures-of-the-past as *historical exercises*, as an assemblage of traces to understand the future that symmetrically resembles how we routinely attempt to understand the past. This is what I mean in saying that the future has a history, and it is surely in the domain of historians to make sense of how these futures worked, both *as* history and *in* various historical epochs.

The Struggle for the Long Term in Transnational Science and Politics: Forging the Future decisively demonstrates the benefits not only of treating the futures of the past as matters for historical rumination, but it exposes to our gaze some very significant characteristics about the history of the future. The topic is so vast and so potentially metaphysical at its extreme edges that there is risk of getting lost in the weeds before one begins. Yet the editors and contributors of this volume have managed to ground a complex story in historical bedrock, and they have done so by creating a rarity in today's historical profession: a genuine collaboration. This volume consists of a diversity of essays ranging around the globe and across decades and methodologies, producing a totality that is significantly greater than the sum of its (already robust and informative) parts. We not only learn a good deal about the history of futures past, we also encounter a striking, emergent argument about the *structure* of the history of the future.

The stakes of this argument are large, and equally significant are its implications. There are, in my reading, three core lessons this volume teaches, and I will take each in turn.

The first is that, although there have surely always been futures past, and even historical futures, The Future blossomed at a particular historical moment. Reading through the essays presented here, it is impossible not to notice how rapidly certain similar approaches to projecting the future—call it future studies, futurology, futurism (not to be confused with the Italian art movement of earlier years)—began to proliferate in quite different parts of the world in the 1960s and 1970s. Whenever a historian sees a phenomenon like this, she is well-advised to cycle through three possible scenarios. First, that it is a coincidence, just a freak happenstance that futurological projects in Ghana happen to resemble conversations taking place in Rome, Moscow, Bucharest, and Cambridge, Massachusetts. This is always a sensible point to entertain, and it serves as the null hypothesis. The historian can always return to this if the other two major explanations do not cohere. The next explanation is diffusion: a model of future studies developed in one site and then spread throughout the world. To be sure, this happened: the Club of Rome's projections into the future, published as *The Limits to Growth* in 1972, were appropriated in multiple different contexts. We see throughout these essays central institutions and pivotal individuals who transmit specific ways of thinking about the future to another node in the network. This cannot, however, explain everything, for the globality of this proliferation (which I will get to in a moment) took place within several different networks, not within just one. The final possible explanation is that something structural was happening worldwide that accounts for the great shift toward future tense.

The essays present a convincing case that the third explanation, the structural approach, was at work regarding the sciences of the future. As becomes clear, there was not a single nucleating cause but several. The key to these types of explanations is always in the timing. Why at the cusp of the 1970s, and why worldwide? One root force was demographic: the generation coming of age at this moment was the first that did not personally experience (or, in any event, remember) the Second World War. The world order that its parents had assembled no longer suited the aspirations of this generation, and ferment emerged from below. It erupted in Paris; in Washington, DC; in Frankfurt; in Hanoi; in Prague; in Beijing under various guises—for example the Prague Spring, the Great Proletarian Cultural Revolution—and is now often called "global 1968."[1] (In this case, too, there was diffusion as activists from around the world communicated across borders; the same was true with "global 1989.")[2] Most of the actors in these pages were older than this generation of disaffected youth, but they responded to its discontent and felt something analogous themselves.

This demographic trend converged, however, with a separate, technological emergence: the transition from the large industrial mainframe computer to the much cheaper minicomputer, accessible to a far wider range of users around the globe. Calculating became easier, and this made dreams of what might be calculated all the more ambitious, even hubristic. (The rise in attention to software, Eglė Rindzevičiūtė shows us, was crucial in this transition.) Finally, widespread recognition of ecological webs, energy trade patterns, and an interconnected economy came to a head as pollution crises, oil embargoes, and linked recessions prompted intellectuals from a diversity of origins to reconceive of the future in the light of new patterns. In this moment, starting from the high Cold War and moving forward, the future became, as Jenny Andersson and Eglė Rindzevičiūtė put it in their introduction, a "powered affair."[3]

Interconnected with this first point is the second core lesson of the volume: the history of this period can only be written from a global perspective. Our present, so we hear endlessly, is globalized.[4] So was the past—and so, this volume insists, were the futures generated in that past. Yet we cannot forget that the postwar world was not just one world but, in the idiom of the time, threefold: a First World, the loose agglomeration of mostly capitalist societies under the vague leadership of the US; a Second World, Marxist polities that traveled with or sometimes against the Union of Soviet Socialist Republics; and an ostensible Third World, which made up the remainder (and, as it happens, the majority of Earth's population). Ironically—or perhaps not?—for the Club of Rome, the über-futurologists chronicled beautifully in the essay by Elodie Vieille Blanchard, there were also three worlds: World 1, World 2, and World 3. (Tellingly, in this case they were computer models.) Despite the penchant of many to split the world, like Gaul, into threes, the geopolitical boundaries were sharp only in the fantasies of Cold War strategic planners; the People's Republic of China, for example, veered between Second and Third according to circumstance. Nonetheless, we find in these essays family resemblances among the projections of the future depending on the sphere of origin, a confirmation of Jenny Andersson's astute observation in her contribution that these scenarios of the future were both manifestations of the Cold War and a means to protest against it. So, whereas Andersson, Vieille Blanchard, and Stefan Cihan Aykut (writing on West Germany) show us mostly a First World perspective, Rindzevičiūtė, Vítězslav Sommer, and Ana-Maria Cătănuş show us the Soviet, Czechoslovak, and Romanian approaches, respectively. Nor is the Third World—caught between American and Soviet spheres of contestation—excluded, as Kevin Baker demonstrates for Nigeria and Jeff Grischow and Holger Weiss for Ghana. And, in the end, Paul Warde and Sverker Sörlin present a comparatively *longue durée* view in this temporally-tight volume by focusing on environmental projections from the 1920s onward. We get, therefore, an international history, a transnational history, and a global history in one.

Last but certainly not least, we find a third set of stakes in the history of knowledge (or, as it is known in German, *Wissensgeschichte*).

Projecting the future was not a matter of random guessing or science-fictional epiphanies—for these actors, it was a *science*, an important and emerging field among the social sciences. In an intriguing pattern, the academic discipline of the history of science—which owed a good deal to Auguste Comte's famous doctrine of positivism, first articulated in the 1830s—has traversed the same path as Comte's much maligned hierarchy of sciences: from physics to chemistry to biology to psychology to, finally, sociology, which Comte considered, *qua* "social physics," as the most important science for the social order. Only now have historians begun to apply the same powerful battery of historical techniques and interpretative frames to the social sciences. This historiography has tended to focus, not coincidentally, on the Cold War era, the moment when the social sciences mushroomed across the social order—on both sides of the Iron Curtain—to cope with the nuclear age.[5]

Yet for all the attention now lavished on the social sciences, the study of the future using social scientific techniques—the highest-stakes, most stimulating, and provocative incarnation of the interdisciplinary social sciences—has eluded attention. That is, until this volume. Again and again across these essays we see interactions between economists and sociologists, operations researchers and demographers, and many others, all trying to make sense of their present by projecting data from the past into the future. As Warde and Sörlin implicitly show, the transition from economic thinking to ecological thinking and back again complicates even the supposedly clear boundary between the social and the natural sciences. What has been missing from the history of the social sciences is now clear: the history of The Future. This was the area that connected all the others, that brought models from one science into the heart of another, and that fused those sciences with the global moment of the 1960s and 1970s, embedded in lines of computer code, all over the world. The future always lies ahead, but its history is what energizes the present. It was true in the postwar moment, and it is true in the post-Cold War moment as well—much as it seems to hold today.

The reader of this volume is in for a treat. What these contributors have assembled is in itself a wonderful interdisciplinary approach to a slice of the past, a moment when the future was not only a playground contemplated by wooly-headed dreamers, but equally a terrain for hard-headed technocrats attempting to shape their present with rigorous knowledge. The future was a serious and gripping affair in the past; the history of the future is no less so.

NOTES

1. See, for example, many of the essays in a similarly wide-ranging collaborative project: Kathrin Fahlenbrach, Erling Sivertsen, and Rolf Werenskjold, eds., *Media and Revolt: Strategies and Performances from the 1960s to the Present* (New York: Berghahn, 2014).
2. On the latter event, see George Lawson, Chris Ambruster, and Michael Cox, eds., *The Global 1989: Continuity and Change in World Politics* (Cambridge, UK: Cambridge University Press, 2010).

3. Historian W. Patrick McCray, focusing on a similar period in the wake of the Club of Rome report, has called these attempts to apply engineering techniques to understanding the future "visioneering," and his analysis displays many resonances with that offered in this volume: *The Visioneers: How a Group of Elite Scientists Pursued Space Colonies, Nanotechnologies, and a Limitless Future* (Princeton: Princeton University Press, 2012).
4. An exemplar of the pundit's-eye-view of this matter is Thomas L. Friedman, *The World Is Flat: A Brief History of the Twenty-First Century* (New York: Farrar, Straus & Giroux, 2005). More sober-minded histories have extended this "globalized" picture back substantially before the dot-com boom, as in Jürgen Osterhammel and Niels P. Peterson, *Globalization: A Short History*, tr. Dona Geyer (Princeton: Princeton University Press, 2005).
5. For example, Paul Erickson, Judy L. Klein, Lorraine Daston, Rebecca Lemov, Thomas Sturm, and Michael D. Gordin, *How Reason Almost Lost Its Mind: The Strange Career of Cold War Rationality* (Chicago: University of Chicago Press, 2013); and Mark Solovey and Hamilton Cravens, eds., *Cold War Social Science: Knowledge Production, Liberal Democracy, and Human Nature* (New York: Palgrave Macmillan, 2012).

REFERENCES

Erickson, Paul, Judy L. Klein, Lorraine Daston, Rebecca Lemov, Thomas Sturm, and Michael D. Gordin. *How Reason Almost Lost Its Mind: The Strange Career of Cold War Rationality*. Chicago: University of Chicago Press, 2013.

Fahlenbrach, Kathrin, Erling Sivertsen, and Rolf Werenskjold, eds. *Media and Revolt: Strategies and Performances from the 1960s to the Present*. New York: Berghahn, 2014.

Friedman, Thomas L. *The World Is Flat: A Brief History of the Twenty-First Century*. New York: Farrar, Straus & Giroux, 2005.

Lawson, George, Chris Ambruster, and Michael Cox, eds. *The Global 1989: Continuity and Change in World Politics*. Cambridge, UK: Cambridge University Press, 2010.

McCray, W. Patrick. *The Visioneers: How a Group of Elite Scientists Pursued Space Colonies, Nanotechnologies, and a Limitless Future*. Princeton: Princeton University Press, 2012.

Osterhammel, Jürgen and Niels P. Peterson. *Globalization: A Short History*. Translated by Dona Geyer. Princeton: Princeton University Press, 2005.

Solovey, Mark and Hamilton Cravens, eds. *Cold War Social Science: Knowledge Production, Liberal Democracy, and Human Nature*. New York: Palgrave Macmillan, 2012.

Introduction

Toward a New History of the Future

Jenny Andersson and Eglė Rindzevičiūtė

By bringing together authors from different historical perspectives around a set of studies on the multifaceted history of prediction, forecasting and futures studies, this book calls for the need to reconsider the power of the idea of the future in historical writing. The authors of this volume share an interest in the ambiguous role played by the future, both for consolidating post-war regimes of power and control and for mobilizing crucial forms of dissent and visions of change. It is, therefore, the ambivalent and fundamentally powered role of the future that is at the center of this volume. Specifically, the chapters in this volume are held together by their interest in how post-war understandings of the future were constituted by particular forms of prediction and future expertise, and in the role played by these in exercising power over time. Drawing from insights in cultural, social, political, environmental and science history, our book thus aims to rethink the future as a historical category, and set the searchlight on the emergence of particular forms of future knowledge that set the future as a distinct temporal field in the post-war period.

We believe that this is an important challenge for historiography, and that it is of interest outside of the historical discipline too, in a number of adjacent fields such as science and technology studies (STS), culture studies, international relations and political sociology. While there are a number of works on the cultural history of time, utopia and apocalypse,[1] and indeed an emerging range of studies around futurity and anticipation in the wider field of social science,[2] we propose a particular historical approach to the way that the future itself became a specific field of scientific and political action in the decades after 1945. Our approach marks a break with a previous wave of historical writing on the future in the area of cultural and conceptual history. Conceptual history approaches understood the future as a conceptual invention of the historic shift from *Ancien Regime* to the Enlightenment. The German school of *Begriffsgeschichte* posited the separation between natural and historical time, and understood the concept of future as a semantic expression of this separation, a precondition for the modern notion of progress as a question of linear change.[3] The decoupling of the idea of the future from the idea of fate and destiny was understood as a precondition for emerging conceptions of democratization, as it resituated the future from the sphere of theocracy to the sphere of scientific reason and human will. While Reinhart Koselleck rejected claims to metahistory, the underpinnings of *Begriffsgeschichte* nevertheless reflected a universalizing historiographic ambition to find the conceptual corner stones of modernity.[4]

The universalizing project of *Begriffsgeschichte* has today been unsettled by a range of historical writings that have brought out a different, global, and much more multifaceted history of time.[5] We know from these studies that calendars and clocks were sites of contestation in the encounters between Europeans and non-Europeans, because time was, and is, a powered affair involving the legitimacy of state and regime. We know, from anthropology and cultural history, that human societies across time and space have very different understandings of time and historicity.[6] The prolific field of science studies and science history has contributed to unpacking contemporary notions of world, Man, or indeed, that of future, and has shown that these are not only conceptual entities but categories that are materially constituted in processes that involve networks of actors, specific forms of study and observation, and the technologies and tools that make things visible and amenable to manipulation. A handful of historical studies have from this perspective started reexamining the role played by different forms of forecasting, predictions, and modeling in shaping specific understandings of the present world.[7]

This book draws on these emergent developments and reexamines the idea of the future as a highly complex and often times contradictory notion that is inherently involved with power and with the claim, from a wide set of arenas, to control social futures. We do so with the wish to resituate the future as an object of study for political and cultural history. The chapters of this book bring out the many paradoxes of the rapidly shifting images of the future of the post-war era, and show that post-war understandings of the future cannot be fully comprehended as a question of succeeding regimes of historicity, as a post-war narrative of progress collapses with the onset of multiple crises of modernity in the 1960s and 1970s.[8] Rather, notions of progress and crisis, apocalypse and utopia, are fellow travelers in future thinking in a wide range of fields across the decades of the post-war period. The chapters of this book also trouble the distinction made by modernist conceptual historians, of a fundamental shift in the idea of the future from the sphere of fate and destiny to the sphere of human action, political will and scientific rationality. While post-war notions of the future were not as such linked to ideas of divine laws, they explicitly invoked the notion of fate and destiny of Mankind. The distinction between an *avenir* shaped by external forces and coming onto societies, and a *future* actively shaped by human beings and human societies, which Lucien Holscher saw as emerging the seventeenth century political thinking, is in many ways reiterated in post-war understandings of futurology and futures research as a scientific quest for laws governing social change, and in contrast, understandings of the future as a question of human choice, values, and moral sentiment.[9] The return of the future on a grand and global scale in the decades from 1945 to the 1970s signal, as Michael Gordin suggests in his Foreword, a global concern with the fate of human civilization, the future of the world and the planet.

While some essays in the book as a virtue of historicization step outside of the post-war time frame, our focus is on the post-war period and indeed on the Cold War era. Certainly the history of prediction is much longer than we account for in this volume.[10] But ideas of a stage driven social development that could be predicted and foreseen, drawn in the writings of Nicolas de Condorcet and Auguste Comte, were in many ways given new air in the attempts, immediately after the World War II, to forge a new future science. Claims to prediction were central elements of the new policy sciences that emerged after the Second World War, and the idea of controlling things, people and, indeed, the future itself was in many ways inherent to them. Different strands in futures research stood in either striking proximity to or critical engagement with modernization theory, which garnered authority in both social science and politics by the early 1960s. Similarly important were emerging postulates of rationality, created with an aim to explain and foretell social developments so that desirable ones could be privileged and undesirable ones avoided. Through such approaches in the social sciences, the future reemerged as a scientific interest, but also as an object of control and intervention.

Prediction was indisputably a Cold War product. Indeed, a literature emanating from sociology, risk and disaster studies has brought forward a Cold War genealogy in the shaping of notions of risk and catastrophe, which in turn brought about an interest in forms of foreseeability and predictability.[11] The challenge here is not to essentialize these as hegemonic aspects of contemporary governmentalities, as is the case in some of this literature,[12] but rather, to historicize and contextualize these notions. Moreover, while Cold War concerns with security and disaster were certainly central to the origins of post-war futurology, the multiple forms of futurity of the post-war era could not be reduced to governmental attempts at control. Certainly, both capitalist and communist systems were inherently interested in the epistemic tools that might allow them a measure of control on what seemed like a dangerously open future. However, the many different constructions of the post-war future resembled in fact an archipelago of contrasting and often conflicting ideas of what this future was, how it could be told and indeed how it could be actively shaped and forged. These reflected the many shifting and contradictory images of the future of the post-war decades, from the futuristic discourses of potentials of the atom, space travel, or ocean mining; to the apocalyptic concepts of ecocide or nuclear holocaust; and the somewhat more optimistic notions of Mankind as a new global community capable of actively reshaping its fate.

The future, we propose, brought together fundamental reflections on the both destructive and constructive nature of human being. To emerging forms of protest and dissent in the 1960s and 1970s, the future took the shape of a global *subject* in the name of which urgent action had to be taken. As a subject, the future had forms of agency on the present; it could literally strike back by unleashing a set of catastrophes that would erode human civilization. It also had a deeply moral quality, calling out for renewed notions of human responsibility and fate, and pleading for a rethinking of the relationship between Man, world and environment.

From this perspective, studying the role of the concept, image, or category of the future is a way of understanding the return, in a sense, of utopian energy in the 1960s and 1970s, as informed indeed by notions of possible different futures and as carried by new forms of transnational mobilization. The historian of international relations, Samuel Moyn, has recently argued that the idea of human rights, propagated as a universal interest by new social movements from the 1970s on, emerged as a fresh utopia substituting the historic utopias proposed by Marxism, liberalism, and communism. This goes in the direction of a revival of the Koselleckian notion of the sphere of expectation as a fundamental mobilizing factor of human action. It also underscores the fact that the future *as such* was a central focus of political imaginaries after the great crises of liberalism after the world wars. As such, the idea of the future was directly connected to both epistemological and political notions of the necessity of deconstruction, reconstruction and alternative. As pointed out in a recent volume by Michael Gordin, Hellen Tilley and Gyan Prakash, utopia (and future) should not only be analyzed in terms of the content of the image or model society that it proposes, but also understood as process and method, as the act indeed of imagining, constructing and constituting possible other worlds.[13]

Utopian visions, as it is widely known, were long used by scholars, intellectuals and politicians as platforms to pose criticisms about present state of being.[14] The role played by the future is that of allowing an estrangement from the present, an estrangement through which the present and its forthcoming consequences appear as amenable to hypothetical changes. This innovative and potentially subversive spirit motivates many of the actors that this volume is concerned with, and underpins, their claim to depict future or long-term developments as a prism through which their present appeared as critical and amendable. They saw the images of the future that they produced—whether it was in the shape of qualitative accounts or models and simulations—as conducive to action and as ways of forging alternative worlds and other presents, thereby attributing a different role to the future itself: embodied in both methods and facts, the future could and did act on the present. Such an understanding enables a move from a previous historiographical understanding of "futures past" to an analysis of the active construction of the future as a category and field constituted by actors through a wide repertoire of instruments, technologies and narratives, held together by their ambition to shape and reshape the modern world.

Science studies have unpacked a box of questions to do with the constitution of scientific objects and the claims to facticity, observation and truth that underlies this process of constitution. Indeed, one of the questions behind this volume was the question of how versions of the future are constructed, and on the basis of whose knowledge and expertise.[15] It is also what has brought the attention in the chapters of the book, from an interest in future images and future narratives derived essentially from cultural and political history, to the much more specific activity of prediction and the manifold forms of future expertise that we can today analyze productively with the tools of the history of science.

Meanwhile, as indicated above, the history of prediction is a very complicated thing. The history of forecasting, futurology or futures studies—activities with unclear boundaries and definitions (which is why we have made the editorial choice to let authors engage differently with these labels in their respective chapters), which proliferated in the 1950s, 1960s and 1970s in diverse intellectual and geographical settings—is a fascinating but virtually uncharted territory. Several chapters in our volume show, indeed, that the future was a key concern of control in both Western and Eastern political systems. But the authors of our volume also show that prediction was a highly varied activity. This goes far beyond questions of scientific representation, and into the dimension of political implication and consequence. Global modeling, for instance, which emerged, as suggested by Elodie Vieille Blanchard, this volume, as a near hegemonic way of imagining the future in the period following onto the publication of *The Limits to Growth* report in 1972, began as a highly contested activity. The visions of the world future produced by the World 1–World 3 models were a virtual battlefield of conflicting images, not only or even mainly around the scientific accuracy of representation, but around inherently political questions of responsibility and consequence, as well as existential notions of human agency versus systems logics. Several authors also show how similar predictive methods can be used to tell and shape very different futures in the hands of actors with different intentions. Methods created to ensure a measure of future control could be given subversive content in other contexts. The scenario technique, for instance, was invented by RANDian Cold War warriors to project nuclear holocaust, but it could also be used by radical social movements trying to conjure up alternative futures, as in Stefan Cihan Aykut's example of energy scenarios and the West German environmental movement.[16] In these processes, the relationship between science and politics is far from straightforward, and our chapters bring out an aspect of the history of the future which has been neglected both by conceptual history and more recent science and technology studies (STS) approaches, namely the political life of prediction, and the role played by the idea of the future as a space for protest and struggle not only over the content of what was coming, but over the forms of science, expertise and technologies with which this future was projected.

PREDICTION AND THE COLD WAR

The Cold War competition was central to the development of futures research and forms of prediction. Meanwhile, predictive knowledge also produced effects that contributed to the unsettling of the Cold War world. The wide field of futures research gave rise to important forms of collaboration and emerging agendas that transgressed Cold War divisions.[17] This transgressive effect of futures thinking can be discerned on the national, transnational and global levels. In the state socialist bloc, forecasting had a long history within the framework of planning, but forms of forecasting gained a new role from the 1960s on as they became tools involved in the relative opening up of the future of the communist system. In the Soviet context, forecasting expertise was developed as part of the broader modernization process, and became a symbol of the advanced economy. If the first long-term plans and the State Planning Committee (Gosplan) in the Soviet Union were initiated in connection to the electrification of Russia,[18] in the 1950s the Soviets developed prediction-based control under the umbrella of cybernetics, which was labeled *the* science of governance.[19] The speed of the scientific-technical revolution, Soviet planners argued, demanded increasingly sophisticated forecasts, capable not only of implementing science as part of the governance of all parts of Soviet society, but also of monitoring changing social structures, possible value revolutions and even future world communist revolutions.

As hinted, prediction was in fact a central site of Cold War exchange, transfer and communication, in particular during the period of détente through the bridge-building efforts of Lyndon Johnson.[20] In the US, the perception that the Soviet Union was gaining the upper-hand in this field spurred the development of long-term forecasting after the Second World War. Indeed, much ground-breaking work was done by émigré scholars, such as the Hungarian economist John von Neuman or the German mathematician Olaf Helmer, whose research resulted in new methods for gaming, simulation and predictive models of international relations. Scholars have detailed the central role of RAND, which gathered scientists who would pioneer a diverse range of predictive techniques, mainly based on mathematical methods and relying on the newly available computer power. These different techniques ranged from statistical methods for stochastic analysis (Monte Carlo), a motley crew of approaches grouped under systems analysis to scenario method and the Delphi method of forging expert consensus about the future events through repetitive opinion questionnaire. In doing this, RAND built an epistemic Cold War arsenal: these techniques were used to know an enemy whose future behavior was to be revealed through forms of virtual experimentation and "synthetic fact" in the absence of conventional knowledge. [21]

But telling the future also had a communicative purpose. Indeed, RANDians also sought to make the Soviets more like the Americans by subscribing them to the same vision of the world as a single, integrated, and predictable system. The ultimate goal of these RANDian techniques, in short, was to make Cold War world foreseeable and therefore manageable from both sides.[22] Soviet experts closely followed Western forecasting, even if the very term and some branches of Western "futurology" would eventually be officially dismissed by the Soviets as bourgeois and ideological (as opposed to the scientific nature of predictions grounded in Marxism-Leninism). Several chapters in this volume indicate that the reception and transfer of forecasting technologies within the East bloc was a complicated story of translation, circulation and reception. The fate of the term "futurology" is indicative of these complex interrelationships between social science and political power: if the Soviets used "futurology" only as a pejorative term to describe liberal Western trends, the Czechoslovaks accommodated "futurology" as a legitimate, Western-oriented approach up until 1968, whereas in Ceaușescu's Romania, "futurology" would remain an accepted term, describing a science embraced by Ceaușescu personally.

Importantly, several chapters in this volume suggest that the activities of modeling, simulation and forecasting had highly unforeseeable effects on regimes of power. Forecasting the future constantly revealed new factors of change and new possible instabilities that then called for new forms of control and action. As Eglė Rindzevičiūtė's chapter shows, even the forecasts commissioned by the Communist Party could in the end produce an understanding of the world that appeared increasingly beyond the reach of any simple forms of governmental control: the more knowledge there were about the future, the more insecure and risky the future appeared to be. The development of climate change science, world energy resources and population modeling prove this point. Thus, forms of prediction produced even at the heart of authoritarian regimes actually contributed to eroding the understanding that long-term control was possible by revealing uncertain and threatening developments. The idea according to which no national future of the environment, economy or population could be managed without appropriate knowledge of wider regional and global trends was part of these subversive effects of predictive knowledge.

This question of the multiple and less known paths which eventually opened up the communist systems to change is laid out also in the chapters by Vítězslav Sommer and Ana-Maria Cătănuș. The question of effect is, to be sure, complicated: if some techniques were used to reinforce the authoritarian system through empowering the surveillance and control of the individuals, other techniques contributed to significant modifications of the communist governance. As demonstrated in particular by Sommer's chapter on the developments of futurology and *prognostika* in Czechoslovakia, Radovan Richta's revisionist version of futurology, proposing possible alternative futures of the system, was an important reformist critique of Marxism-Leninism. The limits of tolerance for such a critique differed, explaining institutional trajectories of futures research and limits to the possibility of using social science and other repertoires to depict possible system change. Although future studies were not strongly associated with dissent in the Soviet context, the growing social networks of future scholars disturbed the Party. The one academic article so far produced on futures studies in the Soviet Union showed that the new association of Soviet forecasters was clamped down upon.[23] Yet futurism did give rise to dissidence in socialist countries where the grasp on the social sciences were slightly less tight. Cătănuş traces the trajectory of Mihai Botez in Ceauşescu's Romania. An accomplished mathematician, Botez was well-integrated in international networks of systems analysis and close to power as the director of research at the Laboratory for Prospective Research in Bucharest. Although the personal route of Botez from a highly positioned scientific expert to a dissident was rather unique, the reasons which led him to dissent were widely shared among the forecasters in the Eastern bloc: the discrepancy between the failing future of the system that they predicted, and the official future told. If Richta in his *Civilization at the Crossroads* (1966) argued that there were many possible ways of future social development, Botez carried out his own, independent study on the preferred futures of the Romanian society at the end of the 1970s.[24] Both Richta and Botez were sanctioned, albeit for different reasons and in different ways: Richta was brought to radically revise his position after the Prague Spring, whereas Botez was severely repressed, excluded from academic life and eventually emigrated in 1987. In both cases the official ideological stance was that there was only one communist future, yet it was acknowledged that there could be many different roads leading to this communist future. These different roads, as the above examples as well as the case of scenarios showing negative growth of Soviet economy discussed by Rindzevičiūtė show, were confined to the specialists' discussions behind the closed doors and never in public. This internal control was, however, accompanied with a certain room for maneuver on the international arena, where East and West forecasters systematically met to incrementally forge a new, global idea of the future.

ENVISIONING THE GLOBAL FUTURE

Several chapters in the volume show how reflections on the future were in fact reflections on the world, and how forms of prediction were centrally implicated in new envisioning and conceptions of a world system that both reenacted and challenged Cold War divisions. Forms of future knowledge performed, as Paul Warde and Sverker Sörlin suggest in their chapter, an act of integration, of bringing together different methods and forms of fact into new images of interrelationships and causal dependencies that permitted viewing the world as a global whole. This act of integration, of using the idea of the future as the symbol for a set of reflection on the fate of human civilization in a postapocalyptic world, was central to the shaping of forms of future thinking from the 1940s on. Warde and Sörlin trace the emergence of what they call a new genre of environmentalist and catastrophist scenario to the rise of ecologism in the interwar period, heavily influenced by demographics and early climate science. Such writings paved the way for the eventual breakthrough of the idea of environmental collapse, well before the Club of Rome report in 1972. They propose that the category of the environment, as opposed to nature, was inherently future oriented, relying on various forms of prediction and prophecy. The categories of the environment and the future seem indeed to go hand in hand, united by ideas of common fate, of shared human responsibility, but also by the tools and technologies that permitted their constitution. An important conductor here was a rise of a new type of knowledge which Warde and Sörlin label metaexpertise, and a new type of scientist, the metaexpert as it were, who easily moved across disciplinary boundaries and between statistical date and literary accounts of disaster.

It is quite possible to extend this notion of metaexpertise to describe a greatly heterogeneous group of future scholars who oftentimes referred to themselves as futurists. Jenny Andersson's chapter discusses the rise of futurism in the decades from the early 1940s on in the body of thought of leading cultural critics such as Lewis Mumford or the economist Kenneth Boulding. Andersson proposes that futurism reenacted a cultural critique of liberalism after 1945 by focusing on the universal values that might be capable of forging a global community and re-enchanting the future. The utopian dimension of futurism was transformed, however, over time, as futurists would also lay claims to metaexpertise, and positioned themselves as experts on world developments and trends.

The act of integration in ways of imaging the future that Warde and Sörlin refer to is also visible in the way that forms of prediction were used to build bridges between different scientific disciplines, governmental sectors and, ultimately, between East and West. It is no accident that the idea of the future, conceived as an object of neutral investigations of natural and mathematical sciences, served as platform for contacts and collaboration in a world ridden by Cold War bipolarity.[25] The scientific tools of prediction and forecasting could be historicized not only as mundane instruments of gathering knowledge about geophysical systems and their deep past and remote future, but also as influential intellectual tools which promoted new forms of global understanding. In particular, from the late 1960s, future studies emerged as a central arena for new ways of thinking about common, indeed global problems and for transnational activity.

In this context, the case of the geophysical sciences, as discussed in Rindzevičiūtė's chapter, is particularly telling: the necessity for prediction in the fields of geophysical sciences opened up the closed Cold War world for

East–West cooperation with an aim of getting to know the long-term future of climate, but also of energy, population and economy. This opening up to the idea of a long-term but also global future entailed the formulation of a governmental agenda of global and "common" problems to which subscribed many prominent scholars and policy makers both from East and West. As a result, the pursuit of predictive knowledge, even if understood as an instrument to advance national interest by the conflicting governments, led to an important shift from the bipolar struggle of the early Cold War. The late 1960s and early 1970s saw the birth of several institutions and informal groups that coordinated international efforts to address the global future. For instance, by the initiative of the US and the Soviet Union, the International Institute of Applied Systems Analysis (IIASA) was established in Austria in 1972. The IIASA was the first international think-tank that facilitated East–West collaboration in data collection about the common problems of humankind. Such common problems, it was suggested, could be solved with the aid of universalizing predictive models that could be applied to the communist as well as to the capitalist world. Another central transnational enterprise was the Club of Rome, discussed by Vieille Blanchard in a chapter that resumes a discussion first examined in the author's PhD dissertation, which is arguably one of the first systematic historic studies of the production of the *Limits to Growth* report.[26]

Forms of prediction were indeed central for emerging attempts to think about the world, and also for giving shape to debates on world development. Indeed, modeling, forecasting and simulation were not confined to the developed world and state socialist bloc. The recent study by Eden Medina on cybernetization of the economic planning in Salvadore Allende's Chile illustrates well this point.[27] We clearly lack historical studies of the role played by future images and forms of future expertise in the global south, which were structurally included in the advancement of the nuclear age.[28] For this reason, the question of whether Western, but also Soviet future studies were emancipatory or colonial projects in the Third World remains to a large extent to be answered. However, two studies in our volume point to the importance played by the future and ways of planning for it or simulating it for opening up understandings of a possibly different African future. These chapters again bring out the powered dimension of the future and the role of future for thinking the hierarchies of world order in the 1960s and 1970s. While both revisionist Marxists and dissidents in the Eastern Bloc tried to model alternative futures of the communist system, the introduction of modeling as part of development economics in Africa address the question of a possible postcolonial future. In their chapter on Ghana, Jeff Grischow and Holger Weiss show how shifting political allegiances from the US to the Soviet Union stimulated the President Kwame Nkrumah to introduce long-term planning based on forecasting. The failure to implement this plan, however, clearly shows the limits of the scientific forecasting as a planning technique, which has to do with both the absence of statistical data and political consensus on the objectives. Similar issues were also revealed in the Nigerian attempts to use computer models for branch development in the 1970s, as discussed by Kevin Baker in this volume. If forecasting-based planning in Ghana was backed by the Soviets, and the Nigerian models by the Americans, both struggled with the issues pertaining to the lack of infrastructure, involvement with local expertise and failure to build a wider consensus as to the envisioned governmental programs.

In sum, the future that is the concern of this volume served different purposes than other traditional categories of fantasy or utopia. The future was constructed as a field of action, intervention, management and protest. By showing how different approaches to future studies intertwined the domains of science and politics, were used for control and change, and erased the dividing line between progress and disaster, this volume brings out a new dimension of research into the history of the future, anchored in political history but introducing insights from science and technology studies, environmental history, cultural history and sociology.

NOTES

1. Ernst Bloch, *The Spirit of Utopia* (1918; reprint, Stanford: Stanford University Press, 2000); John R. Hall, "The Time of History and the History of Times," *History and Theory* 19, no. 2 (1980), 113–131, Stephen Kern, *The Culture of Time and Space, 1880–1918* (Cambridge, MA: Harvard University Press, 2003); Chris Lorenz and Berber Bevernage, *Breaking Up Time, Negotiating the Borders Between Present, Past and Future* (Amsterdam: Vandenhoeck and Ruprecht, 2013), Fredric Jameson, *Archaeologies of the Future: The Desire Called Utopia and Other Science Fictions* (London: Verso, 2005); Mircea Eliade, *The Myth of the Eternal Return: Or, Cosmos and History* (1954; reprint, Princeton: Princeton University Press, 1992).
2. Barbara Adam, *Timewatch: The Social Analysis of Time* (Cambridge: Polity Press, 1995); Nick Brown, Brian Rappert, and Andrew Webster, eds. *Contested Futures: A Sociology of Prospective Techno-Science* (Aldershot, UK: Ashgate, 2000); Barbara Adam and Chris Grove, *Future Matters: Action, Knowledge, Ethics* (Leiden, Netherlands: Brill, 2007); see also American Historical Review Forum, "Histories of the Future" (2012); Elke Seefried, "Steering the Future: The Emergence of 'Western' Futures Research and Its Production of Expertise, 1950s to early 1970s," *European Journal of Futures Research* 2 (2013); and a special issue of *Cahiers du monde Russe* (forthcoming, 2015).
3. Reinhart Koselleck, *Futures Past: On the Semantics of Historical Time* (Cambridge, MA: The MIT Press, 1985).
4. Lucian Holscher, *Die Entdeckung der Zukunft* (Frankfurt: Fischer, 1999).
5. Dipesh Chakrabarty, *Provincializing Europe: Postcolonial Thought and Historical Difference*, (Princeton: Princeton University Press, 2000); Lynn Hunt, "Globalisation and Time," in Lorenz, Chris and Berber Bevernage, eds. *Breaking up Time, Negotiating the Borders between Present, Past and Future* (Amsterdam: Vandenhoeck and Ruprecht, 2013) 199–216.
6. Nancy Munn, "The Cultural Anthropology of Time: A Critical Essay," *Annual Review of Anthropology*, 21 (1992): 93–123;
7. Paul Edwards, *A Vast Machine: Computer Models, Climate Data, and the Politics of Global Warming* (Cambridge: MIT Press, 2010); Jakob Vogel and Heinrich Hartmann, *Zukunftswissen: Prognosen in Wirtschaft, Politik und*

Gessellschaft seit 1900 (Frankfurt and New York: Campus Verlag, 2010); Amy Dahan-Dalmedico and Dominique Pestre, eds. *Les modèles du futur. Changement climatique et scénarios économiques : enjeux politiques et économiques*, (Paris: La Découverte, 2007);

8. Francois Hartog, *Régimes d'historicité, presentisme et experiences du temps* (Paris: Le Seuil, 2003).
9. Lucian Holscher, "History of the Future, the Emergence and Decline of a Temporal Concept," *History of Concepts Newsletter* 5 (2002): 10–15; see also Bruno Latour, "An Attempt at a Compositionist Manifesto," *New Literary History* 41 (2010): 471–490; and Barbara Adam, "History of the Future: Paradoxes and Challenges," *Rethinking History* 14, no. 3 (2010): 361–378.
10. Jenny Andersson, "The Great Future Debate and the Struggle for the World," *American Historical Review* 117, no. 5 (2012): 1411–1430; Ariel Colonomos, *La politique des oracles: raconter le future aujourd'hui* (Paris: Albin Michel, 2014), but also Donald McCloskey, "The Art of Forecasting: From Ancient to Modern Times," *Cato Journal* 12, no. 1 (1992): 23–43.
11. Gregoire Mallard and Andrew Lakoff, "How Claims to Know the Future Are Used to Understand the Present: Techniques of Prospection in the Field of National Security," in Charles Camic, Michèle Lamont, and Neil Gross, eds., *Social Knowledge in the Making* (Chicago: University of Chicago Press, 2010), 339–379.
12. See Claudia Aradau and Rens van Munster, *Politics of Disaster: Genealogies of the Unknown* (Abingdon, UK, and New York: Routledge, 2012).
13. Michael Gordin, Helen Tilley, and Gyan Prakash, eds. *Utopia/Dystopia: Conditions of Historical Possibility* (Princeton: Princeton University Press, 2010).
14. Jameson, *Archaeologies of the Future*; Ruth Levitas, *The Concept of Utopia* (Syracuse, NY: Syracuse University Press, 1990).
15. These questions are at the heart of the research project Futurepol, conducted at Sciences Po in Paris, France. We acknowledge the funding provided to us by the European Research Council within the framework of an European Research Council starting grant.
16. Sharon Ghamari-Tabrizi, *The Worlds of Herman Kahn: The Intuitive Science of Thermonuclear War* (Cambridge, MA: Harvard University Press, 2005); Stefan Cihan Aykut, this volume.
17. The turn to study the Cold War as a period of intense international contacts is expressed in such studies as Odd Arne Westad, *The Global Cold War: Third World Interventions and the Making of Our Times* (Cambridge: Cambridge University Press, 2007); Sari Autio-Sarasmo and Katalin Miklóssy, eds., *Re-Assessing Cold War Europe* (London and New York: Routledge, 2011); György Péteri, ed., *Imagining the West in Eastern Europe and the Soviet Union* (Pittsburgh: University of Pittsburgh Press, 2010).
18. Eglė Rindzevičiūtė, "A Struggle for a Soviet Future: The Birth of Scientific Forecasting in the Soviet Union" (in progress).
19. Slava Gerovitch, *From Newspeak to Cyberspeak: A History of Soviet Cybernetics* (Cambridge, MA: MIT Press, 2002).
20. Francis Gavin and Mark Atwood Lawrence, eds. *Beyond Cold War: Lyndon Johnson and the New Global Challenges of the 1960s* (Oxford: Oxford University Press, 2014).
21. Jenny Andersson, "The Great Future Debate"; Fred Kaplan, *The Wizards of Armageddon* (Stanford: Stanford University Press, 1991); David Jardini, *Out of the Blue Yonder: The RAND Corporation's Diversification into Social Welfare Research, 1946–1968* (PhD diss., Carnegie Mellon University, 1996); Ghamari-Tabrizi, *The Worlds of Herman Kahn.*

22. Here, value-neutral rational choice theory became useful. For more, see S. M. Amadae, *Rationalizing Capitalist Democracy: The Cold War Origins of Rational Choice Liberalism* (Chicago: University of Chicago Press, 2003).
23. Gordon L. Rocca, "A Second Party Within our Midst. The History of the Soviet Forecasting Association," *Social Studies of Science*, 11, no. 2 (1981): 199–247.
24. Radovan Richta, *Civilization at the Crossroads: Social and Human Implications of the Scientific and Technical Revolution* (White Plains, NY: International Arts and Science Press, 1969).
25. Mathematical and statistical methods have historically been purified as relatively apolitical, mainly with an aim to use them for political purpose. Similarly, those fields that used mathematical methods were deemed as most suitable for East–West cooperation. See Alain Desrosières, *The Politics of Large Numbers: A History of Statistical Reasoning* (Cambridge, MA: Harvard University Press, 2002); Theodore Porter, *Trust in Numbers: The Pursuit of Objectivity in Science and Public Life* (Princeton: Princeton University Press, 1996); Slava Gerovitch, "Mathematical Machines of the Cold War: Soviet Computing, American Cybernetics and Ideological Disputes in the Early 1950s," *Social Studies of Science*, 31, no. 2 (2001): 253–287; Eglė Rindzevičiūtė, "The Politics of Governance in an Authoritarian Regime: Hybridization and Purification of Cybernetics in the Soviet Union," *Archiv fur Sozialgeschichte* 50 (2010): 289–309.
26. Elodie Vieille Blanchard, *Les limites à la croissance dans un monde global: modélisations, prospectives, réfutations*, École des Hautes Études en Sciences Sociales, Paris, 2011.
27. Eden Medina, *Cybernetic Revolutionaries: Technology and Politics in Allende's Chille* (Cambridge, MA: MIT Press, 2011).
28. Gabrielle Hecht, ed. *Entangled Geographies: Empire and Technopolitics in the Global Cold War* (Cambridge, MA: MIT Press, 2012).

REFERENCES

Adam, Barbara. "History of the Future: Paradoxes and Challenges." *Rethinking History* 14, no. 3 (2010): 361–378.

Adam, Barbara. *Timewatch: The Social Analysis of Time*. Cambridge, UK: Polity Press, 1995.

Adam, Barbara and Chris Grove. *Future Matters: Action, Knowledge, Ethics*. Leiden, Netherlands: Brill, 2007.

Amadae, S. M. *Rationalizing Capitalist Democracy: The Cold War Origins of Rational Choice Liberalism*. Chicago: University of Chicago Press, 2003.

American Historical Review, "Histories of the Future", 117, no. 5 (2012).

Andersson, Jenny. "The Great Future Debate and the Struggle for the World." *American Historical Review* 117, no. 5 (2012): 1411–1430.

Aradau, Claudia and Rens van Munster. *Politics of Disaster, Genealogies of the Unknown*. Abingdon, UK, and New York: Routledge, 2012.

Autio-Sarasmo, Sari and Katalin Miklóssy, eds. *Re-Assessing Cold War Europe*. London and New York: Routledge, 2011.

Bloch, Ernst. *The Spirit of Utopia*. Stanford: Stanford University Press, 1918/2000.

Brown, Nick, Brian Rappert, and Andrew Webster, eds. *Contested Futures: A Sociology of Prospective Techno-Science*. Aldershot, UK: Ashgate, 2000.

Chakrabarty, Dipesh. *Provincializing Europe: Postcolonial Thought and Historical Difference*. Princeton: Princeton University Press, 2000.

Colonomos, Ariel. *La politique des oracles: raconter le future aujourd'hui*. Paris: Albin Michel, 2014.

Dahan-Dalmedico, Amy and Dominique Pestre, ed. *Les modèles du futur. Changement climatique et scénarios économiques: enjeux politiques et économiques*. Paris: La Découverte, 2007.

Desrosières, Alain. *The Politics of Large Numbers: A History of Statistical Reasoning*. Cambridge, MA: Harvard University Press, 2002.

Edwards, Paul. *A Vast Machine: Computer Models, Climate Data, and the Politics of Global Warming*. Cambridge, MA: MIT Press, 2010.

Eliade, Mircea. *The Myth of the Eternal Return: Or, Cosmos and History*. Princeton: Princeton University Press, 1954/1992.

Gavin, Francis and Mark Atwood Lawrence, eds. *Beyond Cold War: Lyndon Johnson and the New Global Challenges of the 1960s*. Oxford: Oxford University Press, 2014.

Gerovitch, Slava. *From Newspeak to Cyberspeak: A History of Soviet Cybernetics*. Cambridge, MA: MIT Press, 2002.

Gerovitch, Slava. "Mathematical Machines of the Cold War: Soviet Computing, American Cybernetics and Ideological Disputes in the Early 1950s." *Social Studies of Science* 31, no. 2 (2001): 253–287.

Ghamari-Tabrizi, Sharon. *The Worlds of Herman Kahn: The Intuitive Science of Thermonuclear War*. Cambridge, MA: Harvard University Press, 2005.

Gordin, Michael, Helen Tilley, and Gyan Prakash, eds. *Utopia/Dystopia: Conditions of Historical Possibility*. Princeton: Princeton University Press, 2010.

Hall, John R. "The Time of History and the History of Times." *History and Theory* 19, no. 2 (1980): 113–131.

Hartog, Francois. *Régimes d'historicité, presentisme et experiences du temps*. Paris: Seuil, 2003.

Hecht, Gabrielle, ed. *Entangled Geographies: Empire and Technopolitics in the Global Cold War*. Cambridge, MA: MIT Press, 2012.

Holscher, Lucian. "History of the Future, the Emergence and Decline of a Temporal Concept." *History of Concepts Newsletter* 5 (2002): 10–15.

Holscher, Lucian. *Die Entdeckung der Zukunft*. Frankfurt: Fischer, 1999.

Hunt, Lynn. "Globalisation and time." In *Breaking up Time, Negotiating the Borders between Present, Past and Future*, edited by Chris Lorenz and Berber Bevernage, 199–216. Amsterdam: Vandenhoeck and Ruprecht, 2013.

Jameson, Fredric. *Archaeologies of the Future: The Desire Called Utopia and Other Science Fictions*. London: Verso, 2005.

Jardini, David. *Out of the Blue Yonder: The RAND Corporation's Diversification into Social Welfare Research, 1946–1968*. PhD diss., Carnegie Mellon University, 1996.

Kaplan, Fred. *The Wizards of Armageddon*. Palo Alto: Stanford University Press, 1991.

Kern, Stephen. *The Culture of Time and Space, 1880–1918*. Cambridge, MA: Harvard University Press, 2003.

Koselleck, Reinhart. *Futures Past: On the Semantics of Historical Time*. Cambridge, MA: MIT Press, 1985.

Latour, Bruno. "An Attempt at a Compositionist Manifesto." *New Literary History* 41 (2010): 471–490.

Lorenz, Chris and Berber Bevernage, eds. *Breaking up Time, Negotiating the Borders between Present, Past and Future*. Amsterdam: Vandenhoeck and Ruprecht, 2013.

Levitas, Ruth. *The Concept of Utopia*. Syracuse, NY: Syracuse University Press, 1990.

Mallard, Gregoire and Andrew Lakoff. "How Claims to Know the Future Are Used to Understand the Present: Techniques of Prospection in the Field of National Security." In *Social Knowledge in the Making*, edited by Charles Camic, Michèle Lamont, and Neil Gross, 339–379. Chicago: The University of Chicago Press, 2010.

McCloskey, Donald. "The Art of Forecasting: From Ancient to Modern Times." *Cato Journal* 12, no. 1 (1992): 23–43.

Medina, Eden. *Cybernetic Revolutionaries: Technology and Politics in Allende's Chile*. Cambridge, MA: MIT Press, 2011.

Munn, Nancy. "The Cultural Anthropology of Time: A Critical Essay." *Annual Review of Anthropology* 21 (1992): 93–123.

Péteri, György, ed., *Imagining the West in Eastern Europe and the Soviet Union*. Pittsburgh: University of Pittsburgh Press, 2010.

Porter, Theodore. *Trust in Numbers: The Pursuit of Objectivity in Science and Public Life*. Princeton: Princeton University Press, 1996.

Rindzevičiūtė, Eglė. "The Politics of Governance in an Authoritarian Regime: Hybridization and Purification of Cybernetics in the Soviet Union." *Archiv fur Sozialgeschichte* 50 (2010): 289–309.

Rindzevičiūtė, Eglė. "A Struggle for a Soviet Future: The Birth of Scientific Forecasting in the Soviet Union." (In press.)

Seefried, Elke. "Steering the Future: The Emergence of 'Western' Futures Research and Its Production of Expertise, 1950s to early 1970s." *European Journal of Futures Research* 2 (2013): 15–29.

Rocca, Gordon L. "A Second Party Within Our Midst: The History of the Soviet Forecasting Association." *Social Studies of Science* 11, no. 2 (1981): 199–247.

Vogel, Jakob and Heinrich Hartmann. *Zukunftswissen: Prognosen in Wirtschaft, Politik und Gessellschaft seit 1900* (Frankfurt and New York: Campus Verlag, 2010).

Westad, Odd Arne. *The Global Cold War: Third World Interventions and the Making of Our Times*. Cambridge, UK: Cambridge University Press, 2007.

1 Midwives of the Future

Futurism, Futures Studies and the Shaping of the Global Imagination

Jenny Andersson

This chapter traces the rise of futurism in the immediate post war period in the ideas of a number of intellectuals central to its making: the American urbanist Lewis Mumford; the Dutch sociologist Fred Polak; the economist Kenneth Boulding and his equally prominent wife, peace activist Elise Boulding; the German journalist Robert Jungk; the Norwegian international relations theorist Johan Galtung; and the former RANDian, systems theorist Hazan Ozbekhan. In this chapter, I argue that through the ideas of these intellectuals and scientists, the future reemerged as a utopian category.

As a set of utopian (and dystopian) reflections, futurism made use of the notion of the future as an organizing concept for a reflection on what it meant to be human and act with humanity in what futurists perceived as a dangerous, inhumane and irrational world.[1] Futurists understood the future as an imperative for the active reinvention of human civilization; a re-forging, as it were, of "Mankind." They constituted futurism through a range of assumptions and descriptions of human behavior, many of which were in fact continuations of key elements from interwar philosophy and social science. Whereas futurists rejected post-war definitions of rationality as applied, calculable, or predictable, they relied, instead, on fundamental notions of subjectivity, reflexivity and the human imagination. In many ways these notions can be traced back to interwar conceptions of human reason as a question of normative, situated human experience and *being*.[2] These notions ranged, in the thinking of futurists, from romantic and conservative reenactments of a *Kulturkritik* central to liberalism in the interwar period, to emerging forms of radicality that would become associated with the new social movements of the late 1960s and 1970s.

I want to propose that the reemergence of the idea of the future in the immediate post-war decades is indicative of two things: first, prevailing understandings, after 1945, of a fundamental crisis of liberalism; and second, the idea that such crisis could only be escaped through a reinvention of humanity and the human capacity to reimagine the world. The idea, concept and category of the future in this sense embodied a reinvention of utopian energy and structured key debates about the malleability of the world and the scope of human agency. In this specific sense, utopian thinking was an integral element in 1960s and 1970s social movements, fundamentally committed to the possibility of radical system change, and often appealing to interests beyond nation, projected on a universal or global scale.[3] It is in its wider meaning as a signifier of the potential of human action and of the possibilities of reshaping world order that the future gains its relevance in these decades.

The intellectual historian Samuel Moyn has argued that utopianism changed focus after 1945, as World War Two had seemed to fundamentally discredit historical utopias of liberalism and communism. To Moyn, the struggle for human rights is a substitute utopia, a conversion of previous dreams of model societies. By utopia, he means the *project* of human rights, in other words the conception of human rights as a "covenant" of humanity.[4] Many of the "global" social movements of the late 1960s and early 1970s entertained ideas of universalism, global unity and common destiny; futurism is a case in point here. The utopianism of futurism consisted, however, not only in the idea of the future as a possibly better place, but also and specifically, in the active *process* of imagining possibility.[5] They saw this process as one of radical deconstruction, a process that had two steps for futurists. First, the reimagining of the future of the world as something unbound from any form of determinism, including the projections and predictions emanating from the post-war (social) sciences. Second, the use of such imaginings as imperatives for the making of a different world through new forms of activism that mobilized the post-war social sciences and their obsession with modeling, simulation and prediction. These activities, to futurists, were tools for the active imagination.

The world, to futurists, was a system, a holistic entity fraught by a series of antagonistic relationships, between man and nature, between human bodies and technology, between nations and blocs. How could this conflicted system be recreated as a question of balance and harmony? How could man himself be made whole, indeed cured of his pathological drive for destruction? These questions preoccupied intellectuals from a wide range of currents, straddling liberalism, realism and Marxism in the post-war decades and it is not an accident that the idea of the future would bring them together.

The first section of this chapter accounts for how futurists used the idea of the future as a way of addressing these questions. The second part shows how futurists not only imagined a different world future, but invented forms of interrogation capable of conjuring different futures of the world system. Futurism gave rise to a repertoire of forms of knowledge and counterknowledge—construed to undo what futurists understood as hegemonic forms of expertise that contained world futures, and replace these with forms of future consciousness that they saw emanating from an emergent global public. This idea of counterexpertise was central as futurism radicalized into a particular subfield of social science, so-called futures studies, from the late 1960s on. The final part of the chapter discusses how the ambition to create a radical form of expertise for the world future evolved over time, as futurists, willingly or unwillingly, became increasingly caught up in world politics, in particular after 1989. In so doing, futurists abandoned some of the utopian potential embodied in the idea of the future, and they replaced this, I propose, with forms of professionalization and marketization that were not less futuristic but were much less concerned with the utopian activity of opening up alternative worlds.

A COLD WAR SCIENCE?

Seen from within the context of the established intellectual history of the post-war period, futurism and futures studies can seem as oddities or ephemeral projects without lasting consequence to the world. Futurism did not really put a mark on political history, although it concentrated a lot of intellectual focus at a particular point in time. Futures studies never quite entered the canon of the social sciences. I want to argue, however, that futurism and futures studies are of great historical interest, and that they were in fact the site of a key reflection on what Hannah Arendt called the human condition and the role of humankind in a surrounding, socially constituted world.[6] As such, they are also an example of a particular 1960s form of activism that not only straddled the divide between the scientific and the political, but also mobilized religious and eschatological notions alongside arguments derived from social science.

I also want to take futurism and futures studies out of a burgeoning literature that has begun to address it as a particular form of "Cold War science."[7] In a growing body of literature, the origins of new epistemological approaches to prediction, such as modeling, gaming and simulation, in the immediate post-war decades have been interpreted as constituting a particular Cold War knowledge, fundamentally tainted by a governmental, militaristic obsession with foreseeability and predictability.[8] This is clearly true for some aspects of prediction, but futurism, futures research, and futures studies, complicated and interconnected terms, were also fields in which the many tensions and contradictions embedded in "Cold War science" were played out most clearly.[9] Futurism was in many ways dependent on the predictive techniques that postulates of rationality and experiments with modeling and simulation had enabled. But these techniques displayed a great variety of future rationalities, indicative of the radically different imaginaries that forms of prediction could embody. Some of these future shaping techniques projected single futures, as in the case of one move games, whereas some projected multiple, plural futures, as for instance, in games with open outcomes or scenarios that allowed for the projection of standard and alternative worlds. For some futurists, such techniques were tools to contain troubling future developments, but for others, they were aids to the imagination, as they seemed to allow for near visual images of alternative roads ahead.[10] Futures research contained a deeply utopian and romantic strand that would express itself in the rejection of the idea that scientists could postulate and predict rational human behavior and hence foretell the future.[11] The future, futurists argued, was not a question of extrapolation of current trends, nor was it a derivative of the actions of a rational and liberal Cold War subject. The future was an active human construct, a question of normative desires and values. As such, futurism was a central form of protest against the Cold War world order, as well as a fundamental reflection on the disembodied postulates of science, technology and progress.

The problem of the future began, according to futurists, within Man himself, as Man was himself the architect of the great series of unfolding future catastrophes. This Man had to be reformed. Futurism began, therefore, with a fundamental reflection on the human subject, and as such it is directly related to a multitude of existential interrogations in the 1940s and 1950s, many of which were triggered by the use of nuclear weapons. Whereas Lewis Mumford remains the central character in the first generation of post-war futurists, such reflections were constitutive to a much larger field of thinking ranging from Ronald Niebuhr to Herbert Marcuse.[12]

The red thread of futurism was the appeal to the human imagination and the rejection of the idea of science as something distinct from, and superior to, the human imagination. As argued recently by Paul Erickson and others, it was this claim to a disembodied form of rationality that informed the applied and mechanistic turn in the post-war social sciences.[13] This radical claim was much contested, and arguably much less hegemonic than the Cold War science literature has made it out to be. Futurists, some of whom were eminent social scientists, entertained long discussions of whether futures studies were a science or an art, discussions that focused on questions such as disembodied facts versus questions of values, morality and imagination. One of the first terms describing the field of futures research came in the early 1940s from the Jewish émigré Ossip Flechtheim, who used the term futurology to describe a new philosophical approach to the field of the future.[14] He would be criticized by other futurists for the use of such a scientific label.[15] But Flechtheim's use of the term had nothing in common with scientific approaches to prediction; it should be put in the context, rather, of earlier nineteenth and twentieth century philosophy. The term futurology was the result of his attempts to develop the theories of history of Marx and Hegel; the phenomenology of Husserl, with whom he had studied; and the thoughts on the necessity of utopia as opposed to ideology that were put forward in the 1930s by his fellow in exile Karl Mannheim. Flechtheim was also fundamentally inspired by the social psychologist Erich Fromm.[16] This shows the complex connections between futurism and an interwar world of both liberal and Marxist thought that engaged actively with utopia and did not make post-war distinctions between rationality and normativity, observation and subjectivity. Flechtheim's 1940s reflections on humanity's need for a peaceful theory of history developed in the 1960s, drawing extensively on Mannheim, Fromm, and the Marxist revisionist Lezlek Kolakowski, into his idea of futurology as the thorough analysis of societal objectives such as growth or freedom, and the always value laden and normative choice of ideological goals. This activity of scrutiny, deconstruction and construction of alternatives could be performed, Flechtheim proposed, with scientific rigor and reflexivity, but had to retain a fundamentally utopian dimension of the possible construction of alternative societal objectives.[17]

Flechtheim was throughout his life a convinced Marxist whose aversion to Soviet led him instead to a never-folding belief in the possibility of world federation.[18] Other futurists, such as the federalist Denis de Rougemont, or the French political theorist Bertrand de Jouvenel, were conservatives or liberals. Nevertheless, they could also gather around these postulates of the future as the quintessential social construct, the result not of predetermined trends but of the human imagination. De Jouvenel was a deeply conservative thinker, an enigmatic character who travelled, through his life, from an interwar past in fascism to post-war membership in Friedrich Hayek's Mount Pelerin Society, and eventually, into political ecology. His ideas of the future were remarkably adaptive to all of these political strands, to the point of being situated *beyond ideology* and in the domain of neutral expertise. Futures research was, de Jouvenel proposed, an art, distinguished from other social sciences by the awareness of the self-fulfilling prophecy, in other words the constant risk that projections might act performatively on the object of study.[19] This led to a principle of reflexivity (indeed, of social responsibility), shared by many futurists: the idea that because the purpose of futures studies was to conjure better social futures, futurism had to be thought of as a fundamentally value-laden activity. The futurist could not hide behind notions of objectivity, but had the fundamental responsibility of making the assumptions and postulates underlying future projections clear. Such post-positivist stances, necessitated by the theoretical problem of the self-fulfilling prophesy, made futures research distinct from the other social sciences; it could be proposed that it in fact carried a constitutive discussion in the post-war social sciences of the value of objectivity.[20]

Despite this insistence on future research as an art and not a science, futurists, particularly in the 1960s and 1970s, would also claim that futures research belonged in the social sciences. In fact it did. Futures research interested the mainstream social sciences in the period from the mid-1960s on. Several disciplines developed conferences and subcommittees to the themes of prediction, futures research or futures studies. Some sciences, psychology, economics, political science, and international relations saw futures research as a way of pushing their rationality postulates to the test of predictive capacity. Actors such as the Ford Foundation or the National Science Foundation funded and assisted futures research, motivated by the hope that it might represent a new behavioral science in the making.[21] Other disciplines, such as anthropology and sociology, emerging from their earlier focus on strict categories of class or race, took an interest in futures research as an interrogation into culture, values and human diversity.[22] The American Anthropological Association created, in 1972, a subcommittee for future anthropology under the benediction of Margaret Mead.[23] The International Sociology Association created a Research Committee for the sociology of futurology in 1970. It brought together futurists and forecasters from both sides of the Iron Curtain and functioned as an academic extension of the World Futures Studies Federation, created in 1973.[24]

TRANSGRESSING SCIENCE: FUTURISM AND THE HUMAN IMAGINATION

According to most accounts, post-war social science was based on the evacuation of fundamental forms of subjectivity and normativity, and the substitution of these categories with notions of objectivity, applicability and indeed rationality.[25] But there were many dissenting voices in the social sciences against this development. Specifically, a number of intellectuals within the field of futurism would take odds with the premises of behavioralism and argue that value change was not an outcome of the sociotechnical processes of modernization, but an undetermined, conscientious and fundamentally social process. In this they isolated the question of values and possible value change as the hope for system change, and their concept of future was in a sense a derivation of anticipated value revolutions. In the thoughts of futurists, the future was understood as a deeply transgressive object, indeed as one that rendered a number of boundaries of the post war world artificial: the East–West divide, the division between North and South and the emerging Third World, but also the distinction between scientific reason and human imagination and indeed between expertise and militancy. Such boundaries were understood as projections, actively created by the models, forecasts and predictions of applied social science, and as the artefacts of an impoverished view of world development that could only envision the future as an extrapolation of already ongoing trends. The problem of the future, as put by the science fiction writer Arthur C. Clarke, was a problem of a "lack of imagination."[26] The world, futurists argued, was a product of this crippled human imagination, or more precisely, of a schizophrenic separation between the human mind and its capacity for boundless projections on the one hand, and the much more limited scope of bodily sentiment and soul on the other. Such separation could be overcome through the creative powers of imagination, and imagination, as such, was detached from understandings of reality or realism. This appeal to the radical power of the imagination is visible in futurists' appeal to art, journalism, psychoanalysis and science fiction as ways of imagining alternative worlds. Alternative worlds were the worlds that could not be conjured by scientific reason, too circumscribed by logics of facticity, trend extrapolation and prediction. Futurists made reference to science fiction writers such as Arthur C. Clarke (whose book and film *2001,*

A Space Odyssey was discussed as a genuine attempt at futurism) or Isaac Asimov.[27] At several points the limits between science fiction and futures research are more than blurred.[28] This is important: science fiction played a key role for unleashing a fundamental debate on the irrationality of the Cold War, and for thinking about possible consequences of machine technology and nuclear world order. Sharon Gamari-Tabrizi has shown how Cold War technologies such as the scenario method drew actively on the human imagination.[29] Herman Kahn's apocalyptic visions of nuclear war were both derived *from* Hollywoodian scripting techniques, and intended to work *on* the human imagination by triggering a vivid imagery of nuclear holocaust that, or so Kahn argued, would spur defensive action.[30] Gaming drew, too, on important scripted and narrative elements and even hardliner RANDians acknowledged the role played by imagination and subjectivity in producing what was in fact in the world of modeling called "artificial" or "synthetic" fact. In other words, scientific rationality and imagination were not separate fields but intercommunicating notions, which reflects the fact that the boundaries of post-war social science were not those of today. In the 1950s and 1960s, social science research existed in a hybrid field of journalism, activism, think tank activity, and academia, and drew on repertoires from these fields and others. Futurism is a good example of this multitude of repertoires and of the way that post-war social science was open to a number of ways of knowing. In particular, futurism drew on the idea of design as an active process of world-making that began in utopian conceptions of possibility. Futurism would be fundamentally influenced by the ideas of the American urbanist and literary critic Lewis Mumford. Mumford's life-long reflections on machine civilization took an urgently pessimistic turn after 1945, as he wrote several studies devoted to the fundamental irrationality of the post-war age.[31] His tremendous work *The Condition of Man*, first published in 1944, explained the development of human civilization since the Enlightenment as a gradual separation between mind and body, and as a gradual closing in on the mechanistic idea of technical rationality. Baconian and Cartesian science had created a schizophrenic human subject, unable to think, feel, and control technology. Alienated from fundamental aspects of being, modern man had invested machines with the spirit of human civilization. The "religious dogma" of progress had operated a virtual schism in time, a rupture between past and future that had left human beings erring, disorientated, and lost. Separating himself from Marxist ideas of alienation, Mumford argued that human civilization had the potential to re-embed technology in social values, and that this had to be done as a pragmatic and reformist project of re-forging human subjectivity. Salvation began with acting on the values of Man, and only through a systematic reeducation of humanity could the future of human civilization be saved and brought into a restored continuity with history. Utopia, to Mumford, was the creation of a new world consciousness, based on the "goodness of Man."[32]

Like Flechtheim, Mumford anchored his thoughts in the long history of human philosophy. In a way, he is a very romantic writer, imbued with a sense of Hegelian civilizing mission, but also apocalyptic and messianic. The atomic bomb, to Mumford, was the quintessential proof of the dystopian power of technology over the human subject, and the confirmation that the world had gone mad, an idea he expressed in his immediate post-war call for a unified world government of the bomb. [33] Mumford's reflections on the fundamentally destructive effects of science and technology on human civilization were part of a much wider set of ideas in the immediate post-war period that stand as a bridge between continental phenomenology and Heideggerian conceptions of time and being, and Anglo-Saxon transcendental notions of wholeness and emancipation as put forward by Mumford or Raymond Williams.[34] These ideas also anticipated the much more radical reflections emerging in the 1950s and 1960s from the Frankfurt school.[35] It could be proposed that futurism operated a link between these epistemological projects, one rooted in the inter-war period and the other linked, eventually, to emerging system critical social movements. The origins of the Frankfurt school must be found in the observations of the symbiosis between totalitarianism and technology in Nazism and Stalinism that was first interpreted by intellectuals and Jewish refugees such as Walter Benjamin, Hannah Arendt, Gunther Anders and Hans Jonas, the students of Husserl and Heidegger. They would find different answers to the problem of salvation, ranging from propositions of forms of world government for the governance of atomic weapons (Anders), to the creation of epistemological principles for the governance of science and technology and the moral imperative of responsibility (Jonas).[36] They saw the human subject as fundamentally split, and as holding the keys to destruction but also the possibility of salvation. They understood this deeply schizophrenic position as the very the essence of being, and as constituting a profound moral crisis of liberalism. Some futurists had direct links with these thinkers—Flechtheim was a childhood friend of Anders and entertained a lifelong correspondence with him. Mumford cannot be easily placed in this context, but shared with Arendt, Jonas and Flechtheim a critical view of the effects of technology on humanity and the necessity to rethink human rationality. In Mumford's thinking, the act of salvation had to pass through a process of reeducation, which could be both a question of science and education, and a question of spirituality. The final chapter of the *Transformation of Man* drew up the guidelines for such a process of reeducation as a process taking place through social science, pedagogy, and spiritual reform.

Mumford became a founding member of the so-called World Futures Studies Federation. As such, he was a representative of an older, and almost gone, world of liberal thought. Nevertheless Mumford's understanding of the future as an active process of design, built on human values, would have enduring significance on the development of futures studies.[37] Among the other intellectuals who lay the foundations for futures studies we find the Dutch Fred Polak, the American Quaker couple Kenneth and Elise Boulding, and the West German journalist Robert Jungk. Polak was the author of a book that became tremendously influential on the global peace movement, as he argued for the crisis of all mobilizing images of the future of Mankind. In the absence of hopeful images of the future, consumerism, conflict and rivalry had become the mobilizing forces of divided human beings. *The Image of the Future* ended with a dramatic call to humanity to recreate images of the future that could pave the way out of this crisis of human civilization.[38] Polak had spent a few years after the war as a fellow at the Institute for Advanced Studies in the Behavioral Sciences at Stanford. There he met the Bouldings. Elise Boulding was so inspired by Polak's *Image* that she learnt Dutch in order to translate it. Kenneth Boulding dictated, according to his own account, his own book *The Image: Knowledge in Life and Society* in one monologue after discussing with Polak. Boulding's *Image* argued that human beings act according to informed and socially shaped images of the world. *The Image* called for social scientists to understand the motivating factor of love in human agency, and presented a radically different understanding of social organization from the Parsonian view of the social system as a system of exchange relations, published in the same year as Parson's book.[39]

Both of the Bouldings became central to American futurism, Elise in particular as one of a small group of feminist thinkers on the future. European futurism was fundamentally influenced by the thoughts of yet another intellectual, Robert Jungk. Jungk was a prominent figure in the European peace movement following publication of his bestseller *Brighter than a Thousand Suns*, on the atomic scientists of the Manhattan project.[40] In a second book, *Tomorrow Is Already Here*, full of emotional and Christian descriptions of the depravation of American civilization, Jungk described both the super computer at RAND, to which American forecasters had trusted the future, and the jetfighters that propelled the bodies of young male pilots into the atmosphere defying, quite literally, heavenly laws.[41] The future had been colonized, turned away from its state as human fate and destiny, and transformed into a kind of last frontier. Jungk saw the problem of the future as a possible promise of human unity, a way of articulating common fate. Jungk developed his thoughts on futurism from his position as a figurehead of the European peace movement in the mid-1960s. He would then conceptualize the future as a question of the common interest of "Mankind," an understanding that he hoped would be capable of surmounting the division of the international peace movement, split in a liberal and socialist version mirroring the Cold War division of the world.[42]

OVERCOMING WORLD ORDER: FUTURES OF THE SYSTEM

The networks of futurists overlapped with other networks, such as those of the global peace movement in the early 1960s, and the Club of Rome by the end of the decade. The field of futurism organized, from 1968, through a number of meetings and conferences focused on the creation of a World

Federation for Futures Research, or Futures Studies as it would become known. Futures studies were a radicalization of post-war futurism. As indicated, 1950s and early 1960s futurism had a romantic streak, from within which futurism appeared as a project of restoration of a human civilization that was understood as historically constituted but cut loose from past and future by the disruptive forces of war, science and technology. Reconstituting the human experience of being, indeed the project of Man himself, was not as such nostalgic, but imagined as an active process of reconstruction, innovation and design; nevertheless, it drew on a long legacy of liberal and Christian concepts of human existence. The influence of the systems theories of systems analysis, ecologism and revisionist Marxism from the late 1960s on changed these partly conservative notions of futurists such as Polak and Mumford. Futurism developed from its post-war reflections on the future as a reflection on *being* into a critique of the world as a constituted system created, engineered and organized by man. As such, the world was understood as a product of a deeply faulty and unbalanced human value system, and inspired by behaviorism and Parsonian sociology, futurists such as Johan Galtung would embark on large-scale value studies of global populations.[43] As values thus appeared to have a systemic quality, futurists would also radicalize their claim to the necessity of value change, as a process necessarily passing through an act of radical imagination.[44] This second generation of futurists understood the future as not as such lost nor unfettered, but rather *colonized* by the temporal projections of geopolitical, military and business interests. This led them to be very critical of the use of prediction, modeling, and forecasting, which they saw as the scientific tools for the maintaining of fundamental forms of world order.[45] Futures studies were thus conceived of as kind of global counter-expertise, a counter-hegemonic project the purpose of which was to transform the system by means of critical epistemology, acting as the trigger of dormant visions of alternative futures in global publics.

This assumption place quite a formidable cataclysmic power in the hands of futurists themselves. As futurists organized through a series of conferences in the late 1960s and early 1970s, they envisioned themselves as not only the saviors but as the midwives of a future yet unborn but in urgent need of delivery. The introductory statement of the first conference in 1967 emphasized the future as a global common good, a shared concern of peoples living in all parts of the world and under both systems.[46] It rejected the way that futures research had placed a particular power over the future in the hands of the military and industrial establishments, by which was meant RAND and the Organization for European Cooperation and Defense (OECD). The purpose of "establishment futurology," some futurists explained, was to complete the transition to the Marcusian One Dimensional Man, by obliterating any human capacity to think, feel and desire a different and more human future from that projected by the affluent society.[47] Against establishment futurology they posited, therefore, "critical futurology" as part of a "cultural revolution" bringing out the agency of Man, and opposing the way that "dominant cultures" tried to extrapolate themselves and their interests into time.[48] The plural "*s*" in "futures" signified that the future was open and plural, not a singular development. This idea led to some trouble for East European futurists, as the plural form was inacceptable to some of the regimes in the Communist Bloc; but the plural form was defended as a way of allowing for representations of possibly different futures in the two systems.[49] The future was not a pre-given object that could be extrapolated and studied in the halls of science—it was an object to be invented, by acting on the hopes and fears of people.

The democratization of futures research, which would be spread to the world by futurists once these were organized and equipped with tools and data capable of convincing of the merits of the new future art, would be a way of escaping the systemic lock-ins of prediction, and developing new forms of integrated, or "integral" thinking. "Integral studies" was a term from the American futurist John McHale, drawing heavily on Kenneth Boulding's vision of the social sciences as integrated disciplines. To McHale, the social sciences, in their applied and behavioral turn, had made social change into their object of study, but they had not tried to create it, *invent it*. The social sciences represented, therefore, a quintessentially pessimistic project. Established social theory, relying on Auguste Comte, Herbert Spencer, Karl Marx, Emile Durkheim, Sygmund Freud, Max Weber, was about human anxieties and fears, fundamental limitations to human experience and not about the limitlessness of human potential. Theories of postindustrialism, for instance, depicted change as disruptive and dangerous and as a process of breakdown of traditional institutions such as established value systems or the family, and not as a potential question of newness and change. "The future may be literally as we choose to view it, and the conscious degree to which we may materially control our future is quite unprecedented. Our need, therefore is to widen the process and objectives of social and cultural forecasting so as to expand rather than diminish our range of choices and options."[50] Jungk described futures studies as a form of human anthropology, including forms of mental "psychonaut" experimentation that could deliver the human self. Futures research should have the task of imagining radical departures: as in art, it should push for jumps in the human imagination. Such creative imagination would be able to surmount the Cold War world order, and create a new "state of mind" beyond the controversies and conflicts of the Cold War era. "Creative imagination is not content with extending, combining, or negating already existing trends. It attempts, by breaking out of the existing system or counter system, to strike out on a completely new course breaking radically with prevalent concepts. The creative imagination gives rise to a new era."[51]

Following from these premises, futurists tried to design forms of prediction and future interrogation that would allow for futures research to break out of the ontological cage of the present and give voice to yet unarticulated future dreams and desires. This idea lay behind the invention of a number of unorthodox methods, such as Jungk's *Zukunftswerkstätte*, radical future workshops inspired both by the development theorist Paulo Freire's radical social work and by psychoanalysis. It also inspired a critique of the idea of fact as based on past occurrences, hence unsuitable for forms of knowledge with bearing on social futures. For Jungk, data was based on experience and not on future anticipation, and was hopelessly reiterating, therefore, "something that is already dead." Futures research required, instead, metadata, "data-in-being," by which he meant forms of fact capable of discerning unfolding changes in the value structure of societies. Through changes in social values, important transformations could take place and world futures freed.[52] This idea that fact and facticity was a reflection of the past, and helpless in terms of addressing the future, came also from futurists with other political affinities than the leftist Robert Jungk. Jungk's notion of data-in-being was close to de Jouvenel's notion of *futuribles*, the *futurs possibles* that he proposed to be the future-oriented equivalents of fact for the activity that he labeled conjecture. But whereas Jungk saw data-in-being as the hopeful social trends that futurists could discern in order to act on them, to the deeply conservative thinker de Jouvenel, *futuribles* were potentially threatening factoids hiding forms of revolutionary change in the global mass.[53] Jungk would continue to cultivate a great interest in social innovation, community studies, social work, and urban planning.

Futurists' belief in the power of imagination relied on the idea that only by imagining different, and better worlds, would humanity be able to see the absurdity of the world it had actually created and hence be spurred to seek action. In a sense, futurists therefore also relied on the idea of transgression, or integration, also pointed to by Paul Warde and Sverker Sörlin in this volume. It was only through such an act of transgression that the future could gain an independent presence from the present. Whereas, in Warde and Sörlin's case, this refers to the creation of the category of the environment as a crisis concept demanding new imaginative envisionings of the world, for futurists the act of transgression focused on the Cold War world order as the manifestation of a solid crisis of modernity. Futurists shared the conviction that the Cold War was in fact not a struggle between two blocs, but an expression of totalitarian tendencies within authoritarian and liberal systems alike. Flechtheim returned to West Germany in the 1960s. He would remain convinced that Nazism had not disappeared but transformed itself into the state structures of the *Bundesrepublik*. He understood both the economism of the social market economy and the repression of terrorism in the early 1970s as expressions of a deeply authoritarian turn in Western liberalism, similar to how Mumford, in 1946, understood the atomic negotiations as the Nazification of American liberalism. For Flechtheim, a new global and humanistic version of Marxism was the only hope for a common future. He entertained, for this purpose, a desperate correspondence with many leading Marxist revisionists and dissidents, the revisionist groups emanating from the Richta group in the Czechoslovak Academy of Sciences, the Polish Poland 2000 group, and the Yugoslav group Praxis. He gave countless addresses on Radio Free Europe, and in 1968 he launched the journal *Futurum*, which was an important site of East–West circulation before the Prague spring, as it published East European futurists, and particularly the Richta group, in German.[54]

Other chapters in this volume show the importance of the idea of common world futures to forms of systemic and planetary prediction in the 1960s and 1970s. Indeed, the activity of prediction embodied understandings of a unified human civilization exposed to fundamental processes of convergence. But whereas radical futurists embraced the idea of a common future, embodied in notions of the world system or in the idea of common human needs as a notion of universal destiny instrumental to breaking with Cold War rationalities, they rejected modernization and convergence theory. The decolonization process had opened up a plurality of possible futures emanating from the different belief and value systems of human cultures. Futures studies, then, was a way of delivering these emergent world futures and create a new form of global knowledge about unfolding trends.[55]

Johan Galtung was instrumental in spreading futures studies to a number of Western think tanks and the UN system. To Galtung, the objective of futures research was to help all parts of Mankind pursue their images of the future, and also bring out those images that would be capable of uniting humanity, for instance images of a disarmed world. In the mid-1960s he began carrying out large interview studies in order to show how the peaceful future images of people around the world contrasted with the aggressive ones held by world leaders.[56] This fit with his notion of futures studies as a form of "prescriptive social science," which would function like architecture in the active design of new worlds. Futures research, to Galtung, was a reinvented form of political activism, in which the desired changes projected by the futurists should also be actively pursued by them, in "opposition to the predictor removed from social action in the ivory tower of a think tank."[57] As the counterimage of this predictive think tank, clearly modeled on the RAND, Galtung posited the idea of the World Futures Studies Federation as a democratic world federation for the future.

MODELS AS MICROUTOPIAS

The idea of future research as a way of expanding the options of the present so that desirable new societies could be actively created led to a great interest of futurists in the activity of modeling. This is slightly paradoxical. Futurists rejected prediction as an outcropping of Cold War science. But futurists were also fascinated by the technologies, methods and models that they understood as containing potential radical imaginaries and representations of the future. In particular those futurists who would engage in exercises of high technological sophistication, such as Ozbekhan or indeed Galtung, saw models as ways of representing possible alternative worlds, and as providing necessary solidity to debates about the future. Modeling was also understood as embodying a new and more democratic understanding of utopianism. Through modeling, future worlds could be experimented, without having to actually embark on a risky and possibly totalitarian project. Models were, as Elise Boulding would put it at a much later stage, exercises in micro-utopias, future test labs that did not run the risk of running astray.[58] In that sense modeling was, as Galtung repeatedly argued, democratic; a way of offering global public visions of worlds that they could then choose from, provided they got access to methodologically sophisticated futures studies. In other words, it was not the mathematical modeling that was problematic to futurists, but its use as a consolidation of a troubling world future; if put at the service of the creative global imagination, modeling could become a kind of democratic smorgasbord for alternative futures.[59] In fact a young Galtung explained these assumptions to a manifestly perplexed interviewer for a grant with the Rockefeller foundation, stating that he wanted to develop mathematical modeling for the purposes of developing world peace in the footsteps of Gandhi.[60] So he would: Galtung launched, with the Peace Research Institute Oslo as his base, a series of books in the theory and methods of the social sciences, arguing for quantitative value studies as a way of building peace futures.

The idea of modeling alternative worlds was however perhaps more prominently represented by Ozbkehan, a former RANDian forecaster and researcher of the Systems Development corporation in Santa Monica, California. Ozbekhan was also a member of the Club of Rome, and, as described by Elodie Vieille Blanchard in this volume, developed an alternative world model to that of Jay Forester's famous World 1 and World 2 which explicitly incorporated the possibility of value change as the dynamic component of the "world problematique"—in other worlds, as the key to system change. Ozbekhan understood modeling as a way of imagining plural futures in a complex system. Contributing to an OECD report in 1970, Ozbekhan suggested that planning, dominated until the 1960s by forms of extrapolation, must take into account the fundamental factor that the future might be different from the present. "The future must fill the whole, vast and empty canvas with imaginings, with wishes and goals and novel alternative configurations that somehow possess reality and present shared, or at least shareable, values." The planning process, he argued, in this and a number of writings that circulated between the OECD, RAND, the Club of Rome, and the networks of futurists, had to be rethought from that of excavation of law bound developments, to that of an active process of conception and "imaginative futures creation." Modeling was a tool in this active "willing" of the future, in other words, in the active design of system change.[61] The Club of Rome, in its first draft "The Predicament of Mankind," to which Ozbekhan was a key contributor, spoke in very similar terms of the need to reject "positivist" science and develop an ethic of the human condition.[62] Ozbekhan would go on to consider planetary and world problems such as world hunger and the distribution of food resources, problems that he saw as created not by limited resources but by a human failure in envisioning a different organization of world production and distribution, and a problem that could thus be solved by modeling.[63]

EPILOGUE: THE PROFESSIONALIZATION OF FUTURES RESEARCH AS WORLD EXPERTISE

This chapter has argued that futurism and futures studies emerged as a form of utopian thinking in the post-war decades, and that as such, their importance did not lie in the project or actual worlds that they proposed but in their insistence on the fact that the world was a human and social construction that emanated from a set of values that could, in turn, be reflected upon and reconsidered as a question of active subjectivity. This belief, that the world was malleable and could be altered through the mobilization of social science, values, and positive images of tomorrow might seem utopian to us, but we would be well advised not to dismiss it as a fad but to take it rather, as a central conception to many approaches in social science and politics alike in the period discussed. Nevertheless, as all utopian categories, the future mobilized by futurists was full of tensions and reflections of the power structures of the world that it tried to reform. In particular, futurists, in their claim to possess a special link to the future, stepped onto inherently powered ground. Futurists saw themselves not only as scholars of the future, but as midwives of the future. Inspiring, hopeful images of the future, capable of saving the world from disaster, were out there in the minds of people, but they needed to be delivered. As futures did not per se exist, but had to be created through processes of mind release and taunting of the imagination of populations and decision makers, futurists invented a highly prominent role for themselves as the potential saviors of the world.

The motivations for future midwifery changed fundamentally from the utopian, and desperate, perceptions of the future in the 1950s and 1960s, to the professionalization of futurists that took place in the decades after 1989. In the 1990s and 2000s, futures research and futures studies became a consultancy-based practice, embodied in think tanks and institutes catering to public and private clients, oftentimes selling a particular method, model, or software capable of opening up the "box" of the future.[64] These tendencies were arguably there all along. Futurists mistrusted the "establishment" and the boring halls of academia. They saw the think tank model, or indeed the consultancy format, as a guarantee of freedom from political or scientific intervention. From the late 1960s on, futurists created a number of rival powerhouses for the future, including de Jouvenel's Paris-based *Futuribles*, Polak's *Stiftung BeWeTon* in the Netherlands, Flechtheim's *Zentrum fur*

Zukunftsfragen in Berlin, Jungk's Future Library in Salzburg, Olaf Helmer and Daniel Bell's Institute for the future in the US, and Edward Cornish's World Future Society in Washington, DC. The World Futures Studies Federation was an odd collection of futurists of different political color and epistemological shape. This plurality was praised by futurists, because they saw as primordial the purpose of spreading the gospel of futurism to an ailing world. Professionalization and organization was, in the end, more important than epistemological principle. Therefore, the World Futures Studies Federation conceived of itself, not only as the global public future think tank dreamed up by Galtung, but also as a fundamental meeting space for the training of professional futurists and the spread of the methods of futures research. From the axioms on reflexivity and awareness of playing a part in creating global self-fulfilling prophecies, futurists seemed over time to seamlessly slip into an understanding of themselves as a new category of global experts, capable of offering advice on global trends. Over time, futures research seems to turn on itself as a reflection on the methods, models, tools and resources with which to study world futures. Futurists took an early interest in bibliometrical tools, which they saw as essential for a documentation function that would help organize and consolidate futures research. Many of the publications of futurists are, from the mid-1960s on, repositories and information tools pertaining to futures research itself. The many institutes, documentation centers and think tanks devoted to the future produced a sea of newsletters, first on paper and quickly in electronic form, in which a lot of attention was devoted to the methods and tools of futures research, but eventually very little to world problems. Several institutes are organized on the idea of becoming a kind of world data bank of factors contributing to the shaping of world futures and creating forms of metaexpertise (see Sörlin and Warde in this volume) capable of analyzing a kind of catalogue of global factoids. Over time, futurists also became entangled in a number of messy processes of world politics. This was at least partly a consequence of the principles adopted at the moment of the creation of the World Futures Studies Federation in 1973. In order to be able to establish futurism as a bridge across the Cold War divide, futurists accepted a principle of strict political neutrality. But after 1989, the idea of being a bridge between the systems collapsed. It would be the case, interestingly, that the demand for future advice in the post-Cold War decades of the 1990s and 2000s did not predominantly come from democratic and liberal regimes, but from regimes in processes of transition or facing some form of future challenge, oftentimes stemming from the process of democratization. In the 1990s and 2000s, futurists were involved with the Chinese government (Galtung himself, convinced that the Confucianism of Chinese Party officials embodied a particularly harmonious vision of the future) with governments of Morocco, Iran, Egypt, Belarus, Uzbekistan and Mexico. In other words, futurism, from the 1990s on, became directly entangled in new histories of future control, as futurists, equipped with advanced models, online tools and collective intelligence software, restaged themselves as consultants and experts on the future of the emergent world.

NOTES

1. Compare Russell Jacoby, *Picture Imperfect: Utopian Thought for an Antiutopian Age* (New York: Columbia University Press, 2005).
2. Paul Erickson, Judy Klein, Lorraine Daston, Rebecca Lemov, Thomas Sturm, and Michael Gordin, eds., *How Reason almost Lost Its Mind: The Strange Career of Cold War Rationality* (Chicago: University of Chicago Press, 2013).
3. See Akira Irye, *Cultural Internationalism and World Order* (Baltimore: John Hopkins Press, 1997); Matthew Evangelista, *Unarmed Forces: The Transnational Movement to End the Cold War* (Ithaca, NY: Cornell University Press, 2001).
4. Samuel Moyn, *The Last Utopia: Human Rights in History* (Cambridge, MA: Harvard University Press, 2010).
5. Michael D. Gordin, Helen Tilley, and Gyan Prakash, "Introduction: Utopia and Dystopia beyond Space and Time," in Michael D. Gordin, Helen Tilley, and Gyan Prakash, eds., *Utopia, Dystopia: Conditions of Historical Possibility* (Princeton: Princeton University Press, 2010), 11–21.
6. Hannah Arendt, *The Human Condition* (Chicago: University of Chicago Press, 1958/1998).
7. Kaya Tolon, "Futures Studies: A New Science Rooted in Cold War Strategic Thinking," in Mark Solovey and Hunter Craven, eds., *Cold War Social Science, Knowledge Production, Liberal Democracy and Human Nature* (Basingstoke, UK and New York: Palgrave, 2012), 45–63.
8. See Gregoire Mallard and Andrew Lakoff, "How Claims to Know the Future Are Used to Understand the Present," in Michelle Lamont, Charles Camic, Nick Gross, eds., *Social Knowledge in the Making* (Chicago: University of Chicago Press, 2012), 339–379.
9. See Mark Solovey, "Cold War Social Science? Specter, Reality or Useful Concept?," in Mark Solovey and Hunter Craven, eds., *Cold War Social Science* (Basingstoke: Palgrave, 2012), 1–25; Paul Erickson, Judy Klein, Lorraine Daston, Rebecca Lemov, Thomas Sturm, Michael Gordin, eds., *How Reason almost Lost Its Mind: The Strange Career of Cold War Rationality* (Chicago: University of Chicago Press, 2013), 159–181.
10. Rebecca Lemov, "Hypothetical Machines: The Science Fiction Dreams of Cold War Science," *Isis* 101, no. 2 (2010): 401–411; Sharon Ghamari-Tabrizi, "Simulating the Unthinkable: Gaming Future War in the 1950s and 1960s," *Social Studies of Science* 30, no. 2 (2002): 163–223.
11. Jenny Andersson, "The Great Future Debate and the Struggle for the World," *American Historical Review* 117, no. 5 (2012): 1411–1431.
12. I want to thank Rens van Munster and Caspar Sylvest for interesting discussions on this point, see their forthcoming book on nuclear realism.
13. Erickson et al., *When Reason Almost Lost Its Mind*. For an earlier argument, see Phillip Mirowski, *Machine Dreams: How Economics Became a Cyborg Science* (Cambridge: Cambridge University Press, 2002).
14. Ossip Flechtheim, "History and Theodicy, Some Critical Remarks on the Theories of History of Hegel and Marx," *Phylon, Atlanta University Review of Race and Culture*, third quarter (1941): 238–249.
15. Bertrand de Jouvenel dismisses the term, see Bernard de Jouvenel, *L'Art de la conjecture* (Monaco: Editions du Rocher, 1964), 31.

16. Karl Mannheim, *Ideology and Utopia* (New York: Harcourt, 1936); Erich Fromm, *The Sane Society* (New York: Rinehart, 1955).
17. Ossip Flechtheim, *Futurologie* (Berlin: Verlag Wissenschaft und Politik, 1970), 217f.
18. Mario Kessler and Ossip K. Flechtheim, *Politischer Wissenschaftler und Zukunftsdenker 1909–1998* (Köln: Bohlau Verlag, 2007), 125, 220.
19. De Jouvenel, *L'art de la conjecture*.
20. See for instance, de Jouvenel, *L'art de la conjecture*.
21. This explains for instance Talcott Parsons or Paul Lazarsfeld's interest in forms of social forecasting, as well as that of international relations theorists, such as Karl Deutsch's interest, in forecasts of the international system. There were several conferences on forecasting and prediction in the early to mid-1960s, at Yale Law School in 1962, the International Political Science Association in 1964 and American Political Science Association in 1967. Henry David of the National Science Foundation attended the first international conferences of future research.
22. Dorothy Ross, "The Changing Contours of the Social Science Disciplines," in Theodore M. Porter and Dorothy Ross, eds., *The Cambridge History of Science*. Vol. 7: *The Modern Social Sciences* (Cambridge MA: Cambridge University Press, 2008), 205–237.
23. Margaret Mead, "A Note on the Contribution of Anthropology to the Science of the Future," in Proceedings of the American Anthropological Association Symposium on Cultural Futurology, 1971.
24. Records of the International Sociological Association, Research Committee on the Sociology of Futurology, International Institute for Social History, Amsterdam, vol. 695–700.
25. See for instance S. M. Amadae, *Rationalizing Capitalist Democracy:The Cold War Origins of Rational Choice Liberalism* (Chicago: University of Chicago Press, 2003); or Mirowski, *Machine Dreams*.
26. Clarke cited by D. Livingston, "The Study of Science Fiction as Forecasting Methodology," in Japan Society of Futurology, *Proceedings of the International Conference in Future Research* (Kyoto, 1970), 71.
27. Ibid.
28. Several science fiction writers were members of the World Futures Studies Federation and there was also a Futurological Society organized by science fiction writers, not to be confused with Stanislav Lem's satirical rendering of a futurological congress in his 1971 novel, *The Futurological Congress* (translated into English in 1976).
29. Sharon Gamari-Tabrizi, *The worlds of Herman Kahn. On the intuitive science of nuclear war*, Cambridge MA: Harvard University Press, 2005.
30. Ghamari-Tabrizi, *The Worlds of Herman Kahn*.
31. Lewis Mumford, *The Myth of the Machine: Technics and Human Development* (New York: Harvest Books, 1967); Lewis Mumford, *The Pentagon of Power* (New York: Harvest Books, 1976).
32. Lewis Mumford, *The Condition of Man* (New York: Harvest Books, 1944/1956), passim, 236.
33. Lewis Mumford, "Gentlemen, You Are Mad," *The Saturday Review*, 1946; Lewis Mumford, *A Programme for Survival* (Abingdon: Abbey Press, 1946).
34. Richard Wolin, *Heidegger's Children: Hannah Arendt, Karl Lowith, Hans Jonas and Herbert Marcuse* (Princeton: Princeton University Press, 2001).
35. Richard Wolin, *The Frankfurt School Revisited* (London and New York: Routledge, 2006); Anson Rabinbach, *In the Shadow of the Catastrophe: German Intellectuals between Apocalypse and Enlightenment* (Berkeley: California University Press, 2001).

36. Hans Jonas, *Dass Prinzip Verantwortung* (Frankfurt am Main: Insel Verlag, 1979).
37. Lewis Mumford, in *Human Futures*. Published papers from the Rome conference in futures research, Rome, 1970.
38. Fred Polak, *The Image of the Future* (Amsterdam: Elsevier, 1956/1983).
39. Kenneth Boulding, *The Image: Knowledge in Life and Society* (Ann Arbor: University of Michigan Press, 1956). Philippe Fontaine suggests that *The Image* was written as an alternative view to Parson's systems theory, as presented in Parson's 1956 book, *Economy and Society: A Study in the Integration of Economic and Social Theory*. Philippe Fontaine, "Stabilizing American Society: Kenneth Boulding and the Integration of the Social Sciences, 1943–1980," in *Science in Context* 23, no. 2 (2010): 221–265.
40. Robert Jungk, *Brighter than a Thousand Suns* (New York: Harcourt Brace, 1958).
41. Robert Jungk, *Tomorrow Is Already Here* (New York: Simon and Schuster, 1954).
42. Robert Jungk, "The Peace Movements and the Future" (draft, records of the International Confederation for Peace and Disarmament, International Institute for Social History, Amsterdam, 1963–1964).
43. Helmut Ornauer and Johan Galtung, *Images of the World in the Year 2000: A Comparative Ten Nation Study* (Atlantic Highlands: Humanities Press, 1976).
44. Systems theory was a highly complex world of theories. Most studies of it, for instance David Mindell's interesting analysis, focus on systems analysis as the application of an engineer's view on human society. But radical versions of systems analysis, as carried by ecologism, Marxism and the peace movement, were a fundamental influence on the late 1960s and 1970s precisely because they permitted the envisioning of the world as a holistic system built on the interaction between human values, technology and nature. David Mindell, *Between Human and Machine: Feedback, Control and Computing before Cybernetics* (Baltimore: Johns Hopkins University Press, 2002).
45. Johan Galtung, "On the Future of the International System," *Mankind 2000*, 1968, 12–41.
46. Johan Galtung and Robert Jungk, opening address, in, Johan Galtung and Robert Jungk, *Mankind 2000*. Proceedings from the first world conference in future research, Oslo, 1968, 367–378.
47. Herbert Marcuse, *One Dimensional Man: Studies in the Ideology of Advanced Industrial Society* (New York: Beacon Press, 1964).
48. Bart van Steenbergen, "Critical and Establishment Futurology," in Japan Society of Futurology, *Proceedings from the International Future Research Conference*, Kyoto, 1970, 93–103, 96, 97.
49. Interview with Eleonora Masini conducted by Jenny Andersson, January 2012. The Romanian futurist Pavel Apostol's charter for the World Futures Studies Federation first referred to "future studies" in the singular form, presumably because the plural form was inacceptable to Nicolae Ceaușescu, who hosted the first official futures studies conference in Bucharest in 1973. Charter for the World Futures Studies Federation, *Federation Mondiale des Etudes du Futur*, Unesco Archives, 1972/001 A506(498)71. See also Ana-Maria Cătănuş's chapter in this volume.
50. John McHale, "Problems in Social and Cultural Forecasting," in Japan Society of Futurology, *Proceedings of the International Future Research Conference*, Kyoto, 1970, 9–17, 16.
51. Robert Jungk, "The role of Imagination in Future Research," in Japan Society of Futurology, *Proceedings of the International Future Research Conference*, Kyoto, 1970, 1–7, 6.

52. Ibid., 7.
53. De Jouvenel, *L'Art de la conjecture.*
54. Correspondence files, Ossip Flechteim Nachlassung, Deutsche National Bibliothek, Frankfurt; *Futurum*, in particular vol. I and II, 1968.
55. Galtung would launch a series of handbooks in social science methods around this idea.
56. Galtung et al., *Images of the World in the Year 2000*. See Sibylle Duhautois's ongoing PhD project at Sciences Po in Paris.
57. Johan Galtung, "On Future Research and Its Role in the World," In Japan Society of Futurology, *Proceedings of the International Future Research Conference*, Kyoto, 1970, 103–117, 104.
58. Elise Boulding and Kenneth Boulding, *The Future: Images and Processes* (New York and London: Sage, 1995).
59. Galtung, "On Future Research," 107.
60. Grant interview with Johan Galtung, November 15, 1956, Rockefeller Foundation, Rockefeller Archives Center, RF R6 22 GC 1957.
61. Hazan Ozbekhan, "Toward a General Theory of Planning," Organisation of Economic Development and Cooperation [OECD], *Perspectives on Planning*, Paris, 1970, 47–155.
62. Hasan Ozbekhan, *The Predicament of Mankind: Quest for Structured Responses to Growing Wold Wide Complexities and Uncertainties. A Proposal.* Club of Rome, New York, NY, 1970; Ozbekhan, "Toward a General Theory."
63. Hazan Ozbekhan, "The Role of Goals and Planning in the Solution of the World Food Problem," in Johan Galtung and Robert Jungk, *Mankind 2000. Proceedings of the first world conference in future research*, Oslo 1968, 117–150.
64. See for instance the Millennium Project, a global but Washington, DC-based think tank with nodes all over the world offering advice to policy makers and business leaders.

ARCHIVES

Records of the International Sociological Association, Research Committee on the Sociology of Futurology, International Institute for Social History, Amsterdam.

Ossip Flechteim Nachlassung, Deutsche National Bibliothek, Frankfurt.

World Futures Studies Federation, *Federation Mondiale des Etudes du Futur*, Unesco Archives, Paris.

Rockefeller Foundation Grant Files, Johan Galtung, Rockefeller Archives Center.

REFERENCES

Amadae, S. M. *Rationalizing Capitalist Democracy:The Cold War Origins of Rational Choice Liberalism*. Chicago: University of Chicago Press, 2003.

Andersson, Jenny. "The Great Future Debate and the Struggle for the World." *American Historical Review* 117, no. 5 (2012): 1411–1431.

Arendt, Hannah. *The Human Condition*. Chicago: University of Chicago Press, 1958/1998.

Boulding, Elise and Kenneth Boulding. *The Future: Images and Processes*. New York, London: Sage, 1995.

Boulding, Kenneth. *The Image: Knowledge in Life and Society*. Ann Arbor: University of Michigan Press, 1956.
de Jouvenel, Bernard. *L'Art de la conjecture*. Monaco: Editions du Rocher, 1964.
Erickson, Paul, Klein, Judy, Daston, Lorraine, Lemov, Rebecca, Sturm, Thomas, Gordin, Michael. *How Reason Almost Lost Its Mind: The Strange Career of Cold War Rationality*. Chicago: University of Chicago Press, 2013.
Evangelista, Matthew. *Unarmed Forces: The Transnational Movement to End the Cold War*. Ithaca, NY: Cornell University Press, 2001.
Flechtheim, Ossip. *Futurologie*. Berlin: Verlag Wissenschaft und Politik, 1971.
Flechtheim, Ossip. "History and Theodicy, Some Critical Remarks on the Theories of History of Hegel and Marx." *Phylon, Atlanta University Review of Race and Culture*, third quarter (1941): 238–249.
Fontaine, Philippe. "Stabilizing American Society: Kenneth Boulding and the Integration of the Social Sciences, 1943–1980." *Science in Context* 23, no. 2 (2010): 221–265.
Fromm, Erich. *The Sane Society*. New York: Rinehart and Company, 1955.
Galtung, Johan. *On Future Research and Its Role in the World. Japan Society of Futurology*, Proceedings of the International Future Research Conference, Kyoto, Japan, 1971: 103–117.
Galtung, Johan. "On the Future of the International System." In Galtung, Johan and Jungk, Robert. *Mankind 2000*, Proceedings of the first world conference in future research, Oslo, 1968: 12–41.
Galtung, Johan and Robert Jungk. Opening address. In Galtung, Johan and Jungk, Robert. *Mankind 2000*, Proceedings of the first world conference in future research, Oslo, 1968: 367–378.
Ghamari-Tabrizi, Sharon. "Simulating the Unthinkable: Gaming Future War in the 1950s and 1960s." *Social Studies of Science*, 30, no. 2 (2002): 163–223.
Gamari-Tabrizi, Sharon. *The worlds of Herman Kahn. On the intuitive science of nuclear war*, Cambridge MA: Harvard University Press, 2005.
Gordin, Michael D., Helen Tilley, and Gyan Prakash. "Introduction: Utopia and Dystopia beyond Space and Time." In *Utopia, Dystopia: Conditions of Historical Possibility*, edited by Michael D. Gordin, Helen Tilley, and Gyan Prakash, 11–21. Princeton: Princeton University Press, 2010.
Human Futures. Needs, societies, technologies. Based on papers presented at the special world conference in futures research, Rome, 1970.
Irye, Akira. *Cultural Internationalism and World Order*. Baltimore: John Hopkins Press, 1997.
Jacoby, Russell. *Picture Imperfect: Utopian Thought for an Antiutopian Age*. New York: Columbia University Press, 2005.
Jonas, Hans. *Dass Prinzip Verantwortung*. Frankfurt am Main: Insel Verlag, 1979.
Jungk, Robert. *Brighter than a Thousand Suns*. New York: Harcourt Brace, 1958.
Jungk, Robert. "The Peace Movements and the Future." Draft, records of the International Confederation for Peace and Disarmament, International Institute for Social History, Amsterdam, 1963–1964.
Jungk, Robert. "The Role of Imagination in Future Research." In Japan Society of Futurology. *Proceedings of the International Future Research Conference*, 1–7. Kyoto, 1970.
Jungk, Robert. *Tomorrow Is Already Here*. New York: Simon and Schuster, 1954.
Kessler, Mario. *Ossip Flechtheim. Politischer Wissenschaftler und Zukunftsdenker 1909–1998*. Köln: Bohlau Verlag, 2007.
Lemov, Rebecca. "Hypothetical Machines: The Science Fiction Dreams of Cold War Science." *Isis* 101, no. 2 (2010): 401–411.
Livingston, D. "The Study of Science Fiction as Forecasting Methodology." In Japan Society of Futurology. *Proceedings of the International Conference in Future Research*, 71–79. Kyoto, 1970.

Mallard, Gregoire and Lakoff, Andrew. "How Claims to Know the Future Are Used to Understand the Present." In *Social Knowledge in the Making*, Lamont, Michelle, Camic, Charles, Gross, Nick. Chicago: University of Chicago Press, 2012: 339–379.
Mannheim, Karl. *Ideology and Utopia*. An introduction to the sociology of knowledge. New York: Harcourt, Brice and Company,1936.
Marcuse, Herbert. *One Dimensional Man: Studies in the Ideology of Advanced Industrial Society*. New York: Beacon Press, 1964.
McHale, John. "Problems in Social and Cultural Forecasting." In Japan Society of Futurology. *Proceedings of the International Future Research Conference*, 9–17. Kyoto, 1970.
Mindell, David. *Between Human and Machine: Feedback, Control and Computing before Cybernetics*. Baltimore: Johns Hopkins University Press, 2002.
Mirowski, Phillip. *Machine Dreams: How Economics Became a Cyborg Science*. Cambridge, MA: Cambridge University Press, 2002.
Moyn, Samuel. *The Last Utopia: Human Rights in History*. Cambridge, MA: Harvard University Press, 2010.
Mumford, Lewis. *The Condition of Man*. New York: Harvest Books, 1944/1956.
Mumford, Lewis. "Gentlemen, You Are Mad." *The Saturday Review*, vol. 1946: 5–6.
Mumford, Lewis. *The Myth of the Machine: Technics and Human Development*. New York: Harvest Books, 1967.
Mumford, Lewis. *The Pentagon of Power*. New York: Harvest Books, 1976.
Mumford, Lewis. *A Programme for Survival*. London: Secker and Warburg, 1946.
Ornauer, Helmut and Galtung, Johan. *Images of the World in the Year 2000: A Comparative Ten Nation Study*. Atlantic Highlands: Humanities Press, 1976.
Ozbekhan, Hasan. "The Role of Goals and Planning in the Solution of the World Food Problem." In Galtung, Johan and Jungk, Robert. *Mankind 2000*, Proceedings of the first world conference in future research, Oslo, 1968: 117–150.
Ozbekhan, Hazan. "Toward a General Theory of Planning." Organisation of Economic Cooperation and Development [OECD], *Perspectives on Planning*, 47–155. Paris: OECD, 1968.
Ozbekhan, Hazan and the Club of Rome. *The Predicament of Mankind: A Quest for Structured Responses to Growing World Wide Complexities and Uncertainties. A Proposal*. 1970.
Polak, Fred. *The Image of the Future*. Amsterdam: Elsevier, 1956/1983.
Rabinbach, Anson. *In the Shadow of the Catastrophe: German Intellectuals between Apocalypse and Enlightenment*. Berkeley: California University Press, 2001.
Ross, Dorothy. "The Changing Contours of the Social Science Disciplines." In *The Cambridge History of Science. Vol. 7: The Modern Social Sciences*, edited by Theodore M. Porter and Dorothy Ross. Cambridge, MA: Cambridge University Press, 2008.
Solovey, Mark. "Cold War Social Science? Specter, Reality or Useful Concept?" In Solovey, Mark and Craven, Hunter, *Cold War Social Science. Liberal democracy and human nature*, Basingstoke, UK and New York: Palgrave, 2012:1–25.
Steenbergen, Bart van. "Critical and Establishment Futurology." In Japan Society of Futurology, *Proceedings from the International Future Research Conference*, 93–103. Kyoto, 1970.
Tolon, Kaya. "Futures Studies: A New Science Rooted in Cold War Strategic Thinking." In *Cold War Social Science, Knowledge Production, Liberal Democracy and Human Nature*, edited by Mark Solovey and Hunter Craven. Basingstoke, UK and New York: Palgrave, 2012.
Wolin, Richard. *The Frankfurt School Revisited*. London and New York: Routledge, 2006.
Wolin, Richard. *Heidegger's Children: Hannah Arendt, Karl Lowith, Hans Jonas and Herbert Marcuse*. Princeton: Princeton University Press, 2001.

2 Expertise for the Future

The Emergence of Environmental Prediction c. 1920–1970

Paul Warde and Sverker Sörlin[1]

THE ENVIRONMENTAL TIPPING POINT

In 1948, the president of the American Association for the Advancement of Science, Edmund Sinnott, declared, "Man's command over nature has grown more rapidly than his mastery of himself. Man, not nature, is the great problem today."[2] Yet if "man," in Sinnott's view, was "the problem," this was not because he had identified some new characteristic or behaviour displayed by the species. What was new was not the message about the inner nature of man: but about nature, planet, and in a new way: *the environment*.

The year 1948 had seen the publication of two books, William Vogt's *Road to Survival* and Fairfield Osborn's *Our Plundered Planet* that heralded what was, within a decade, to become a literary genre that remains buoyant and marketable today: *concerned reports on the global condition*. "The world is sick," wrote William Vogt in the foreword that he penned in August 1948 for the British edition. As for any patient the next question must be: what is the prognosis?[3]

The planet had just survived a world war and humanity was busy erecting a new framework of global institutions to manage relations between nations, the financial order, development, and culture. As revealed in the work of Thomas Robertson, Osborn and Vogt had begun to formulate their works during the war itself and in part as a direct response to the carnage, events that set the nature of "man" in a rather dim light.[4] It is perhaps not so surprising that Vogt and Osborn thought in terms of the whole Earth (although "planet" remains striking as evoking the singular nature of our world in the solar system), which itself was, of course, not new. It is more worthy of remark that the survivors now warned that their very survival was at stake, not just for a particular nation, society, or location, but for humanity: that the future was a place of ever-greater risk. Both Osborn and Vogt were Americans trained in ecology and had strong conservationist leanings. But they departed from the classical conservationist paradigm which was national or local and built instead on reports from many corners of the world, including work Vogt had carried out on missions in Latin America.[5]

In Vogt we find something else new, certainly in a book that would be read by many millions: an idea now so entrenched that it is a surprise to realize that it emerged so recently: the modern usage of the idea of *the* environment. Not an environment of any particular place, but expressed as a global singularity, with its all-encompassing definite article, a catch-all through which human destiny could be expressed and mapped; something that, like us, had a fate. "Environment" itself was not a new word by any means, as we will see. But Vogt linked "*the* environment" not just to the local and its influence on the individual organism, but to the global. On a broad canvas he painted the world as an increasingly unified place, starting with the age of exploration and culminating in the modern integrated understanding: "that we live in one world in an ecological—an environmental—sense."[6] This was, as Osborn framed it, a space of shared risks and fate, as the rising numbers of human newborns became "a living part of the environment into which each of them has come. Its strength will be theirs, and its weaknesses."[7]

It is important that the idea of *the environment*, and the prediction of a whole global system falling into degradation, emerged together: co-determined, or bound in a kind of double-helix that it is an important challenge to unravel. We would propose that "the environment" bore an orientation towards the future in it from its beginnings, just as, from the same time (if not before), one could hardly say "nature" without evoking a certain nostalgia.

We also propose that "the environment" emerged historically as a crisis concept, closely wedded to prediction, and fears of the future. In turn, these predictions were also based on *techniques*, ways of gathering, processing, and visualizing data. The concept of the environment became a tool by which many things that, in retrospect, appear obviously and unquestionably linked, became subject to attempts to develop a new integrated understanding. In other words, their connection was not self-evident at all, and this integration required work. What we will seek to do in this chapter is suggest why this future-orientated environmental moment coalesced in the years around 1950, and examine some of the numerous strands that became entwined in its creation. As befits an all-embracing and global idea, these are very numerous and can only be glossed here, very much with the intention to prompt discussion and further investigation. These strands had long histories of varying lengths, but the pace of change, scale of alarm, and possibilities for integration seem to have accelerated from another post-war decade, the 1920s. In turn, we would like to speculate a little about the aftermath of the integrative act, too: how the new paradigm was developed and operationalised in new stories, techniques, and disciplines over the post-war decades.

The new environmental outlook of 1948 was first and foremost a work of integrative imagination, of combining a set of already existing issues and problems on various scales into a meaningful whole, which was at the same time harbouring a set of serious concerns that became visible with and through the new concept "the environment." Vogt and Osborn established a proto-version of what we could call the *modern environmental problem catalogue*. It included, but was not limited to: population growth (by far the number one issue at the time), water scarcity, soil erosion, overconsumption, overgrazing, overfishing, resource depletion, pests, industrial wastes, the retarding productivity of soils, and species loss. Only later would they be joined by "climate change," although climate had played a greater or lesser role in discourses about the nature and fate of civilizations since Antiquity. Indeed, part of the reason that climate change could rapidly become a problem conceptualized as belonging to the whole world was the groundwork in preparing the idea of "the environment" even before computer climate simulations raised alarm about greenhouse gas emissions among scientists in the 1970s.[8]

At this point, we note a significant shift in the meaning of the word "environment." This concept had been popularized by Herbert Spencer in the middle of the nineteenth century, and previously had denoted the sum-total of factors and properties *working on* an organism or a set of organisms, often to which they adapted in evolutionary narratives.[9] There were of course also notions, especially in ecology, of the mutual interaction of organisms and their environment, but these were more usually characterized as communities, a biocenosis, or an equivalent term.[10] More commonly, in both ecology and geography "environment" denoted a cause and explanation of differences rather than an object of change—and prediction—itself. Now, in the 1940s, it came to stand for a limited (albeit large) number of features of nature that were subject to change through human action, in a knowable way: their change was quantifiable and measurable. After the emergence of "the environmental" as a global and systemic reality, every subsequent problem that could be related to the catalogue—pesticides, climate—found itself integrated into this broader story of planetary dynamics, an inextricably connected "web," although sometimes with change attributed to an ultimate single cause: "the causal chain of deterioration is easily followed to its source. Too many cars, too many factories, too much detergent, too much pesticide . . . too little water, too much carbon dioxide—all can be traced easily to *too many people*."[11]

The integration of "the environment" brought a new quality to understanding of the Earth, of humanity as a species, and of their inextricably bound futures. It was an act of imagination, but one that required and prompted new metrics and literary and visual genres of its own. Despite being closely associated with environmental data, however, these new genres retained their storytelling quality: Vogt's work provided the model for many more jeremiads on the state of the environment. Vogt began his work with fictionalised vignettes of scattered individuals facing up to "environmental resistance" in a globalised world; and indeed much of *Road to Survival* has the quality of an impassioned narrative, with little scientific data presented. Fourteen years later, Rachel Carson would also strikingly and famously employ the literary mode, in a fictionalised vision of the damage caused by chemical pollution in the famous preface to *Silent Spring*, although the subsequent work provides detailed discussions of scientific results from the field.[12] She had used sophisticated literary techniques in her earlier non-fiction books on maritime environments as well.

Yet it was also the case that after the 1940s, no one would take seriously a work of predictive (or "science" fiction) simply because of the qualities of literary imagination displayed, as had been the case in the past. By 1970, when Loren Eiseley remarked that the future had become "our primary obsession," it was scientists who had become "soothsayers," expected to deal with "questions involving the destiny of man over prospective millions of years."[13] In other words, "the environment" also evoked a new kind of expertise, as well as a reconfiguration of some old ones: experts who had the extraordinary and polymathic task of having the skills to predict the future of the interaction of nearly everything, and the capacity to articulate this through a range of repertoires.

The cornerstone of this approach was to turn nature into numbers, the changes in which were assumed to have an almost inevitable direction towards decline. This production of knowledge simultaneously generated and mandated a specific kind of predictive environmental expertise in the collation and interpretation of environmental information. In turn, this was instrumental in the forging of the environment as an integral or "world" problem. The problem was essentially one of expectation, or prognostication, which is why the emergence of predictive environmental expertise was so emblematically a part of the general orientation of attitudes towards futures. Whereas the declensionist narratives of ongoing degradation favoured in environmental thought had a long and deep past, what was new here were the claims by scientists that the *rates of change* were not only knowable but also predictable, and indeed, collectively so for many, many different things. This differentiated this kind of work from earlier phases of declensionism associated with, for example, the writings of George Perkins Marsh, or the analysis of changing landscapes and climates in specific regions such as the Russian steppes.[14] The possession of numbers became an integral part of a futures narrative, and one's plausibility in pronouncing it.

But, importantly, it was not obvious how entwining of futures and environment was supposed to be interpreted. The potential of the story was directed in two quite different directions. One pointed to the soon emerging computer-based large-scale predictions associated with very large programs of data assembly such as the International Biological Program. This ran parallel to a certain strand of scenario-building, sometimes involving the military-funded super science of applied mathematics. Another absorbed the creative and synthetic energies of thinkers and philosophers that were deeply concerned with the "fate" of "the planet" and "humanity." It is against this backdrop that we ask how and why "the environment" arrived at this central position in the futures narrative.

ACCOUNTING THE CONDITIONS OF THE EARTH

The approaches that coalesced around post-war environmental thought drew on a wide range of disciplines: demography, geography, resource management and economics, ecology, and soil science. The practitioners of these disciplines were not yet "environmental" experts, but they employed methods and concepts that would turn out to be apposite for the kind of integrative thinking that the new post-war understanding of the environment demanded. Importantly, they would offer various roads to understanding why "the environment" became so profoundly attached to the future. Scientists in many disciplines, but often with a mathematical or at least strongly quantitative orientation as a common denominator, offered a route through which later environmental thinkers could claim expertise. Ironically, this inheritance would itself close down the theoretically limitless elements that might be included in the environment to those that could be measured and analysed by a limited number of widely accepted techniques. There was, undoubtedly, a supply-side element to attempts at integrative thinking, in that the emergence of time series and statistical datasets had from quite an early date prompted investigations into their relationships, such as W. S. Jevons's attempt to link crop yields and thus economic performance to sunspot cycles in the 1880s (later expanded by his son H. S. Jevons in 1910), or immediately after the First World War, William Beveridge's use of series of barometric pressure readings as an explanation (and hence predictor) of trade cycles, which brought a rapid rebuff from the head of the Meteorological Office in the UK.[15]

There was plenty of expertise around in the first half of the twentieth century, and plenty of environmental thinking, but it was more often retrospective rather than predictive, and in its particular forms embraced only parts of what became the problem catalogue of post-war thought. Those areas and methods that would prove most influential were approaches which could easily be *scaled*—that is, work at the local and global levels, and indeed allow some sense of integration between the two. Through this route expertise in one, often very localised field, in the sense both of the a limited set of phenomena being studied, and an area of study being a limited space on the surface of the earth, could open up avenues to analysing global-scale phenomena. This would particularly be the case with formalised "systems" approaches, that were influential in, for example, studies of animal population dynamics.[16] We will now trace some of these in a little more detail.

One fundamental discipline was demography, and especially among those following the gloomy prognostications of Thomas Malthus. Theirs was an argument from design, because according to Malthus, the power of population would overwhelm resources irrespective of how productive the latter became. This brute generative power became a crucial part of the Darwinian argument for natural selection, and would underpin varieties of evolutionary thought and the expectation and fear that "nature" was constantly at risk of being thrown out of balance. Malthusians were certainly interested in empirical data, but found themselves having to constantly refine their predictions in the light of new information (as did Malthus himself). After the First World War, the issue of population emerged in a new form, conceived as possibly the first truly global issue, inasmuch as the population growth that put a pressure on agricultural capacities happened virtually all over the world and was recognized as needing concerted action. This was the idea behind the first world population conference held in Geneva in 1927, which assembled population statisticians, economists, and other experts from many countries, including notables such as J. M. Keynes.[17] The focus of the agenda was basically Malthus writ large: would the lands suitable for agricultural production suffice to feed the growing number of people? The issue was attacked with a combination of calculations of land size, agricultural productivity, and demographic data. Predictions of population change turned essentially into a case of applied human ecology, dealt with by a set of newly installed experts.[18] The carrying capacities of the world's agricultural producers were contrasted with empirical predictions of population growth and emerging understanding of the demographic transition.[19]

The significance of population studies to the emerging category of the environment is how the *conditions of the earth* are reinterpreted as facts of significance in a *narrative of humanity's relation to the earth and presented as numbers*. As census data improved, the debate increasingly employed *actual* numbers, such as the Australian work that led to George Knibbs's *Mathematical Theory of Population* (1917), and after the Geneva conference, his *The Shadow of the World's Future or the Earth's Population Possibilities and the Consequences of the Present Rate of Increase of the Earth's Inhabitants* (1928).[20] But there were other tools of this trade that could be applied where data was lacking: a logistic curve and a sense of where a population might be on it would do just as well.[21] Frequently, leading figures in interwar population theorizing had spent time in a range of disciplines that later came together as "environmental"; one such example is Alexander Carr-Saunders, author of *The Population Problem* of 1922, who trained as a zoologist under leading statistician Karl Pearson.[22] Other significant figures were population biologists Raymond Pearl, the great booster for the logistic curve, or Edward Murray East, men who had also been given administrative responsibility in securing food supplies during World War I and its aftermath.[23]

A shared set of techniques and axioms about population behaviour and its "environment" allowed, it seemed, easy translation across disciplines via mathematical techniques; Alfred J. Lotka for example, exported mathematical methods and statistics from age distribution to epidemiology to bibliometrics.[24] Lotka, in the typical fashion that eroded the divide between natural and social systems, studied both predator–prey behaviour and the behaviour of scientists (e.g., Lotka, "The Frequency Distribution of Scientific Productivity," 1926), [25] paving the way for the more comprehensive algorithms of ecology that the brothers Eugene and Howard T. Odum launched after the war and that would become a standard textbook approach drummed into students for decades afterwards.[26] A certain kind of environmental numbers were thus created and drawn into the social realm.

The immediate post-World War II period accelerated these trends. Census data was now much more widely available, especially from the developing world. Population was found to be rising rapidly (some countries experiencing increases in rates of life expectancy of over one year per year!), even in the US.[27] Ideas that could trace their origin back to Malthus's striking formulations, but that had become "social facts"[28] belonging to a widely accepted measurable problem in the 1920s, now became a central aspect to the environmental problem catalogue, even as a more sophisticated understanding of demographic transition and a more differentiated view of population questioned the simple pieties of earlier predictions.[29] Simultaneously, the post-war era saw a major preoccupation with the other side of the equation from population dynamics, that of raw materials (including food). This too was in part a direct consequence of the challenge of global logistics during the war. Attempts to predict the longevity of such resources have a history stretching back to early modern times, but they acquired new form with the development of mathematical and geometrical techniques for assessing rates of consumption and depletion. Concerns about resource availability became a kind of race between rates of consumption, replacement, and technological change on a global scale.[30] As resource use increased at faster rates with industrialization, economic geography became dominated by the distribution of resources and their availability as a predictor of the level of economic prowess, or even "civilization." The idea was a widespread trope propagated not least as a deterministic relation between the availability of cheap energy and material and cultural "progress."[31]

In the first decades of the twentieth century, the prospect of resource exhaustion had largely been considered in terms of individual resources. The logistical effort of World War Two helped broaden horizons to consider pressure on resources as a general rather than a particular phenomenon. The British think tank Political and Economic Planning was explicit in trying to counter "a sense of frustration with the fragmentary, specialized or distorted views beyond which even well-informed individuals find it difficult to proceed when tackling so huge and complex a problem."[32] In the immediate post-war years, there were a flurry of surveys and attempts to predict the longevity of resource use, notably the President's Commission on Material Resources (also known as the Paley Commission) that published *Resources for Freedom* in 1952; the left-leaning report *World Population and Resources* in the UK; and efforts such as Palmer Putnam's *Energy in the Future* (1953, at the behest of the Atomic Energy Commission), Samuel Ordway's *Resources and the American Dream* (1953), Harrison Brown's *The Challenge of Man's Future* (1954) and Georg Borgström's *The Hungry Planet* (1953).[33] These arrayed for the first time the full panoply of pressures, with a particular concern for metals and water, extending the traditional nexus of population and food resources, and the particular worries that had applied to timber and coal. The inclusion of metals was significant because they called into question the possibility of resolving long-term resource pressures by simply increasing reliance on the technological wonders of modern industrial society. The Paley Commission launched the Resources for the Future institute that would make major subsequent contributions to resource economics and produce large and systematic resource audits and future projections by the 1960s.[34]

At around the same time the experience of individual American oilfields and high levels of petroleum demand during the war lead to attempts to more widely inventory oil reserves and predict future rates of consumption among petroleum geologists. Leading figures among these were C. L. Weeks, Eugene Ayres, and most famously M. King Hubbert. The latter was influenced by the population biology of Pearl and Carr-Saunders, and in a series of famous articles from 1949 onwards came to formulate the model of "peak oil," a concept that originated at this time and was referred to as "peak petroleum." Some analysts predicted that this moment would come for America even before 1960.[35] The post-war period also brought to the fore evidence that consumption of resources in industrial societies generally was advancing far more rapidly than population. The inferred implication was that even with improvements in efficiency, global development would follow much the same path; in other words, the limited data on a few countries, above all the US, was treated as the basis for a universal theory about the future.[36]

As we have already noted, the integration presented by Vogt and Osborn was strongly steered by their training in ecology. In his famous paper of 1935, Arthur Tansley defined the "ecosystem" and showed how it could be used both in natural communities and in those modified by human activities, such as grazed landscapes, in which plants are modified by animals and the relationship between the two is closely monitored by humans. Tansley wrote of "the whole complex of physical factors forming what we call the environment of the biome—the habitat factors in the widest sense,"[37] which is in a somewhat traditional sense where the "environment" plays the role of conditioning natural selection. In the earlier *Animal Ecology* (1927), Charles Elton, a student of Carr-Saunders, focused on *food chains* and the *food cycle* as part of the preconditions for life itself that demanded the study of an entire interlinked system of relationships to be able to understand the fortunes of any one organism.[38] Elton's definition of *niche*, or the place of the animal in its community, was also important in understanding the way the community functions as a whole; the progress of life could only be understood relationally. It was in the study of animal populations and rangelands that the ecological idea of "carrying capacity" emerged, an idea central to Vogt's writing, and that he would have heard used regularly in exchanges with his friend, tutor Aldo Leopold.[39] Leopold, like Tansley, frequently employed the term "environment" as a way of integrating the factors affecting a particular animal population in a particular place: "every environment carries not only characteristic kinds of animals, but characteristic *numbers* of each. . . ."[40] In turn, for some ecologists it became almost a habitual tick to attribute the political disasters of the 1930s and 1940s to overpopulation.[41] Modern man had become like the famous benighted and starving deer of the Kaibab Plateau, and further industrialisation could only accentuate this process: as Leopold argued in 1944, "the impending industrialization of the world means that any conservation problems heretofore local will shortly become global."[42]

Influential as these strands of ecology, zoology, and indeed ornithology (Vogt's specialism) were, there were other strands in thinking about the conditions of the Earth with longer histories that also accelerated in influence from the 1920s to the 1940s. We can see increasing interest in the unified Earth as a relevant biophysical category, such as Vladimir Vernadskii's 1926 book *Biosfera*, soon (1929) translated into French, but available in English only in 1998.[43] In 1948, Vernadskii's ideas on the biosphere (the zone around the Earth that supported life) entered wider English-language circulation via Yale ecologist Evelyn Hutchinson. Modern man was eroding his own "survival" by wasting "parts of the *biosphere* which provide the things that Homo sapiens as a mammal and as an educable social organism needs or thinks he needs. The process is continuously increasing in intensity as population expands."[44] Of particular influence would be the use of energy budgets, revived in the 1920s as a way of quantifying the physical in what came to be called the ecosystem, again with the pervasively creative Lotka in a pivotal role.[45] Indeed, to a certain extent this brought intellectual development full circle, because the development of equilibrating ideas in ecology by which nature or local ecosystems were supposed to settle into a steady state which humans then "disturbed" itself derived by analogy from an interest in heat budgets and energy flows in Royal Chapman's 1928 paper, "The Quantitative Analysis of Environmental Factors," and in the case of Arthur Tansley's "ecosystem," from an attraction to Freud's ideas of equilibrating functions in the mind that drew him towards the notion that similar processes might exist in local ecologies that tended towards a settled state.[46]

Lotka was also, as we recall, a significant theorist in population dynamics. Conceiving of ecological processes as multi-scalar energy flows would become pervasive both within ecology and beyond, with the publication and long life of the Odums' 1953 textbook, *Fundamentals of Ecology*, which wedded such thinking to a symbolic notation borrowed from electrical engineering. Thus developments across population theory, economics, and ecology (among other disciplines) opened a way by which the very old notion of the "balance of nature" could be reformulated in terms of dynamics and measureable flows to be modelled within an Earth system; an early seed of fully fledged "global change" thinking that only emerged more than a half century after Lotka's equations.

Of course, we must acknowledge that other routes had also been leading to "thinking as a globe," as we might say, for a long time. Geophysical and biological data had been accumulating: Alexander von Humboldt's iso-lines, Heinrich Wilhelm Dove's global isotherms from 1853, the international geodetic services in the middle decades of the nineteenth century.[47] Oceanography had a similar need to assemble worldwide datasets.[48] These were some early appearances, but normal, systematic, everyday *global* data collection about the *entire* earth's physical conditions and how they change is by and large a twentieth century phenomenon. The point here is not to privilege any one discipline, in the way that ecology did once truly seem as the environmental discipline par preference.[49] Rather, we claim that in the period from the 1920s on, environmental phenomena across their widest possible spectrum, especially through depictions of their alarmingly rapid rates of change, were turned into social facts. At the same time, they were unified as facts, by becoming environmental, by their link to a common denominator labelled the environment. But we also make the observation, perhaps even more fundamental, that what emerged in this crucial phase during the interwar decades was a transformation from the gathering of facts about the state of the earth, its plants, animals, and atmosphere to a mathematical understanding of its dynamics and directions. This had not yet turned into the futures narrative of the environment, but in essence we can see most of these proto-performers of predictive environmental expertise busily at work on their equations across the whole ensemble of disciplines, forging *avant la lettre* the new understanding that would so drastically coalesce after the war.

FROM "MAN'S ROLE IN CHANGING THE FACE OF THE EARTH" TO "FUTURE ENVIRONMENTS OF NORTH AMERICA"

In the middle of June 1955, 73 researchers gathered at the Princeton Inn, New Jersey, for the symposium Man's Role in Changing the Face of the Earth. The conference was the brainchild of William L. Thomas, director of the Wenner-Gren Foundation on Anthropological Research that provided the funding. He called in three men to preside over the mix of academics and figures from policymaking and industry: Americans geographer Carl Sauer, entomologist Marston Bates, and polymathic historian, philosopher, and planning theorist Lewis Mumford. Of the 73 participants in this early attempt to convene global environmental expertise there was only one woman, the Indian plant geneticist Edavaleth Jamaki Ammal. Forty percent of the attendees came from the earth sciences, twenty-eight percent from the biological sciences, twelve percent the social sciences and humanities, and twenty percent from applied fields such as planning. The nineteenth century

American conservationist and author of *Man and Nature* (1864), George Perkins Marsh, was evoked as a kind of patron saint of the proceedings, but seven years after Vogt and Osborn had been published, this was something more novel: an attempt to aggregate the necessary expertise to comprehend the new global challenge.[50]

Marsh's work provided a lodestar as one of the earliest works to offer a generalized critique of how human behaviour led to damage in the environment (a term he did not use); a "declensionist" narrative that provided a trajectory if "man" did not amend his ways. His major preoccupation since his youth had been damage to the soil, and here he stood in line with a number of nineteenth century works that explicitly linked the fate of civilizations to their management of the soil.[51] In the 1920s, soil quality and erosion became issues of tremendous importance in debates about potential regional food supplies, often directly related to population growth.[52] The American Dust Bowl erosion disaster in the 1930s made it an issue of world interest, but every continent could provide examples;[53] classifying what he perceived as rapidly deteriorating soils in South America was one of the jobs undertaken by Vogt. Soil science was an early integrative discipline that drew in forestry, hydrology, ecology, chemistry, and meteorology. As a discipline, it also highlighted problematic time horizons: once the erosion was observed, the forces that had led to it had already long been in place and were possibly irreversible. The time horizon of "renewable" became much longer; one could not make judgements about environmental processes from observing them over short periods of time, and actions now might have consequences that became manifest only far into the future.[54]

The anxious and animated gathering of experts at the Princeton Inn in 1955 was not, as yet, very "environmental." People still thought of themselves as "conservationists" rather than "environmentalists." What was more important was a new sociability of experts that defined the emerging field; predominately natural scientists, with a smattering of social scientists and humanists to explain what in human culture was driving what we now, in retrospect, can see as the "great acceleration" of human influence on the planet. Even if integrative techniques were not yet widely on display, the supposition was that integration was necessary and possible. In fact, it was an exercise in imaginative connection: "we did not want it to run into shoals of statistics," commented Thomas.[55] Most of the presentations looked *historically* at human impacts, suggesting that we were on a collective trajectory, but one that led to destabilisation and crisis. At the same time, Charles Elton was moving his work beyond the functions of individual ecosystems to considering their interaction on global scale, writing in *The Ecology of Invasions*, "we are living in a period of the world's history when the mingling of thousands of kinds of organisms from different parts of the world is setting up terrific dislocations in nature."[56] The fundamental problem catalogue of the environment could be scaled from the smallest niche to the planetary. In the published conference proceedings that appeared the following year, the "prospect" section was short, taking up only 171 pages out of a total of 1,152.[57] Nevertheless, the implication was clear: all of the participants had a shared interest in how their work related to the future.

For knowledge to become a social fact, it must be accepted as such by others, and this rests on matters of trust and procedure that are as much to do with the society which harbours and cherishes these facts as the raw force of the data or observations. How then was "environmental" knowledge made? One simple method employed by doomsayers of the post-war era was to put the statistics alongside each other: if the graphs all pointed upwards, we are going to be in trouble (or, conversely, if resource stocks all went downwards, then we were also in trouble). Much as aggregated local statistics about "conditions of society" (from the nineteenth century onward) "invented" society (often understood as a national entity), measurable and quantitative "conditions of the earth" invented "environment" as a global phenomenon. Gathering numbers and statistics became crucial, and the same "trust in numbers" that Theodore Porter demonstrated was necessary to build credibility around engineering and social projects in the nineteenth century was invoked in building the environment as a global issue.[58] This is, in essence, how expertise and environment are co-produced: as a series of efforts to establish predictive authority based on aggregating local data sets to suggest a global condition undergoing dangerous, quantifiable change.

One response was to see an increasing tendency for expertise in one "environmental" field mandating one's "expertise for the future"—an authority to make apparently informed predictions—across the whole of the "environmental problem catalogue" that had become integrated into a set of mutually interacting dangers. This did not mean that the core work in local disciplines was abandoned, of course, and the challenge remained to provide meaningful and widely accepted tools of integration. Much of the early post-war integration of "the environment" remained rhetorical, as we have seen—an imaginative act, and perhaps to a lesser degree a research project rather than a distinct method. The language of Vogt and his peers was frequently forthright, and medicalised metaphors, including viewing the causes of environmental problems as "cancers," were widespread in the genre.[59] Indeed, the uptake of the very term "environment" might be political in the sense of a negotiation of competing interests. In Britain, Solly Zuckermann, a scientific advisor tasked in 1959 by the new Ministry of Science with reviewing possibilities for a new research council to assist in managing what were called "natural resources," was impressed by the coming together of sciences in the International Geophysical Year of 1957. He searched for a way of collectively naming them and came up with "environmental sciences." By 1962 this was in print, and in 1964 Britain had a "Natural Environment Research Council"; the use of "Environment" was a compromise among what we now call the environmental sciences, because none of them would accept the more familiar alternatives suggested by others![60] In the US, the policy impact of the idea was framed above all by the intervention of political scientist and planner Lynton Caldwell, whose article "Environment: A New Focus for Public Policy?" set in train a policy agenda that would be mimicked in European countries, with agencies established Sweden in 1967, and the US and Britain in 1970.[61]

Although in practice no one could reasonably claim deep knowledge of the full range of fields that became part of the environmental sciences, the conviction that one could contribute to such a vast work of integration, along with a certain commonality of technique, made it possible for butterfly experts to proclaim on demography, or for ornithologists and game managers to rail about the state of humanity. This tendency for expertise in one field to mandate expertise for the future across the whole of the environmental problem catalogue had already, as we have seen, emerged among the statisticians, zoologists, and fruit fly experts populating the population and eugenics debates of the 1920s. But this also raised the issue of *validation*: many of these environmental inter-relationships were not in themselves testable or measurable; as with many models, one could establish the inputs, and measure the outputs, but only infer as to how the two were connected with each other.[62] "The environment" is not easily warranted by particular experts; it is an *integrative project* in its own right, which gives an unusually wide range of specialists the opportunity to claim that they can reasonably contribute, and the very precondition of their contribution is the conceptual unity provided by the word itself.

We want to suggest that "the environment" as a research project (in varying forms) depended on a set of integrative acts and techniques that required a new kind of specialist, the meta-specialist, who could speak for the unfolding future. These specialists took two fundamental forms. First, there were individuals who had authority in the institutions and productions of their own discipline, as technicians and analysts, but they also transcended such disciplinary boundaries to interact with others, draw data from other fields into comparison with their core field of work, and in many cases, wrote strikingly for a wider public or started appearing in broadcast media. In the environmental arena, they often emerged from one of the key forerunner disciplines that we have identified, and on the basis of such authority, pronounced as polymathic prophets on a very wide range of issues—as did Paul Sears, William Vogt, Lewis Mumford, Georg Borgström, Kenneth Boulding and, slightly later, Rachel Carson, Bernhard Grzimek, Paul Ehrlich, and Barry Commoner, to name just some of the most influential.[63] These individuals, among others, played major roles in defining environmental problems and a "*world problematique*" during the 1950s and 1960s, culminating in a wave of legislative and popular political action that explicitly evoked the vulnerability of the environment in a range of countries around 1970.[64]

The second kind of meta-specialist presented a more profound and perhaps enduring shift in patterns of research and the generation of social facts: the emergence of researchers and research specialisms that made *the environment* the direct object of their study, and hence shifted the ground of what made a contribution to environmental understanding from a more synthetic achievement, drawing on different disciplines, to a constitutive one. This step would partly be made possible by the analytical possibilities offered by computers, which created the opportunity of *simulating* an environment or climate even when data remained far too scarce to build a rigorously empirical view of global dynamics, as well as the demand for new research that the environmental crisis itself generated. Examples of this new kind of expert emerged early in climate science, although climate was not considered a crisis variable in the post-war years, rather as an economic and military issue.[65].The paradigmatic case of this emerging expertise is found in the attempt to model and predict global dynamics in the *Limits to Growth* report of 1972, produced by a group based at the Massachusetts Institute of Technology (MIT) using the "World3" model. Although it was immediately controversial, it pointed the way to the future.[66]

Ten years after the Man's Role conference, America's Conservation Foundation (which boasted Fairfield Osborn as President) held a gathering in Warrenton, Virginia, deliberately imitating its predecessor.[67] Many of the same people were there, and there was still only one woman. This was an event where there were, as its chair, the British ecologist Frank Frasier Darling observed, "a lot of ideas about the future." They came from participants such as William Vogt, Kenneth Boulding, Raymond Dasmann, Lewis Mumford, Max Nicholson, M. King Hubbert and Lynton Caldwell. Most strikingly of all, the conference was entitled *Future Environments of North America*, and the contributions were full of comments on "public policy," "environmental impacts," or "weather modification." In contrast to the previous decade, "the environment" was everywhere in discourse, but bound with a tight leash to human needs and government action. There could be wide assent for the aim of Pierre Dansereau: "a valid imaginary reconstruction of our world is now our greatest task. It may even be the condition of our survival."[68] If the aggregated expertise at Princeton had, almost accidentally, emerged as a "rich resource of coordinated knowledge and quickened thinking," the *Future Environments of America* aimed to be a "gathering of skills [to] ameliorate, prune, amplify and possibly validate the thoughts of each," as Darling put it, where a common thread to all the papers was "management."[69] Equally, that same year, the American Association for the Advancement of Science's Committee on Science in the Promotion of Human Welfare noted, "The entire planet can now serve as a scientific laboratory."[70]

CONCLUDING REMARKS: CRISIS AND "THE ROLE OF MAN"

Of course, all this environmental change was being caused by something, and that thing was writ large in the 1955 Princeton conference: "man." A kind of paradox lay at the heart of the new environmental paradigm, related to the two kinds of expertise we have identified above. One the one hand, environmental crisis was identified and legitimated through quantification and the authority of scientific metrics. On the other, the root cause of the crisis that emerged so soon after catastrophic global war was identified by many of the most prominent environmental thinkers as a malaise in "man," an inner crisis that was both the result of the deep history of human evolution, and the instrumentalising ethos of modern industrial society (see also Jenny Andersson in this volume).[71] Indeed, in the years we have covered, it was commonplace for all kinds of social ills to be connected to the environmental malaise.[72]

What constituted "man" in a global sense? And what kind of expertise might usefully prise open that black box? Vogt opened *Road to Survival* with his individualised tales of people on their own roads: a peasant woman in Mexico, a recently elected Labour MP, a commercial logger, a boatload of refugees running the gauntlet of border guards, a research biologist. And yet, whereas their experience is diverse, their destiny and attitudes are not. Vogt bewails "the ignorance that runs through the entire social structure of the world," as much composed of "the countryman's limited comprehension" as the "urban man's unawareness." In this dismal parade of "idiotic" ideas, it is only the biologist who "sees his role more clearly."[73] The scientist, then, but above all the meta-specialist, is a kind of soothsayer who sees where humanity is headed as a population. It was certainly not Vogt who first ruthlessly employed the trope that he was reason personified, whereas those who opposed his views were "thoughtless and selfish . . . apt to stigmatise all reasoning about the future as absurd and chimerical." The latter words come from W. S. Jevons's *The Coal Question* of 1865, and are themselves an echo of Malthus.[74] And yet, whereas Vogt indicated a few particularly egregious examples of the human thinking that had gotten the planet into trouble, such as capitalism or Aristotle, extraordinarily wrongheaded thinking was for him a shared property of the whole of humanity; it was inexplicable, aside from a generalised avarice, weakness for advertising, and "'love'" stories and "'comics'"(cultural forms so debased that even Vogt had to put them in inverted commas) and ignorance. In short, the cause of the global trouble is not susceptible to any detailed analysis itself.[75]

With technical roots in electrical engineering (in its systems form) and population ecology, it is perhaps not surprising that much environmental thinking could conceive of the crisis as a kind of engineering problem, an idea that looked forward to Buckminster Fuller's metaphor of the *Operating Manual for Spaceship Earth* (1969).[76] Within this, humans themselves were a de-individualised mass, or at best a distribution: a population. This thinking did not ask the question, "is 'humanity' the relevant level for our analysis?" And yet, humanity appeared to be the problem, driving all of the other indicators of system health. "People have disgusting habits," wrote Kenneth Boulding by way of a joke at the Man's Role conference in 1955: "The evolutionary plan / Went astray by evolving man."[77] Humans were a species like any other, but not a species that was capable of managing the world it had itself created. The perhaps most often repeated trope of the modern environmental age—"Man is the only organism known that lives by destroying the environment indispensable to his survival"—also stemmed from Vogt and his ominous year 1948.[78]

These were not universal views. As much as experts now became united by a systemic understanding of "environment" and by a numbers-driven approach to predicting its future direction, they remained dis-united by other factors. Although the environment had a future possible to predict, it might also be possible to change. This was not only, or not even primarily, a science issue. It had as much to do with politics. It may be useful, therefore, to disentangle the environmental futurologists that started to appear in the post-war decades in ever growing numbers. This is an analytical task of a greater magnitude that this paper can address, but as already suggested above it is at least possible to identify two strands, one that is more technocratic and quantitative, which stems from operations research and statistical data, and a second one that is, to use those words, philosophical and utopian. On the one hand, the systems analysis and integrative thinking of, say, the teams commissioned by the Club of Rome, and on the other the concerned reflexivity on the human condition—to use Hannah Arendt's words—as it was conceived by visionaries with a heart, like Buckminster Fuller, or Lewis Mumford, although these divisions should not be writ too large. Nevertheless, their outputs tended to diverge into different literatures. Not yet even considered as environmental expertise were the humanities, certain strands of the social sciences, and theology, despite the clear religious influence on many of those involved in debates around the environment, and future studies more widely.[79] The focus on numbers effectively circumscribed their opportunities, and they played a minor role, at least before 1970.

Nevertheless, it can be startling how frequently the solutions proposed for environmental problems by leading conservationists of the 1950s and 1960s was simply a renouncement of material desires, evoking a hotchpotch of the wisdom of the ancients, or in the case of the visionary planner Lewis Mumford, rounding up proceedings at the Man's Role conference, people had "to fight against the worldwide tendency towards mechanical uniformity and human utility by trying to reserve some small segment of our life for an individuated development. . . . Must we not cultivate a force that came late even in man's conception of godhood—the force that Henry Adams prophetically summoned up in opposition to the dynamo? I mean the force of love."[80] The solution to the ever-quickening mass of human population was to be found hidden in the recesses of human heart (just as the problem, somewhat mysteriously, seemed to lie in human desire). It might seem extraordinary, in retrospect, that so huge an ambition as the modelling of future global change could offer as its flipside the solution to the crisis as a mass act of individual will, or that the newly diagnosed ailments afflicting the whole globe could be resolved by something so simple. And of course, other treatments were soon suggested, although none so far very effective.

NOTES

1. The authors would like to thank Libby Robin, our colleague and friend in the project, "Expertise of the Future," out of which this paper has come, and without whose many contributions it could not have been written.
2. Cited in H. Nichols, "Greed Held Check to Stretching Natural Resources," *Christian Science Monitor*, 15 September 1948, p. 9.
3. William Vogt, foreword to *The Road to Survival* (London: Victor Gollancz, 1948), hereafter, page numbers from London edition; Fairfield Osborn, *Our Plundered Planet* (Boston: Little, Brown & Co, 1948.)
4. Thomas Robertson, *The Malthusian Moment. Global Population and the Birth of American Environmentalism* (New Brunswick, NJ: Rutgers, 2012), 36–60; Thomas Robertson, " 'This is the American Earth': American Empire, the Cold War, and American Environmentalism," *Diplomatic History* 32, no. 4 (2008): 565; Thomas Robertson, "Total War and the Total Environment: Fairfield Osborn, William Vogt and the Birth of Global Ecology," *Environmental History* 17 (2012): 336–364.
5. Robertson, *Malthusian Moment*, 41–45.
6. Vogt, *Road to Survival*, 14–15; see also 142–143.
7. Osborn, *Our Plundered Planet*, 4.
8. Paul Warde, Sverker Sörlin, and Libby Robin, *The Environment: A History* (forthcoming), Chapter 6.
9. Paul Warde, "The Environment," in Peter Coates, David Moon, and Paul Warde, eds., *Local Places, Global Processes* (Oxford: Windgather, 2015).
10. Robert P. McIntosh, *The Background of Ecology: Concept and Theory* (Cambridge, UK: Cambridge University Press, 1985); Gregg Mittman, *The State of Nature: Ecology, Community, and American Social Thought, 1900–1950* (Chicago: University of Chicago Press, 1992); Lynn K. Nyhart, *Modern Nature: The Rise of the Biological Perspective in Germany* (Chicago: University of Chicago Press, 2009); Lucien Febvre, *La Terre et l'Évolution humaine* (Paris: Albin Michel, 1922), translated into English as *A Geographical Introduction to History*, 2nd ed. (London: Kegan Paul, Trench, Trubner & Co, 1932).
11. Paul Ehrlich, *The Population Bomb* (New York: Ballantine Books, 1968), 67; see also Paul Ehrlich, interview by *Mother Earth News*, 1974. www.mnforsustain.org/ehrlich_paul_interview_1974.htm.
12. Rachel Carson, *Silent Spring* (London: Penguin, 1965).
13. Loren Eiseley, *The Invisible Pyramid* (Lincoln, NE: University of Nebraska Press, 1970), 105.
14. George Perkins Marsh, *Man and Nature: Or, Physical Geography as Modified by Human Action* (London: Murray, 1864); David Moon, *The Plough that Broke the Steppes* (Oxford: Oxford University Press, 2013); Jean-Baptiste Fressoz, *L'Apocalypse joyeuse: Une histoire du risque technologique* (Paris: Le Seuil, 2012)
15. William Stanley Jevons, *The Coal Question: An Enquiry Concerning the Progress of the Nation, and the Probable Exhaustion of Our Coal Mines* (London: Macmillan, 1865); H. S. Jevons, *The Sun's Heat and Trade Activity* (London: King and Sen, 1910); William H. Beveridge, "British Exports and the Barometer," *Economic Journal* 30 (1920): 13–25; William H. Beveridge, "Wheat Prices and Rainfall in Western Europe," *Journal of the Royal Statistical Society* 85, no. 3 (1922): 412–475.
16. Charles C. Adams, *Guide to the Study of Animal Ecology* (New York: Macmillan, 1913); Alfred J. Lotka, "Contribution to the Energetics of Evolution," *Proceedings of the National Academy of Sciences of the USA* 8 (1922): 147–151; Alfred J. Lotka, *Elements of Physical Biology* (Baltimore: Williams

and Watkins, 1925); Sharon E. Kingsland, *Modeling Nature: Episodes in the History of Population Ecology* (Chicago: University of Chicago Press, 1985).

17. Matthew Connelly, "To Inherit the Earth: Imagining World Population, from the Yellow Peril to the Population Bomb," *Journal of Global History* 1 (2006): 299–319; Alison Bashford, "Nation, Empire, Globe: The Spaces of Population Debate in the Interwar Years," *Comparative Studies in Society and History* 49, no. 1 (2007): 170–201; Alison Bashford, *Global Population: History, Geopolitics, and Life on Earth* (New York: Columbia University Press, 2014).
18. Bashford, *Global Population.*
19. D. Kirk, "Demographic Transition Theory," *Population Studies* 50 (1996): 361–387; W.S. Thompson, "Population," *American Journal of Sociology* 34 (1929): 959–975; Matthew Connelly, *Fatal Misconception: The Struggle to Control World Population* (Cambridge, MA: Harvard University Press, 2008).
20. George H. Knibbs, "Mathematical Theory of Population, of its Character and Fluctuations, and of the Factors Which Influence Them," Appendix A of *Census of the Commonwealth of Australia, 1911. Vol. 1: Statistician's Report* (Melbourne: Government Printer, 1917): 455–456; George H. Knibbs, *The Shadow of the World's Future, or the Population Possibilities and the Consequences of the Present Rate of Increase of the Earth's Inhabitants* (London: Ernest Benn, 1928).
21. The biologists Raymond Pearl and Lowell J. Reed used census data to try and fix growth to the logistic curves they developed in studying fruit fly populations. Putnam assessed their predictions in 1953, finding the error in regard to the US very small (0.3 percent), but very big in regard to the Philippines (122 percent), and found an underestimate of 27 percent for the whole world. Palmer C. Putnam, *Energy in the Future* (New York: van Nostrand, 1953), 40. See also Kingsland, *Modeling Nature.*
22. Alexander M. Carr-Saunders, *The Population Problem: A Study in Human Evolution* (Oxford: Clarendon, 1922); Peder Anker, *Imperial Ecology: Environmental Order in the British Empire, 1895–1945* (Cambridge, MA: Harvard University Press, 2001), 86–93.
23. Robertson, *Malthusian Moment*, 16–23; Kingsland, *Modeling Nature.*
24. Ariane Tanner, "Publish and Perish: Alfred James Lotka und die Anspannung in der Wissenschaft," *NTM Zeitschrift für Geschichte der Wissenschaften, Technik und Medizin* 21 (2012): 143–170.
25. Alfred J. Lotka, "The Frequency Distribution of Scientific Productivity," *Journal of the Washington Academy of Sciences* 16 (1926): 317–323.
26. Eugene P. Odum, in collaboration with Howard. T. Odum, *Fundamentals of Ecology* (Philadelphia: W. S. Saunders, 1953).
27. For example, Connelly, *Fatal Misconception*, 116.
28. M. Poovey, *A History of the Modern Fact: Problems of Knowledge in the Sciences of Wealth and Society* (Chicago: University of Chicago Press, 1998).
29. See Putnam, *Energy*; Political and Economic Planning [PEP], *World Population and Resources* (London: PEP, 1955); Paul Demeny, "Demography and the Limits to Growth," *Population and Development Review* 14 suppl. (1988): 213–244.
30. For example, on petroleum reserves see W. Pratt, *Oil in the Earth* (Lawrence: University of Kansas Press, 1942); Eugene Ayres and C. A. Scarlott, *Energy Sources—The Wealth of the World* (New York: McGraw-Hill, 1952); Marion King Hubbert, "Energy from Fossil Fuels," *Science* 109 (1949): 103–108.
31. Jevons, *Coal Question*; Gunnar Andersson, *Kraft och kultur: Kol, petroleum, vattenkraft och vind i människans tjänst* (Stockholm: Geber, 1920); Sydney Frank Markham, *Climate and the Energy of Nations* (London: Milford, 1942).
32. PEP, *World Population*, x.

33. The President's Materials Policy Commission, Resources for Freedom, A Report to the President by the President's Materials Policy Commission, 5 vols (Washington D.C.: U.S. Government Printing Office, 1952); PEP, *World Population*; Putnam, *Energy in the Future*; Samuel H. Ordway, *Resources and the American Dream* (New York: Ronald Press, 1953); Harrison Brown, *The Challenge of Man's Future* (New York: Viking 1954); Georg Borgström, *The Hungry Planet: The Modern World at the Edge of Famine* (Swedish original 1953, English translation New York: MacMillan, 1965).
34. Robertson, *Malthusian Moment*, 570; H. H. Landsberg, L. L. Fischman, and J.L. Fisher, *Resources in America's Future: Patterns of Requirements and Availabilities 1960–2000* (Baltimore: John Hopkins, 1963); B.C. Netschert, *The Future Supply of Oil and Gas: A Study of the Availability of Crude Oil, Natural Gas and Natural Gas Liquids in the United States in the Period through 1975* (Baltimore: Resources for the Future, John Hopkins University Press, 1958).
35. Ayres and Scarlott, *Energy Sources*; C. L. Weeks, "The Next Hundred Years Energy Demand and Sources of Supply," *Journal of the Alberta Society of Petroleum Geologists* 9, no. 5 (May 1961): 141–157; Hubbert, "Energy."
36. For example, PEP, *World Population*, xix.
37. Arthur Tansley, "The Use and Abuse of Vegetational Concepts and Terms," *Ecology* 16 (1935): 299.
38. Charles S. Elton, *Animal Ecology* (London: Sidgwick & Jackson, 1927).
39. Robertson, *Malthusian Moment*, 43.
40. Aldo Leopold, *The River of the Mother of God and Other Essays by Aldo Leopold*, Susan L. Flader and J. Baird Callicott, eds. (Madison, WI: University of Wisconsin Press, 1941/1991); N.F. Sayre, "The Genesis, History and Limits of Carrying Capacity," *Annals of the Association of American Geographers* 98 (2008): 120–134.
41. Robertson, *Malthusian Moment*, 26, 43–46; Robertson, "American Empire," 572–573; Robertson, "Total War," 344–345.
42. Robertson, *Malthusian Moment*, 34.
43. Vladimir Vernadsky, *La Biosphère* (Russian original 1926, French translation Paris: Félix Alcan, 1929); A.V. Lapo, "Vladimir I. Vernadsky (1863–1945), Founder of the Biosphere Concept," *International Journal of Microbiology* (2001): 47–49; Jonathon Oldfield and D.J.B. Shaw, "V.I. Vernadsky and the Noosphere Concept: Russian Understandings of Society–Nature Interaction," *Geoforum* 37, no. 1 (2006): 145–154; K.E. Bailes, *Science and Russian Culture in an Age of Revolutions: V. I. Vernadsky and His Scientific School, 1863–1945* (Bloomington, IN: Indiana University Press, 1990).
44. Evelyn G. Hutchinson, "On Living in the Biosphere," *Scientific Monthly* 67, no.6 (1948): 393–394; V.I. Vernadsky, "The Biosphere and the Noösphere," *American Scientist* 33 (1945): 1–12.
45. Lotka, "Energetics of Evolution."
46. Royal N. Chapman, "The Quantitative Analysis of Environmental Factors," *Ecology* 9 (1928): 111–122; Anker, *Imperial Ecology*; McIntosh, *Background of Ecology*, 171–178, 193–213; Paul N. Edwards, *A Vast Machine: Computer Models, Climate Data, and the Politics of Global Warming* (Cambridge, MA: MIT Press, 2010); Sven Widmalm, "A Commerce of Letters: Geodetic Networks in the 19th Century," *Science Studies* 5, no. 2 (1992): 43–58.
47. Alexander von Humboldt, De distributione geographica plantarum (Paris: Libraria Graeco-Latina-Germanica, 1817); Alexander von Humboldt, "Sur les lignes isothèrmes," *Annales de Chimie et de Physique* 5 (1817): 102–112; Michael Dettelbach, "Humboldtian Science," in Nicholas Jardine, J. A. Secord, and E. Spary, eds., *Cultures of Natural History* (Cambridge and New York: Cambridge University Press, 1996): 287–304.

48. James Rennell, *Currents of the Atlantic Ocean, and of Those Which Prevail between the Indian Ocean and the Atlantic* (London: J. G. and F. Rivington, 1832); Helen M. Rozwadowski, *The Sea Knows No Boundaries: A Century of Marine Science under ICES* (Seattle, WA: University of Washington Press, 2002).
49. Donald Worster, Nature's Economy: A History of Ecological Ideas, 2nd ed. (Cambridge and New York: Cambridge University Press, 1994).
50. William. L. Thomas (ed.), *Man's Role in Changing the Face of the Earth* (Chicago: University of Chicago Press, 1956).
51. Paul Warde, "The Invention of Sustainability," *Journal of Modern Intellectual History* 8, no. 1 (2011): 153–170; Steven Stoll, *Larding the Lean Earth: Soil and Society in Nineteenth-Century America* (New York: Hill & Wang, 2002); Benjamin Cohen, *Notes from the Ground: Science, Soil and Society in the American Countryside* (New Haven, CT: Yale University Press, 2009); Marsh, *Man and Nature*; Justus von Liebig, *Chemistry in Its Application to Agriculture and Physiology* (London: Taylor & Walton, 1842).
52. Bashford, *Global Population.*
53. Paul B. Sears, *Deserts on the March* (Norman: University of Oklahoma Press, 1935); W. Gordon East, *The Geography behind History* (London: T. Nelson Publishers, 1938); E. Mitchell, *Soil and Civilization* (Sydney: Angus and Robertson, 1946); Donald Worster, *Dust Bowl: The Southern Plains in the 1930s* (New York: Oxford University Press, 1979); Moon, *Plough.*
54. H. L. Shantz, *Conservation of Renewable Natural Resources* (Philadelphia: University of Pennsylvania Press, 1941).
55. Thomas, Introductory to *Man's Role,* xi–xxviii.
56. Charles S. Elton, *The Ecology of Invasions by Animals and Plants* (London: Methuen, 1958).
57. Thomas, *Man's Role.*
58. Theodore M. Porter, *Trust in Numbers: The Pursuit of Objectivity in Science and Public Life* (Princeton: Princeton University Press, 1995).
59. Vogt, *Road to Survival*; Robertson, *Malthusian Moment,* 149.
60. John Sheail, N*ature Conservation in Britain: The Formative Years* (London: Stationary Office, 1998), 160–166; John Sheail, *The Natural Environment Research Council—A History* (Swindon, UK: Natural Environment Research Council, 1992).
61. Lynton K. Caldwell, "Environment: A New Focus for Public Policy?," *Public Administration Review* XXIII, no. 3 (September, 1963): 132–139; Kai F. Hünemörder, *Die Frühgeschichte der globalen Umweltkrise und die Formierung der deutschen Umweltpolitik (1950–1973)* (Stuttgart: Steiner, 2004); However, the Swedes still used the old name *Naturvårdsverket*, suggesting that environment needed times to become established, with 1972, the year of the Stockholm "Only One World" conference, as a second bracket after 1948.
62. See for comparison M. Heymann, "Modelling Reality: Practice, Knowledge, and Uncertainty in Atmospheric Transport Simulation," *Historical Studies of the Physical and Biological Sciences* 37 (2006): 49–85.
63. Björn-Ola Linnér, *The Return of Malthus: Environmentalism and the Post-War Population-Resource Crisis* (Harris, UK: White Horse Press, 2003); Michael Egan, *Barry Commoner and the Science of Survival* (Cambridge, MA: MIT Press, 2009); Robertson, *Malthusian Moment*; Sandra Chaney, *Nature of the Miracle Years: Conservation in West Germany, 1945–1975* (New York: Berghahn, 2008), 176–177, 190; Ute Hasenöhrl, *Zivilgesellschaft und Protest. Eine Geschichte der Naturschutz- und Umweltbewegung in Bayern 1945–1980* (Göttingen: Vandenhoeck & Ruprecht, 2011).
64. Robertson, *Malthusian Moment*; Chaney, *Nature*, 176–204; F. Charvolin, "L'année clef pour la définition de l'environnement en France," La revue pour

l'histoire du CNRS, 4 (2001), http://histoire-cnrs.revues.org/3022; F. Charvolin, *L'invention de l'environnement en France* (Paris: La Découverte, 2003); Jens Ivo Engels, *Naturpolitik in der Bundesrepublik. Ideenwelt und politische Verhaltenstile Naturschutz und Umweltbewegung 1950–1980* (Paderborn: Schöningh, 2006).

65. Jacob D. Hamblin, *Arming Mother Nature: The Birth of Catastrophic Environmentalism* (Oxford: Oxford University Press, 2013).
66. See the chapter of Elodie Vieille Blanchard in this volume; Donella H. Meadows, Dennis L. Meadows, Jørgen Randers, and William L. Behrens, *The Limits to Growth: A Report for the Club of Rome Project on the Predicament of Mankind* (New York: Universe Books, 1972); Donella Meadows, J. Richardson, and G. Bruckmann, *Groping in the Dark: The First Decade of Global Modeling* (New York: John Wiley, 1982).
67. Frank Fraser Darling and John P. Milton, eds., *Future Environments of North America* (Garden City, NY: Natural History Press, 1966).
68. Pierre Dansereau, "Ecological Impact and Human Ecology," in Frank Fraser Darling and John P. Milton, eds., *Future Environments of North America* (Garden City, NY: Natural History Press, 1966): 453.
69. Frank Fraser Darling, "Introduction," in Frank Fraser Darling and John P. Milton, eds., *Future Environments of North America* (Garden City, NY: Natural History Press, 1966): 1–7.
70. Lynton Caldwell, *Environment: A Challenge for Modern Society* (New York: Natural History Press, 1970), 244.
71. Osborn, *Plundered Planet.*
72. Meadows, et al., *The Limits to Growth*, 10–11; Vogt, *Road to Survival.*
73. Vogt, *Road to Survival*, 15, 56, 81.
74. Jevons, *Coal Question*, 2.
75. Vogt, *Road to Survival.*
76. The metaphor was employed by Boulding at the April 1965 conference; the same metaphor was used by Adlai Stephenson in July of that year. Kenneth Boulding, "Economics and Ecology," in Frank Fraser Darling and John P. Milton, eds., *Future Environments of North America* 232–234; R. Buckminster-Fuller, *Operating Manual for Spaceship Earth* (Carbondale, IL: Southern Illinois University Press 1969).
77. Thomas, *Man's Role*, 1087.
78. Vogt, *Road to Survival*, 95.
79. Sverker Sörlin, "Reconfiguring Environmental Expertise," *Environmental Science and Policy* 28 (2013): 14–24. See also Andersson in this volume.
80. Meadows, et al., *Limits to Growth*, 10–11; Thomas, *Man's Role*, 1151–1152; Samuel H. Ordway Jr., "Possible Limits of Raw Material Consumption," in William L. Thomas, ed., *Man's Role in Changing the Face of the Earth* (Chicago: University of Chicago Press, 1956): 987–1009.

REFERENCES

Adams, Charles C. *Guide to the Study of Animal Ecology*. New York: Macmillan, 1913.

Andersson, Gunnar. *Kraft och kultur: Kol, petroleum, vattenkraft och vind i människans tjänst*. Stockholm: Geber, 1920.

Anker, Peder. *Imperial Ecology: Environmental Order in the British Empire, 1895–1945*. Cambridge, MA: Harvard University Press, 2001.

Ayres, Eugene and C. A. Scarlott. *Energy Sources—The Wealth of the World*. New York: McGraw-Hill, 1952.

Bailes, K. E. *Science and Russian Culture in an Age of Revolutions: V. I. Vernadsky and His Scientific School, 1863–1945*. Bloomington, IN: Indiana University Press, 1990.

Bashford, Alison. *Global Population: History, Geopolitics, and Life on Earth*. New York: Columbia University Press, 2014.

Bashford, Alison. "Nation, Empire, Globe: The Spaces of Population Debate in the Interwar Years." *Comparative Studies in Society and History* 49, no. 1 (2007): 170–201.

Beveridge, William H. "British Exports and the Barometer." *Economic Journal* 30 (1920): 13–25.

Beveridge, William H. "Wheat Prices and Rainfall in Western Europe." *Journal of the Royal Statistical Society* 85, no. 3 (1922): 412–475.

Borgström, Georg. *The Hungry Planet: The Modern World at the Edge of Famine*. Swedish original 1953, English translation New York: MacMillan, 1965.

Boulding, Kenneth. "Economics and Ecology." In *Future Environments of North America*, edited by Frank Fraser Darling and John P. Milton, 232–234. Garden City, NY: Natural History Press, 1966.

Brown, Harrison. *The Challenge of Man's Future*. New York: Viking, 1954.

Buckminster Fuller, R. *Operating Manual for Spaceship Earth*. Carbondale, IL: Southern Illinois University Press, 1969.

Caldwell, Lynton K. *Environment: A Challenge for Modern Society*. New York: Natural History Press, 1970.

Caldwell, Lynton K. "Environment: A New Focus for Public Policy?" *Public Administration Review* XXIII, no. 3 (September 1963): 132–139.

Carr-Saunders, Alexander M. *The Population Problem: A Study in Human Evolution*. Oxford: Clarendon, 1922.

Carson, Rachel. *Silent Spring*. London: Penguin, 1965. (Originally published 1962)

Chaney, Sandra. *Nature of the Miracle Years: Conservation in West Germany, 1945–1975*. New York: Berghahn, 2008.

Chapman, Royal N. "The Quantitative Analysis of Environmental Factors." *Ecology* 9 (1928): 111–122.

Charvolin, F. "L'année clef pour la définition de l'environnement en France." *La revue pour l'histoire du CNRS* 4 (2001), http://histoire-cnrs.revues.org/3022.

Charvolin, F. *L'invention de l'environnement en France*. Paris: La Découverte, 2003.

Cohen, Benjamin. *Notes from the Ground: Science, Soil and Society in the American Countryside*. New Haven, CT: Yale University Press, 2009.

Connelly, Matthew. *Fatal Misconception: The Struggle to Control World Population*. Cambridge, MA: Harvard University Press, 2008.

Connelly, Matthew. "To Inherit the Earth: Imagining World Population, from the Yellow Peril to the Population Bomb." *Journal of Global History* 1 (2006): 299–319.Dansereau, Pierre. "Ecological Impact and Human Ecology." In *Future Environments of North America*, edited by Frank Fraser Darling and John P. Milton, 425–462. Garden City, NY: Natural History Press, 1966.Demeny, Paul. "Demography and the Limits to Growth" in *Population and Development Review* 14 suppl. (1988): 213–244.

Dettelbach, Michael. "Humboldtian Science." In *Cultures of Natural History*, edited by Nicholas Jardine, J. A. Secord, and E. Spary, 287–304. Cambridge and New York: Cambridge University Press, 1996.

East, Gordon W. *The Geography behind History*. London: T. Nelson Publishers, 1938.

Edwards, Paul N. *A Vast Machine: Computer Models, Climate Data, and the Politics of Global Warming*. Cambridge, MA: MIT Press, 2010.

Egan, Michael. *Barry Commoner and the Science of Survival*. Cambridge, MA: MIT Press, 2009.

Ehrlich, Paul. Interview by *Mother Earth News*, 1974. www.mnforsustain.org/ehrlich_paul_interview_1974.htm

Ehrlich, Paul. *The Population Bomb*. New York: Ballantine Books, 1968.
Eiseley, Loren. *The Invisible Pyramid*. Lincoln, NE: University of Nebraska Press, 1970.
Elton, Charles S. *Animal Ecology*. London: Sidgwick & Jackson, 1927.
Elton, Charles S. *The Ecology of Invasions by Animals and Plants*. London: Methuen, 1958.
Engels, Jens Ivo. *Naturpolitik in der Bundesrepublik. Ideenwelt und politische Verhaltenstile Naturschutz und Umweltbewegung 1950–1980*. Paderborn: Schöningh, 2006.
Febvre, Lucien. *La Terre et l'Évolution humaine*. Paris: Albin Michel, 1922. English transl., *A Geographical Introduction to History*, 2nd ed. London: Kegan Paul, Trench, Trubner & Co, 1932.
Fraser Darling, Frank. "Introduction." In *Future Environments of North America*, edited by Frank Fraser Darling and John P. Milton, 1–7. Garden City, NY: Natural History Press, 1966.
Fraser Darling, Frank and John P. Milton, eds. *Future Environments of North America*. Garden City, NY: Natural History Press, 1966.
Fressoz, Jean-Baptiste. *L'Apocalypse joyeuse: Une histoire du risque technologique*. Paris: Le Seuil, 2012.
Hamblin, Jacob D. *Arming Mother Nature: The Birth of Catastrophic Environmentalism*. Oxford: Oxford University Press, 2013.
Hasenöhrl, Ute. *Zivilgesellschaft und Protest. Eine Geschichte der Naturschutz- und Umweltbewegung in Bayern 1945–1980*. Göttingen: Vandenhoeck and Ruprecht, 2011.
Heymann, M. "Modelling Reality: Practice, Knowledge, and Uncertainty in Atmospheric Transport Simulation." *Historical Studies of the Physical and Biological Sciences* 37 (2006): 49–85.
Hubbert, Marion King. "Energy from Fossil Fuels." *Science* 109 (1949): 103–108.
Humboldt, Alexander von. *De distributione geographica plantarum*. Paris: Libraria Graeco-Latina-Germanica, 1817.
Humboldt, Alexander von. "Sur les lignes isotherms." *Annales de Chimie et de Physique* 5 (1817): 102–112.
Hünemörder, Kai F. *Die Frühgeschichte der globalen Umweltkrise und die Formierung der deutschen Umweltpolitik (1950–1973)*. Stuttgart: Steiner, 2004.
Hutchinson, Evelyn G. "On Living in the Biosphere." *Scientific Monthly* 67, no. 6 (1948): 393–394.
Jevons, H. S. *The Sun's Heat and Trade Activity*. London: King and Sen, 1910.
Jevons, William Stanley. *The Coal Question: An Enquiry Concerning the Progress of the Nation, and the Probable Exhaustion of Our Coal Mines*. London: Macmillan, 1865.
Kingsland, Sharon E. *Modeling Nature: Episodes in the History of Population Ecology*. Chicago: University of Chicago Press, 1985.
Kirk, D. "Demographic Transition Theory." *Population Studies* 50 (1996): 361–387
Knibbs, George H. "Mathematical Theory of Population, of Its Character and Fluctuations, and of the Factors Which Influence Them." Appendix A of *Census of the Commonwealth of Australia, 1911. Vol. 1: Statistician's Report*, 455–456. Melbourne: Government Printer, 1917.
Knibbs, George H. *The Shadow of the World's Future, or the Population Possibilities and the Consequences of the Present Rate of Increase of the Earth's Inhabitants*. London: Ernest Benn, 1928.
Landsberg, H. H., L. L. Fischman, and J. L. Fisher. *Resources in America's Future: Patterns of Requirements and Availabilities 1960–2000*. Baltimore: John Hopkins, 1963.
Lapo, A. V. "Vladimir I. Vernadsky (1863–1945), Founder of the Biosphere Concept." *International Journal of Microbiology* 4 (2001): 47–49.

Leopold, Aldo. *The River of the Mother of God and Other Essays by Aldo Leopold*, edited by Susan L. Flader and J. Baird Callicott. Madison, WI: University of Wisconsin Press, 1941/1991.

Liebig, Justus von. *Chemistry in Its Application to Agriculture and Physiology*. London: Taylor & Walton, 1842.

Linnér, Björn-Ola. *The Return of Malthus: Environmentalism and the Post-War Population-Resource Crisis*. Harris, UK: White Horse Press, 2003.

Lotka, Alfred J. "Contribution to the Energetics of Evolution." *Proceedings of the National Academy of Sciences of the USA* 8 (1922): 147–151.

Lotka, Alfred J. *Elements of Physical Biology*. Baltimore: Williams and Watkins, 1925.

Lotka, Alfred J. "The Frequency Distribution of Scientific Productivity." *Journal of the Washington Academy of Sciences* 16 (1926): 317–323.

Marsh, George Perkins. *Man and Nature: Or, Physical Geography as Modified by Human Action*. London: Murray, 1864.

Markham, Sydney Frank. *Climate and the Energy of Nations*. London: Milford, 1942.

McIntosh, Robert P. *The Background of Ecology: Concept and Theory*. Cambridge: Cambridge University Press, 1985.

Meadows, Donella H., Dennis L. Meadows, Jørgen Randers, and William L. Behrens. *The Limits to Growth: A Report for the Club of Rome Project on the Predicament of Mankind*. New York: Universe Books, 1972.

Meadows, Donella, J. Richardson, and G. Bruckmann. *Groping in the Dark: The First Decade of Global Modeling*. New York: John Wiley, 1982.

Mitchell, E. *Soil and Civilization*. Sydney: Angus and Robertson, 1946.

Mittman, Gregg. *The State of Nature: Ecology, Community, and American Social Thought, 1900–1950*. Chicago: University of Chicago Press, 1992.

Moon, David. *The Plough that Broke the Steppes*. Oxford: Oxford University Press, 2013.

Netschert, B. C. *The Future Supply of Oil and Gas: A Study of the Availability of Crude Oil, Natural Gas and Natural Gas Liquids in the United States in the Period through 1975*. Baltimore: Resources for the Future, John Hopkins University Press, 1958.

Nichols, H. "Greed Held Major Hurdle to Stretch Vital Resources." *Christian Science Monitor* 15 (September 1948): 9.

Nyhart, Lynn K. *Modern Nature: The Rise of the Biological Perspective in Germany*. Chicago: University of Chicago Press, 2009.

Odum, Eugene P., in collaboration with Howard. T. Odum. *Fundamentals of Ecology*. Philadelphia: W. S. Saunders, 1953.

Oldfield, Jonathan and D.J.B. Shaw. "V. I. Vernadsky and the Noosphere Concept: Russian Understandings of Society–Nature Interaction." *Geoforum* 37, no. 1(2006): 145–154.

Ordway, Samuel H., Jr., "Possible Limits of Raw Material Consumption." In *Man's Role in Changing the Face of the Earth*, edited by William L. Thomas, 987–1009. Chicago: University of Chicago Press, 1956.

Ordway Jr., Samuel H. *Resources and the American Dream*. New York: Ronald Press, 1953.

Osborn, Fairfield. *Our Plundered Planet*. Boston: Little, Brown & Co, 1948.

Political and Economic Planning [PEP]. *World Population and Resources*. London: PEP, 1955.

Poovey, Mary. *A History of the Modern Fact: Problems of Knowledge in the Sciences of Wealth and Society*. Chicago: University of Chicago Press, 1998.

Porter, Theodore M. *Trust in Numbers: The Pursuit of Objectivity in Science and Public Life*. Princeton: Princeton University Press, 1995.

Pratt, W. *Oil in the Earth*. Lawrence: University of Kansas Press, 1942.

Putnam, Palmer C. *Energy in the Future*. New York: van Nostrand, 1953.

Rennell, James. *Currents of the Atlantic Ocean, and of Those Which Prevail between the Indian Ocean and the Atlantic*. London: J. G. and F. Rivington, 1832.

Robertson, Thomas. "'This Is the American Earth': American Empire, the Cold War, and American Environmentalism." *Diplomatic History* 32, no. 4 (2008): 561–584.

Robertson, Thomas. *The Malthusian Moment: Global Population and the Birth of American Environmentalism*. New Brunswick, NJ: Rutgers, 2012.

Robertson, Thomas. "Total War and the Total Environment: Fairfield Osborn, William Vogt and the Birth of Global Ecology." *Environmental History* 17 (2012): 336–364.

Rozwadowski, Helen M. *The Sea Knows No Boundaries: A Century of Marine Science under ICES*. Seattle, WA: University of Washington Press, 2002.

Sayre, N. F. "The Genesis, History and Limits of Carrying Capacity." *Annals of the Association of American Geographers* 98 (2008): 120–134.

Sears, B. *Deserts on the March*. Norman: University of Oklahoma Press, 1935.

Shantz, H. L. *Conservation of Renewable Natural Resources*. Philadelphia: University of Pennsylvania Press, 1941.

Sheail, John. *The Natural Environment Research Council: A History*. Swindon, UK: Natural Environment Research Council, 1992.

Sheail, John. *Nature Conservation in Britain: The Formative Years*. London: Stationary Office, 1998.

Sörlin, Sverker. "Reconfiguring Environmental Expertise." *Environmental Science and Policy* 28 (2013): 14–24.

Stoll, Steven. *Larding the Lean Earth: Soil and Society in Nineteenth-Century America*. New York: Hill & Wang, 2002.

Tanner, Ariane. "Publish and Perish: Alfred James Lotka und die Anspannung in der Wissenschaft." *NTM Zeitschrift für Geschichte der Wissenschaften, Technik und Medizin* 21 (2012): 143–170.

Tansley, Arthur. "The Use and Abuse of Vegetational Concepts and Terms." *Ecology* 16 (1935): 284–307.

Thomas, William. L., ed. *Man's Role in Changing the Face of the Earth*. Chicago: University of Chicago Press.

Thompson, W. S. "Population." *American Journal of Sociology* 34 (1929): 959–975.

Vernadsky, Vladimir. *La Biosphère*. Russian original 1926, French translation Paris: Félix Alcan, 1929.

Vernadsky, Vladimir. "The Biosphere and the Noösphere." *American Scientist* 33 (1945): 1–12.

Vogt, William. *The Road to Survival*. London: Victor Gollancz, 1948.

Warde, Paul. "The Environment." In *Local Places, Global Processes*, edited by Peter Coates, David Moon, and Paul Warde. Oxford: Windgather, 2015.

Warde, Paul. "The Invention of Sustainability." *Journal of Modern Intellectual History* 8, no. 1 (2011): 153–170.

Warde, Paul, Sverker Sörlin, and Libby Robin. *The Environment: A History*. Forthcoming.

Weeks, C. L. "The Next Hundred Years Energy Demand and Sources of Supply." *Journal of the Alberta Society of Petroleum Geologists* 9, no. 5 (May 1961): 141–157.

Widmalm, Sven. "A Commerce of Letters: Geodetic Networks in the 19th Century." *Science Studies* 5, no. 2 (1992): 43–58.

Worster, Donald. *Dust Bowl: The Southern Plains in the 1930s*. New York: Oxford University Press, 1979.

Worster, Donald. *Nature's Economy: A History of Ecological Ideas*. 2nd ed. Cambridge and New York: Cambridge University Press, 1994.

3 Energy Futures from the Social Market Economy to the *Energiewende*

The Politicization of West German Energy Debates, 1950–1990

Stefan Cihan Aykut

This chapter analyzes the history of *energy forecasts and scenarios* in West Germany in the decades 1950–1990. It shows that forecasting techniques were crucial in structuring the emerging field of energy policy and analyzes them as *sociotechnical objects* that defined boundaries between scientific and political questions in German energy discourse. The first part of the chapter analyzes how forecasting techniques were introduced into energy debates at the global level and later in national policy-making. I also point to some scholarly debates about how to characterize the functions and effects of such techniques. The second part of the chapter deals with the role of energy forecasts from the late 1950s to the beginning of the 1970s. In this period, characterized by steady economic growth in what has been called the German "*Wirtschaftswunder*" ("economic miracle"), energy modeling techniques were concentrated in the hands a few economic research institutes, energy utilities and state administrations, and forecasts were mainly used as instruments directly intended for the policy process.[1] They represented an attempt to create forms of foreseeability about evolutions in the energy field, and reflected a general trend to the "rationalization of politics." But forecasts were also part of a negotiation game that took place between government and energy utilities, as well as between different groups within government, over public investments in energy technologies, energy infrastructure and power plants. In fact, forecasting reenacted a post-war social contract based on steady economic growth and associated energy demand, and helped to forge a political compromise on energy policy comprising a simultaneous commitment to the dominant free-market ideology and to the protection of the domestic coal sector. Forecasts also encapsulated optimistic visions about the future potential of nuclear energy, constituting the cornerstone of an emerging "economy of techno-scientific promise."[2]

The role of forecasting changed, however, by the end of the 1970s. The third part of the chapter discusses the emergence of energy turnaround (*Energiewende*)[3] scenarios in West Germany. These scenarios appeared in the context of increasing controversies over energy policy after the two oil crises (1973 and 1979), but also in a situation of growing civil society resistance against nuclear energy. In other words, by the early 1980s, some of the cornerstones of post-war energy consensus in Germany crumbled. Elaborated by research institutes with close links to the anti-nuclear movement, energy turnaround scenarios constituted strategic devices in these controversies, and they used the scenario technique specifically in order to allow for the representation of contrasting alternative energy futures, thereby re-politicizing the energy debate. In particular, the chapter analyzes the first transition scenario elaborated by the newly founded *Öko-Institut* (Ecological Institute), in 1980, and shows how this scenario questioned, challenged and destabilized central elements of the post-war consensus.

The fourth part of the chapter looks at how the scenario technique was used in two German parliamentary commissions (*Enquetekommission*), in 1979–1980, and in 1987–1990. These commissions were organized at two very particular moments in time, in which fundamental understandings of energy policy were "unsettled"[4] and energy futures hotly debated. The first commission was created after the Three Miles Island accident, the second one after the Chernobyl catastrophe and first public alerts in West Germany about an imminent "climate catastrophe." In the context of these two commissions, scenario methodology played a key role in the process of negotiating an energy future for West Germany. Scenarios were used to foster consensual recommendations about long-term goals and short-term actions, and allowed the commissions to present different conflicting future visions compatible with these goals. The commissions thus avoided taking an explicit stance in the controversy on nuclear energy, and helped alternative energy scenarios that had emerged out of the radical environmental debate of the 1970s to become accepted and included into official expertise. Scenarios of energy futures were also central, I propose, in a redrawing of boundaries between what should be considered as a "scientific" representation of the future, in other words one that could be delegated to experts—and what should be regarded as open for political discussion in the energy field.[5]

In the concluding part, I point to how this analysis helps to understand in later developments, when the *Energiewende* became official government policy after 1998 (when a coalition government between Social Democrats and Greens decided to phase out nuclear and engage an energy transition) and 2011 (the re-affirmation of the energy transition by a liberal-conservative government after the Fukushima accident).

FORECASTING TECHNIQUES AND ENERGY POLICY: FROM THE GLOBAL TO THE NATIONAL LEVEL

The role of forecasting techniques changed dramatically in the decades from the 1950s to the 1980s that are the focus of this chapter. "The future" became an object of study in the 1950s and 1960s in a majority of industrialized countries. The first forecasting studies originated at the US RAND Corporation in a context of Cold War nuclear confrontation,[6] whereas French planning circles developed and theorized "strategic forecasting" or "*la prospective*" as a method to rationalize decision-making and investment decisions for big industrial projects in a quickly changing sociotechnical environment.[7] Whereas this first line of forecasting techniques were based on a belief in science and the "knowability" of the future,[8] growing environmental consciousness, debates about possible *Limits to Growth*[9] and the emancipatory impetus of the peace, civil rights and environmental movements inspired new approaches in the 1970s, including "doomsday scenarios" aiming to raise consciousness of the unsustainability of current lifestyles and production and consumption patterns,[10] the use of the scenario technique to represent alternative futures,[11] and participatory future-making practices.[12]

Forecasting and scenario techniques were used extensively in the energy field throughout the analyzed period. The first global estimations of future energy demand are even older. They originated from the meetings of the World Power Conferences (the precursor of the World Energy Council), and were issued already in the 1920s. Several decades later, global studies like the International Institute of Applied Systems Analysis' (IIASA) *Energy in a Finite World*[13] modeled the global energy system and sketched possible future developments in energy consumption and production as a response to growing worries about limited energy resources.[14] Since the 1990s, the International Energy Agency, an international organization founded in 1973 after the first oil crisis, has released annual forecasts called *World Energy Outlook* that constitute the global reference point for national debates about the evolution of energy systems.

At the national level, forecasting techniques were introduced in energy policy and used by energy utilities in the aftermath of Second World War, as national reconstruction and rapid economic growth, combined with a tendency toward increasing size in production units (electricity, heat) and the need to build new and ever more heavy and costly infrastructure (mines, refineries, pipelines, transmission lines, energy research, etc.), created a situation where return on investment was often only achieved in the medium or long-term. Public decision-makers and energy companies therefore depended—and still depend—on projections of the evolution of energy demand, prices of different resources and technology developments for their investment decisions. But as this chapter will show, forecasting tools were used very differently from the 1950s to the 1980s, in the hands of different actors with different conceptions of energy futures that mobilized different forms of future-related expertise.

Forecasting techniques emerged in the 1950s and 1960s in Germany and elsewhere with the promise to "rationalize" energy policy-making through a clarification of available policy options and a structured discussion of their foreseeable effects. This was part of a wider trend to science-based policy advice in the legislative process.[15] But a look at their history and the controversies and struggles of which they were objects suggests the "rationalization" paradigm describes rather poorly what forecasts actually do.[16] Instead, the use of forecasts in the post-war period can be seen as part of a general trend to quantification in public policy that has been analyzed in great detail by scholars like Theodore Porter[17] and Alain Desrosières.[18] Porter suggests that "quantitative technologies used to investigate social and economic life work best if the world they aim to describe can be remade in their image . . . . Quantification is simultaneously a means of planning and of prediction."[19] Porter's argument echoes an older claim about self-fulfilling prophecies in the social sciences[20] and has been further developed in a literature on the performativity of economics.[21] Following these insights, this chapter is interested in forecasts and scenarios not as mere representations but as agents of change. In the case of energy forecasts, the performativity claim is backed by the fact that official forecasting practices are clearly aimed to shape reality, rather than just describe possible or probable evolutions. Energy forecasts generally associate major public and industrial actors, and they include strong hypotheses on the future evolution of energy policies. Thomas Baumgartner and Atle Midttun show, in a comparative study on forecasting practices in Western Europe, that energy forecasts are closely associated with planning traditions.[22] Their predictive power therefore stems less from sophisticated methods than from the way that they are used by actors and networks, in particular state administrations and energy companies that have a strong influence on the setting of prices, tariffs, on investments decisions, and so on. Other studies have argued that energy forecasting is essentially a means to legitimize political decisions made independently of modeling outcomes.[23] Wynne, for example, shows in his in-depth study of IIASA's first global energy study *Energy In A Finite World* that energy system modeling emerged as a tool that granted an aura of objectivity to energy forecasts and to the policy recommendations that could be drawn from them.[24] From this point of view, politics is infused in forecasts through the institutional context of modeling exercises,[25] and science is essentially a tool to back up policy decisions.

Forecasts had indeed performative effects and also a legitimizing function. But both of these paradigms finally stipulate the superiority of one set of actors and practices—either scientists and expertise, or politicians and ideology—over the other, and miss therefore the more subtle interplay between science and politics in the making of energy futures. I aim to show in the following analysis that forecasts and scenarios were part of a complex and contentious coproduction[26] process in which the shaping of scientific methods of foreseeability and the negotiation of a societal future for West Germany were closely intertwined and took place at the same time.

GOVERNING THE FUTURE: ENERGY FORECASTS AND THE *SOZIALE MARKTWIRTSCHAFT*

The German state did not have a federal energy policy before 1973. In accordance with Germany's post-war ideology of market liberalism, market forces were to drive the development of energy infrastructure and the construction of new power plants. The absence of an explicit federal energy policy did not mean, however, that the state was not engaged in multiple ways and on different levels in the energy field: through subsidies and funding of energy-related research, through legislative or administrative rule-making affecting the building of transmission lines, power plants, and the extraction of resources, and as a market-actor through publicly owned companies on municipal (*Stadtwerke*) and state (*Länder*) levels. This entanglement between the regulator and regulated companies, together with a longstanding tradition of corporatism in what has been called the *Deutschland AG*,[27] created a complex terrain for energy policy, whose main actors were the federal State, the *Länder*, municipalities, energy utilities, and big industrial consumers.[28]

Four developments transformed, however, energy policy and representations of energy futures, as well as imaginaries and cosmologies in the post-war decades: spectacular economic growth and increased energy demand; quickly rising oil consumption; development of nuclear energy and what was known in German as *Atomeuphorie* or "atom euphoria"; and the opening of the German market to imported coal in 1956, when the creation of the European Community of Coal and Steel put an end to regulated prices. Together, these developments presented an unprecedented challenge to the domestic coal industry[29] and provoked a spectacular rise in energy dependency from eight percent in 1960 to sixty percent in 1977.[30] The German coal industry entered a phase of decline, with production dropping from 103 million tons in 1950 to 80 million in 1973 and the overall share of coal in German energy production falling from ninety percent in 1950 to approximately thirty percent in 1970. As the coal industry employed some 433,000 workers in 1950, these evolutions were of high political significance.[31]

As investments in new power plants exploded, with the development of ever bigger units in coal and gas, and later in the nuclear sector,[32] forecasts were first used as planning tools by energy companies in their struggle to convince public and private funders of the necessity to keep investing in new projects (power plants, energy research, etc.).[33] At the federal level, they were embedded in a different context, as the crisis in the German coal industry spurred heated debate in the government. In this debate, protectionists led by Chancellor Adenauer pledged for governmental support of the mining industry, whereas market liberalists around the minister of the economy Ludwig Erhard refused government help. Political tensions were exacerbated by the fact that Germany had for the first time become a net importer of energy resources in 1955. In this context, Adenauer created a parliamentary commission on energy ("*Energie-Enquête*"[34]) in 1959, with the objective to calculate future energy demand and sketch expected evolutions of German coal. The commission produced a study on the basis of a ten-year forecast made by a consortium composed of major German economic institutes.[35] The advanced methodology used in the study would become the standard in energy forecasting for the years to come.[36] Two features of the forecast were particularly innovative: contrary to most earlier studies that had used an aggregated approach in which the national economy was analyzed as a whole, it disaggregated the economy in three major sectors—industry, transportation and households—and went into further detail in the industry and transportation sectors (detailing three subsectors of each). This was combined with a detailed analysis of substitution processes between different forms of final energy ("modal mix": electricity, heat, movement) and of primary energy sources ("technology mix": coal, gas, oil, nuclear, hydro, etc.). This new approach reflected the political mandate of the commission, as the substitution method allowed more detailed estimations of future energy mixes, and especially of evolutions in the coal sector. The report also included, for the first time in the history of German energy forecasts, a discussion of uncertainties and of plausible alternative evolutions to the forecasted developments. The sectorial approach highlighted the ongoing substitution process between domestic coal and imported oil and gas, whereas the discussion of alternative evolutions gave them a political dimension. Through its method and its results, the report thus backed the call for government to take protective measures that would stop the decline of the German coal sector.

The report prepared a rare infringement to market ideology in German post-war politics. Several measures, intended to protect the coal industry, were introduced by the conservative government on the basis of the report, starting with the Electricity Production Act in 1965.[37] In fact, the *Energie-Enquête* was the first of several studies that directly paved the way for new forms of public intervention in the energy sector: such forecasts of future demand would be produced as a basis for the elaboration of a federal energy policy after the first oil shock, and each of the several updates of the federal energy law after 1973 relied on an actualized forecast.

However, forecasts had been used before in West German energy debates. In the mid-1950s, a debate erupted on the energy gap (*Energielücke*). This was fuelled by a report of the Research Centre of the Energy Industries at the Technical University in Karlsruhe. Published in 1955, the report forecasted an energy crisis beginning in 1975 and peaking in 2000, with dramatic consequences for German industry.[38] The assumption of energy shortages turned out to be wrong by the end of the 1950s, as cheap oil flooded energy markets. Nevertheless, forecasts in the 1960s still nourished fear of future shortages. Based on systematically overestimated projections of energy demand,[39] forecasts would continue to call for massive public investments in the energy sector. At a first glance, the overestimations might appear to be a result of the technical assumptions that underpinned them: forecasts up until the end of the 1970s assumed a stable relationship between economic growth and energy demand, simply extrapolating future energy demand from past developments. They also assumed that energy prices were going to fall as a result of the ever-increasing availability of oil and the expansion of cheap nuclear energy. But both of these assumptions were highly political. Taken together, they reflected central features of the German post-war consensus: German *soziale Marktwirtschaft* (social market economy) was built on the promise of steady and stable growth. Such growth would give a growing middle class access to wealth and affluence and prevent social conflict. The federal government actively promoted economic growth. A central instrument was the Economic Stabilization Act (*Stabilitätsgesetz*), introduced in 1967 by the first coalition government of the Christian democrats (CDU/CSU) and the Social Democrats (SPD) by the minister of the economy Karl Schiller (SPD). It concretized the concept of *Globalsteuerung* "global steering" through the definition of four objectives for economic policy: stable prices, high employment, foreign trade equilibrium, stable economic growth. As trade-offs between them made their simultaneous pursuit almost impossible, these four goals were also called the "magical square" of economic policy.[40] This had important implications for energy policy. In the eyes of decision-makers, growth required access to cheap energy like oil and, in the future, nuclear power. But developing such cheap energy conflicted with two of the other policy objectives, as imported oil already unbalanced the foreign trade equilibrium and both oil and nuclear would increase the crisis in domestic coal production, causing massive unemployment. In this context, the political desire to support a nascent nuclear energy program, while also protecting the struggling domestic coal sector emerged as part of an economy of promise[41] that permitted to apparently reconcile conflicting policy goals through a fundamentally optimistic vision of the West German economic and energy futures. This economy of promise—the German research literature coined the notion "surplus of hope" (*Hoffnungsüberschuss*)[42]—was implicated in forecasting exercises through the systematic overestimation of future shares of nuclear and coal in the energy mix, combined with an underestimation of imported oil and gas.[43]

These post-war forecasts of the German energy market were generally produced by economic institutes with close ties to industry and government, and directly involved in the production of economic data and macroeconomic modelling.[44] *Energiewirtschaftliches Institut* (EWI), founded in 1943 in Cologne, for example, that had a leading role in the *Energieenquete*, was (and is) financed by a consortium of the main energy companies and actors in the energy field, including *Rheinisch-Westfälisches Elektrizitätswerk* (RWE) and the highly coal-dependent land of North-Rhine Westphalia. *Ifo Institut für Wirtschaftsforschung* (Ifo, established in Munich, 1949) and *Rheinisch-Westfälisches Institut für Wirtschaftsforschung* (RWI, established in Essen, 1943), two other economic institutes with energy divisions, have close ties to industry.[45] On the global level, this was later completed by the foundation of IIASA in Laxenburg, close to Vienna, an international institute that specialized in the production of global energy forecasts and was directed from 1975 to 1981 by Professor Wolf Häfele, a nuclear physicist and father and most prominent advocate of the German fast-breeder technology.

These institutes developed econometric models that over time were increasingly sophisticated concerning energy production and consumption.[46] By introducing complex modeling methodology that could only be understood and thus only be challenged by a handful of actors, these energy forecasts were a fundamental source of expertise for the West German state. Indeed, they served the federal government and its administration by legitimizing energy policy objectives of the social market economy. Dominant models were embedded in powerful actor networks and they reproduced those actors' views.[47] This changed in the aftermath of the oil crises, as predictive models failed, but also with the emergence of new actors and methods as a result of the controversy over nuclear energy.

POLITICIZING THE FUTURE: ENERGY SCENARIOS IN THE NUCLEAR CONTROVERSY

The two oil crises (1973 and 1979) were a fundamental shock to this projected, stable future of increasing growth and energy consumption. The failure of dominant econometric models, based until now on the tight coupling of economic growth and energy demand, to predict and properly explain a situation where rising prices led to increased energy conservation, efficiency and a diminishing overall energy demand, triggered the development of new types of models and forecasting practices.[48] A central economic controversy in this context was about the "elasticity" of energy demand, a notion that refers to the reaction of energy consumers to the variability of energy prices. Economic models were built on the premise that overall demand wasn't affected by prices, a dogma that was increasingly contested in the 1970s.[49] This made it possible to think about ways to decouple growth from energy demand through more efficient use of resources. But forecasts also became objects of important forms of political contestation. This struggle led to a multiplication of contrasting forecasts and scenarios, which would no longer be elaborated only by state administrations, research institutes and energy companies, but also by experts close to the nascent environmental movement.[50] In the wake of the oil crisis, energy forecasting thus evolved from an instrument of energy policy-making that underpinned the promises and ideological foundations of the German social market economy to a weapon in sociotechnical controversies, essentially over nuclear energy.

The development of alternative forecasts and their progressive establishment in the energy debate was the result of the growing strength of the German environmental movement. The construction site for a nuclear power plant at Wyhl, in the Rhine valley near Freiburg, was occupied by local farmers and activists in 1975 (it stayed occupied until 1977), and public opposition soon accompanied other planned constructions in the late 1970s, especially in Brokdorf (Schleswig Holstein), Grohnde (Lower Saxony) and Kalkar (North Rhine-Westphalia). A first peak of anti-nuclear protests was reached in 1977, when 40,000 people protested at Kalkar and 15,000 in Grohnde, in what would be depicted by the media as the "battle of Grohnde." Inspired by the Wyhl protests, activists tried to occupy other construction sites, leading to fierce clashes with the police. The formerly peaceful demonstrations thus more and more culminated in violence. Despite this increasingly explosive situation and the radicalization of parts of the movement—fuelled through the fact that anti-nuclear protests in French Malville in 1977 had caused a first dead amongst the activists—the German anti-nuclear movement continued to mobilize broadly around topics like the planned nuclear waste repository in Gorleben. In 1977, the federal government (Social Democrats) and the regional government (Christian Democrats) of North Rhine-Westphalia had decided to explore the possibility of storing nuclear waste at the former salt mine of Gorleben. The decision provoked local anger and national resistance and more than 100,000 opponents to atomic energy demonstrated in Hannover, the regional capital, on March 31, 1979. This was the largest demonstration the land—and the German anti-nuclear movement—had seen so far. A long series of legal cases against the construction of new nuclear power plants for safety reasons accompanied the direct militant actions in demonstrations and occupations, and gave experts close to the environmental movement a public forum. Soon, spontaneously created grassroots, popular education institutions like *VHS Wyhler Wald* (created in 1975), and independent research institutes like *Öko-Institut* (in 1977), went from providing information on the dangers of nuclear energy to informing about alternative energy sources, like solar collectors, photovoltaic panels or wind turbines.

Many ideas and new concepts used by the anti-nuclear movement came from the US, where President Carter (1977–1981) had initiated, as a response to the oil crisis and the taking of US hostages in Iran, a support program for renewable energy through favorable regulations and tax incentives that led to a (temporary)[51] boom in renewable energy technologies. This was especially true for California, where incentives led to a rapid development of wind[52] and solar[53] energy. Along with these experiences came a new way of thinking the relationships between energy, environment, and the economy, embodied in new scientific approaches. Howard T. Odum, after having introduced system's thinking in ecology, now applied his "eco-energetics" (the modeling of energy flows in a system) to American agriculture and shocked with the provocative thesis that "growing potatoes comes from fossil fuel."[54] Alan Knees and Robert Ayres,[55] among others, proposed to re-introduce material flows (from resources to production and waste) in economic analysis, laying the basis of what would later become the fields of ecological economics[56] and industrial ecology.[57] Amory Lovins, a physician working for Friends of the Earth International, published several path-breaking studies on energy in which he laid out a "soft path" in energy policy, that is, an energy strategy based on renewables and energy savings instead of centralized energy systems based on nuclear and fossil fuel power plants. He showed it to be technically feasible and economically viable,[58] and presented his results in official committees in several European countries, including as an invited expert in the German Enquete Commission on nuclear energy. These studies had in common the way that they re-introduced materiality, expressed in an idiom of cycles and flows derived from systems thinking and scientific ecology, in an energy discourse dominated by the disincarnated language of neoclassical economics.[59] They were translated and transposed in the German debate through the work of a few pioneers like Klaus Michael Meyer-Abich, a physicist and philosopher who had done his dissertation with Carl Friedrich von Weizsäcker, a well-known German atomic physicists and one of the founders of peace studies. Meyer-Abich specialized in energy questions and was one of the first Germany researchers to adopt Lovins's proposal to consider energy savings as an actual source of energy.[60] He also proposed a research program on the "social compatibility" (*Sozialverträglichkeit*) of energy systems that built on Lovins's distinction between hard and soft energy paths.[61]

The foundation of the *Öko-Institut* in 1977 in Freiburg by actors who had participated in the protests in Wyhl marked a milestone, as it provided the environmental movement with a source of alternative energy expertise. Financed by a philanthropic association, the Institute soon established itself as an independent voice in German energy debates. Its "energy turnaround scenario,"[62] published in 1980, introduced a major innovation in forecasting techniques, as it proposed to use the scenario technique as a way of projecting possible alternative energy futures. Based on a bottom-up approach inspired by engineering expertise rather than economic theory, it allowed for more accurate representation of energy-saving potentials. It also modeled the possible contribution of decentralized renewables to energy production, and the impact of technological innovation in these fields.[63] Politically, the energy turnaround scenarios were an answer of the environmental movement to the energy gap discourse that had informed post-war energy forecasts and had legitimized the massive investments into nuclear energy.

The three authors of the energy turnaround study were Florentin Krause, a chemist; Hartmut Bossel, an engineer and philosopher; and Karl-Friedrich Müller-Reissmann, a theologist and computer scientist. Together, they managed to assure support for their study from a variety of civil society actors, like the International Friends of the Earth Foundation (founded in 1969 in San Francisco, California, by former members of the Sierra Club), the *Max-Himmelheber-Stiftung* (a foundation of a German entrepreneur converted to the ecological cause, created in 1970 in Reutlingen, near Stuttgart) and the *Forschungsstätte der Evangelischen Studiengemeinschaft* (a protestant research institute founded in 1957/1958 in Heidelberg). Krause, who had just obtained his PhD from the University of Berkeley, and Bossel, who had worked in Southern Californian as a mechanics teacher, had participated in numerous energy projects and also held a PhD from Berkeley, were also sufficiently aware of the American debate to translate some of its recent theoretical and practical developments into the German energy debate.[64]

The *Energiewende* report was designed rhetorically as an argument of guerilla warfare. Following a common slogan on guerilla warfare that suggests to "draw back your fist before you strike,"[65] the authors first made concessions to the dominant German energy discourse. In the subtitle, "growth and prosperity without oil and uranium," the authors indicated that their vision of a radical energy turnaround did not entail material sacrifices, and did not imply a radical departure from Germany's post-war ideology combining market liberalism and a social contract based on growth. Thus, the authors explicitly stressed—and regretted—that their scenario was not as radical as would have been an actual soft path scenario for Germany. They highlighted that they based their scenario on official growth prognosis, that it did not entail deep structural changes in the economy (e.g., deindustrialization or service economy,[66] which would have been radical departures from industrialism), nor drastic changes in energy consumption patterns (for instance through life-style changes) or infrastructure (e.g. modal switch from road to rail). The authors thus qualified their method as a "technical fix scenario," aiming at satisfying projected energy needs even of "overtly growth-euphoric forecasts."[67] But they then operated a series of *strategic displacements* in the report, by redefining some basic notions and concepts from the established energy debate. Thus, they proposed a bottom-up approach based on "energy services"[68]—heat, light, kinetic force, transportation, and so on—instead of aggregated demand, explicitly criticizing the domination of the energy discourse by economists and arguments derived from neoclassical theory. The main argument was that customers don't ask for particular energy carriers (coal, gas, oil) or kilowatt-hours of electricity, but for specific services. This made it possible to reconceptualize energy efficiency as a possible source of energy, and hence to popularize one of the main insights of Lovins' and Meyer-Abich's work in a way that was understandable for the general public.[69] The report was also carefully designed to create alliances with major actors in the West German energy debate. Thus, despite the well-known ecological and health damages caused by coal, the report justified continued coal use, highlighting its quality as a domestic and German resource ("*heimische Kohle*"), thus building argumentative bridges to the trade unions and their historical ally, the SPD, as well as to the coal industry. After thus redefining the terms of the debate, the study concentrated its attacks on three particular elements of the German energy debate: first, the assumption of a tight coupling between economic growth and energy demand; second, the concentration of energy policy on the production-side; third, the reliance on oil and nuclear as basic pillars of the energy system. In other words, it focused on the very bulk of assumptions of post-war forecasts. The scenario-technique played a central role in order to set this argument out.

Most of the report was consecrated to a detailed analysis of future energy needs and corresponding energy services. On the basis of this analysis, the report proposed three scenarios.[70] The first one, a "business-as-usual" scenario[71] assuming unchanged production and consumption patterns, was dismissed by the authors as economically, ecologically and politically unrealistic.[72] The second one, labeled "Coal and Gas," was a concession to those who "don't give much on renewables."[73] These two scenarios were set up, in fact, to privilege the third one, "Sun and Coal," which clearly had the favor of the authors and was qualified as "most realistic." The authors argued that increasing reliance on oil was not an option, as debates about "peak oil" suggested the resource was scarce and prices prone to increase sharply in the future. They also dismissed its substitution with nuclear energy because of the hidden costs and risks associated with this form of energy (risk of major accidents, unresolved nuclear waste storage, etc.). Both oil and nuclear energy, as well as gas, were also criticized for being imported resources, hence undermining German energy autonomy, whereas the "sun and coal" scenario had the unique advantage of granting near self-sufficiency to German energy needs.

The report did not call for de-growth or zero growth. This was advocated in the American debate by figures like Kenneth E. Boulding, a heterodox economist, system scientists and dedicated Quaker who theorized in "The Economics of the Coming Spaceship Earth"[74] an economy with minimal resource use and waste production, or Herman E. Daly, one of the founders of ecological economics, who advocated a "steady-state economy,"[75] criticizing "uneconomic growth" and the modernist belief that technological progress would always make up for the scarcity of natural resources. Such radical voices, however, weren't audible in the mainstream of the German public debate where growth was considered a necessity after the deprivations of the war and the years of reconstruction. Thus, by concentrating on the decoupling of growth from energy demand instead of zero growth, the energy futures expressed in the *Energiewende* report stayed compatible with dominant themes in German post-war debate and were rendered acceptable for a larger part of the German population.

The first reactions to the report were nonetheless hostile: the study received little media attention,[76] and a group from Kernforschungszentrum (KFA) Jülich, a major nuclear power research institution, issued a counter-pamphlet with the similar title "*Energiewende*?" to refute the arguments of the Öko-Institut.[77] But this attack backfired: entering grounds that were unfamiliar to them, the nuclear researchers themselves made miscalculations that they had to correct before finally withdrawing their counter-report altogether.[78] If anything, the KFA Jülich report finally resulted in increased publicity to the theses of the *Energiewende* report.

The long-term influence of the *Energiewende* scenario on German energy discourse can hardly be overestimated. It certainly succeeded in its claim to re-politicize the energy debate by redefining what is an objective practical constraint („Sachzwang“) and what should be seen as a mere constraint to thinking („Denkzwang“) due to the dominant energy gap discourse.[79] For the authors of the study, official forecasts that consistently overestimated future energy demand were technocratic instruments unduly narrowing down the range of possible energy futures, whereas such expertise should instead be opening up spaces for political decision-making and public debate. The success of the report was not only due to the originality of the study, but at least as much to the political context of the late 1970s and early 1980s. After the Three Mile Island accident in 1979, 100,000 people demonstrated in Bonn against nuclear energy. In 1983, the Greens entered the German Bundestag, putting a rapid end to nuclear power on the agenda, and using the reports of the *Öko-Institut* to back this claim. The nuclear accident of Chernobyl in 1986 finally popularized the theses of the report widely, and Öko-Institut's call to create local *Energiewende* committees in German towns and municipalities was followed by hundreds of local citizen's initiatives. These brought the main topics of the report—focusing on energy savings and renewables—to the attention of rural communities and local decision-makers.

Perhaps the most important effect of the *Energiewende* scenario was to re-open the future horizon of the German energy debate. By contesting official energy futures and proposing an alternative vision, the political dimensions and effects of these futures could also be contested on new grounds. The *Energiewende* report gave anti-nuclear militants a "grasp on the future,"[80] and the future scenarios that it comprised were an important resource in their struggle against the nuclear establishment. But it also illustrated and reinforced a growing schism within environmentalism, between more radical interpretations of the limits discourse by de-growth activists mostly in the US and in France, and moderate accounts in Germany that claimed that a deep transformation of the energy system could be accomplished within a reformed Social market economy. Among the proponents of this second vision, Hans Christoph Binswanger, a Suisse economist from the University of Sankt Gallen, was very influential. Binswanger published, together with two German economists—Holger Bonus and Manfred Timmermann from the University of Cologne—a book called *Economy and Environment*, in which he called for a more efficient use of resources through an ecological tax reform.[81] In this book and in follow-up publications, he coined the idea of a "double dividend", claiming that an ecological tax reform would benefit both the employment and the environment. Two other representatives of this line of thought, Joseph Huber[82] and Martin Jänicke,[83] went even further and proposed to reconcile growth-based economic policy and environmentalism through what they called "ecological modernization."[84] The transformations in the energy sector laid out in the *Energiewende* report were a central argument in their reconceptualization of environmental policies as industrial policies.

NEGOTIATING THE FUTURE: SCENARIOS IN PARLIAMENTARY ENQUETE COMMISSIONS

The alternative visions expressed in the *Energiewende* scenario would be progressively institutionalized in the German energy debate, as the *Öko-Institut* was asked to contribute scenarios to two parliamentary Enquete Commissions—the first one on "future nuclear energy policy" (Zukünftige Kernenergiepolitik, 1979–1982), and the second on "precaution for the protection of the Earth's atmosphere" (Vorsorge zum Schutz der Erdatmosphäre, 1987–1990). The scenarios elaborated by the institute on these occasions were halfway between the two energy paths, "Coal and Gas" and "Coal and Sun," of the original *Energiewende* report. Through these two commissions, the radical *Energiewende* scenarios were transformed into an official option for the West German energy future.

These two commissions are of particular interest because they profoundly influenced the German energy debate and because of the particular use the commissions made of the scenario instrument. Organized in the aftermath of major nuclear accidents, the commissions formulated different "energy paths" that described energy futures with and without nuclear energy. The first commission's influence was mainly discursive, in the sense that its use of the scenario technology and its working procedures set a methodological standard for parliamentary energy expertise and because its results were taken up in subsequent commissions. The second commission influenced policy-making more directly, as it was instrumental in the preparation of an ambitious German climate change strategy. Furthermore, recommendations of the second commission were issued unanimously, whereas the Christian democrats withdrew their support from the conclusions of the first commission just before the publication of the results. Both, however, participated in a progressive "democratization of energy policy" and enabled the Bundestag to emancipate itself from government energy policy.

Following a parliamentary reform in 1969, Enquete Commissions became a particular feature of German parliamentary life. Since then, their mandate is larger than classical parliamentary enquiry.[85] Used by parliament (Bundestag or regional parliaments) to prepare decisions on "complex and important subjects,"[86] Enquete Commissions came about as part of an attempt to make parliament more independent from ministerial expertise and increase public accountability in a context of growing critique of technocracy and demands to increase transparence and accountability in public policy-making.[87] These commissions are composed in a particular way: half of their members are members of parliament, the other half are experts, named by the political parties represented in parliament. Thus, experts are not only called in for hearings, they are full members of the commission (including the right to vote) and work together with the parliamentarians throughout the commission's duration. A corollary of this practice is that in politically sensitive topics, the choice of experts is a highly disputed process, and experts are generally chosen not only because they are specialists of a particular topic, but also because they represent a particular social force, political position, or viewpoint. Sheila Jasanoff has described this particular German way of organizing policy knowledge, or public expertise, as a "civic epistemology,"[88] marked by the principles of inclusion, representation and rational dialogue between relevant viewpoints.[89] The flip side of this way of organizing expertise is the risk of corporatist cooption.[90] In the two commissions mentioned earlier, invited experts were mainly chosen depending on their stance on nuclear energy—experts close to the industry or nuclear research institutes were counterbalanced with others having close ties to the environmental movement and to alternative energy research. Significantly, both commissions included a member from the newly founded Öko-Institut.

The 1979 commission[91] was created at a crucial moment: the Wyhl construction site had been occupied by the antinuclear movement for almost two years; there had been violent clashes in Brokdorf, and heated controversy over the planned Kalkar *Superphoenix*—a fast-breeder reactor constructed since 1973 that was to make Germany less dependent on importations of nuclear material and solve its nuclear waste problem, but that was criticized by opponents as far too expensive and even riskier than regular nuclear plants. In neighboring Austria, a popular vote had been organized on the "Zwentendorf" nuclear power plant. Although the plant had already been built, the Austrian population rejected nuclear power in an extremely close vote (fifty-one percent to forty-nine percent). In German politics, growing division and publicly voiced discontent on nuclear energy within the two parties of the governmental coalition, the Social Democrats and the liberals of the *Freie Demokratische Partei*, was further exacerbated by the highly publicized nuclear accident at Harrisburg in 1979. In this heated atmosphere, the stated aim of the commission was to canalize the "emotional" nuclear controversy into "rational" and "reasoned" debate.[92] The fifteen commission members believed this could be attained through the systematic clarification of different options and the implications of these options in political, social and economic terms. This led them to propose four energy paths, two with and two without nuclear energy.

The first of these scenarios corresponded roughly to the earlier social market economy consensus and reflected the vision of the energy utilities. It projected a doubling of energy demand until 2030, most of which was to be satisfied through nuclear energy. The second scenario expressed a more moderate view, commonplace amongst energy experts close to the nuclear industry. It projected a fifty percent increase in energy demand and a more moderate increase in nuclear energy. In the third scenario, demand was stabilized and nuclear energy was progressively phased-out. The fourth scenario built on the *Öko-Institut* report and projected a rapid nuclear phase-out, combined with heavy energy savings. A central feature of the commission's methodology was the establishment of a list of "common criteria" to evaluate the different scenarios. These criteria included economic viability, environmental impact, international compatibility, and social acceptability. This emphasis on the possibility of systematic evaluation of the scenarios was what enabled the commission to make authoritative policy recommendations.

The compromise proposed by the 1979 Enquete Commission was intended to give both the "hard path" (based on nuclear and fossil fuels) and the "soft path" (based on energy savings and renewables) a fair chance. It advocated to follow a "parallel approach" in energy policy for a 10 year period beginning in 1980: pursuing the construction of the Kalkar fast-breeder reactor on the one hand, but implementing an extremely stringent energy saving policy on the other. The members of the commission justified this parallel approach by the fact that key factors like the rapidity of structural changes in the economy and the evolution of public acceptance of nuclear energy, as well as the impact of energy saving measures and the final economic and technical feasibility of the fast-breeder technology, could simply not been foreseen properly. The final decision on which path to choose was therefore postponed to 1990, when the consequences of these developments would be clearer.

By elaborating several different energy paths, establishing a set of evaluation criteria, and forging consensual recommendations for the next ten years, the commission delegated the final decision on the structure of the German energy system to the political system. The fact, however, that it considered a nuclear phase-out to be a viable option at all, was an important success for the anti-nuclear movement.

Following the Chernobyl accident in 1986, the German energy debate was again re-politicized. Chernobyl gave a new dynamic to the anti-nuclear movement and accelerated the German social democrat's turnaround on nuclear energy.[93] But the year 1986 also marked the beginning of a broad public debate on climate change. The initial impetus for this debate came from a short pamphlet issued by the energy working group of the German physician's association, which warned of an "impending man-made climate catastrophe," a message that was taken up by the news magazine *der Spiegel* in a title story shortly after.[94] This was the starting point of the German climate debate, with its characteristic alarmist tone.[95] Chancellor Kohl publicly expressed his worries about climate change and his Christian Democratic party proposed to create a parliamentary commission on ozone and climate change, thus covering the two closely related new global environmental threats and sharpening the ecological profile of the party. Other political powers—especially the strongly anti-nuclear Greens and the Social Democrats—were more skeptical of the climate topic at the beginning, as they feared it might overshadow the debate on the risks of nuclear energy, and as they suspected, not without reason, that this was the main reason for Kohl's conservative party to push for it.

A second parliamentary commission was nevertheless created in 1987. It was composed of eighteen members, with nine scientists coming from different branches of the atmospheric sciences (four members), energy and economical sciences (four members) and international law (one member). It issued a report that comprised a scientific analysis of the ozone hole and climate change problems, a discussion of the international energy situation, and the respective responsibilities of different countries, as well as a blueprint for an international climate convention. The heart of the report, however, was constituted by 420 pages on energy policy, including a reference scenario and three scenarios that responded to a fixed objective of thirty percent reductions of CO_2 emissions. The difference between these three scenarios was to be found mainly in the projections concerning the use of nuclear energy. The first scenario, "Energy policy," was based on the assumption of a stable development of nuclear energy, while market forces were to drive the composition of the rest of the energy mix. The second energy path, "Nuclear expansion," aimed at meeting the reduction essentially through an industrial energy policy whose objective was to substitute nuclear to coal. The third scenario finally, "Nuclear phase-out," proposed a progressive substitution of nuclear by renewables, gas and energy savings until 2005.[96] Like the 1979 commission on nuclear energy, the climate commission used the scenario methodology to issue consensual recommendations while explicitly delegating the question of nuclear energy to the political system. This was expressed in the commission report as follows:

> The commission is convinced that the dangers that result of the enhanced greenhouse effect can only be avoided if the political system endorses the leading role that democracy assigns to it . . . for the commission, this results in the task to adopt a discursive working method for decision-making, in which political agreement as well as factual discord are laid out, and currently consensual and divergent standpoints are made transparent.[97]

This "discursive working method" comprised the inclusion of divergent opinions of commission members in the core of the report, as well as an active communication strategy towards the general public and the participation of the commission members to public hearings. This led to a wide discussion of the three energy scenarios and their implications.

The two Enquete Commissions led to a second discursive decoupling: just as the original *Energiewende* study had contributed to separating the debate about nuclear phase-out from radical growth critique, the Enquete Commissions' use of the scenario technique resulted in a progressive separation of the nuclear and climate change debates in German energy discourse. In other words, the Enquete Commission reports made it difficult for proponents of nuclear energy to argue that nuclear energy was necessary to combat climate change, because the energy paths of the second Enquete Commission officially showed that it was possible to pursue an ambitious CO_2 reduction target and shut down German atomic energy. The situation therefore was very different for example from neighbor France, where climate change became a major argument in the debate on a hypothetic "nuclear re-launch."[98] Again, this displacement of the debate was the result of a partial re-negotiation of the science—policy border in the energy debate, as the experts of both commissions had refused to recommend particular energy technologies in response to the climate change threat.

THREE MOMENTS OF "DECOUPLING" AND THE INSTITUTIONALIZATION OF THE *ENERGIEWENDE*

The analysis in this chapter shows that the "future" in energy forecasts is not something outside the present. It is on the contrary closely tied to the political, social and scientific debates of the time when the forecasts were made.[99] In other words, the future of scenarios tells us more about the society that made those scenarios than about the future itself. In West Germany, energy forecasts reflected fundamental values and normative assumptions about economic growth, social stability and the desirability of specific energy technologies that were embedded in forecasting practices, economic theory and modeling techniques. As a consequence, the evolution of West German society and politics was reflected in and shaped by the forecasts and scenarios we discussed here.

This has implications for the study of foresight practices and social movements: what some regretted as a progressive "scientization" of the ecological movement through increasing reliance on expert knowledge has indeed led to an opening up of energy futures the West German energy debate. The future became political[100] in the sense that social movements used the instrument of scenarios to engage in energy controversies. Alongside the occupation of construction sites, mass demonstrations and the blocking of nuclear transports, scenarios emerged as a part of the contentious repertoire[101] used by the anti-nuclear movement to make its voice heard and influence German energy policy.

Another important conclusion is that the effects of forecasts and scenarios depend on the uses that actors make of them, the practices attached to them and the political struggles they are embedded in. Just as Harro van Lente and Arie Rip conceptualize expectations in technological development as "prospective structures to be filled with in by agency,"[102] *Energiewende* scenarios influenced West German energy policy because they were appropriated by local actors and through their progressive institutionalization in official parliamentary expertise.

I suggest that this institutionalization has been facilitated by three moments of "discursive decoupling" between different discourses, operated through the different scenarios and forecasts in each of the analyzed periods: first, economists and energy specialists discussed in the 1970s the thesis that economic growth and energy demand could be *decoupled*, meaning that in energy efficient countries, strong economic development could be attained without comparable increases in overall energy demand. Second, the authors of the *Energiewende* report built on this argument and increased public acceptance of their energy vision by separating their argument for the necessary transformation of the German energy system from more radical critiques of economic growth. Finally, the members of the 1987 parliamentary Enquete Commission used the scenario technique to separate the debate on the future of nuclear energy from the debate on climate change.

Today, the historical window in which a pluralization of energy futures took place may be closing again, as the official energy vision expressed in the *Energiekonzept* "energy concept" for the Bundesrepublik, a plan defining goals and a schedule until 2050 for the evolution of the German energy system concerning the nuclear phase-out, the increase of renewables and of energy efficiency, as well as emission reduction targets, emerges as a unique official future designed to put the actors of the German energy turnaround to work. Similar to the late 1970s, however, contestation by social movements of the direction taken by the *Energiewende* may well result in the emergence of new alternative future visions that re-politicize parts of the energy debate.

NOTES

1. On "policy instruments," see Pierre Lascoumes and Patrick Le Galès, "From the Nature of Instruments to the Sociology of Public Policy Instrumentation," *Governance* 20, no. 1 (2007): 1–21.
2. Technoscientific innovations have increasingly become embedded in future-oriented narratives expressed in forecasts, roadmaps, scenarios, and so on that aim to convince investors and the general public of the necessity and positive effects of the innovation. See Pierre-Benoît Joly, "On the Economics of Techno-Scientific Promises," in Madeleine Akrich, Yannick Barthe, Fabian Muniesa, Philippe Mustar, eds., *Débordements. Mélanges Offerts à Michel Callon* (Paris: Presse des Mines, 2010), 203–222.
3. The German notion "Energiewende," derived from "Wende," can best be translated as "U-turn" (in sailing or more generally conducting); it describes a more radical policy change than the English term "energy transition."
4. On public controversies as moments "unsettling" routines, convictions and actor-coalitions, see Mike Hulme, "After Climategate . . . Never the Same," in *Exploring Climate Change through Science and in Society: An Anthology of Mike Hulme's Essays, Interviews and Speeches* (Abingdon, UK: Routledge, 2013), 252–264.
5. A similar argument is made by Christian Dieckhoff, *Modellierte Zukunft. Zur Theorie und Praxis von Energieszenarien im Kontext Wissenschaftlicher Politikberatung* (PhD diss., Karlsruher Institut für Technologie, 2014).
6. Founded by the US Air Force with the support of the Ford Foundation, RAND has been one of the pillars of post-war American scientific and intellectual hegemony. Herman Kahn developed the scenario methodology in his studies on nuclear war at RAND. On the role of the RAND Corporation in futurology, see Jenny Andersson, "The Great Future Debate and the Struggle for the World," *American Historical Review* 117, no. 5 (2012): 1411–1431.

7. Gabriele Hecht, *The Radiance of France: Nuclear Power and National Identity after World War II* (Cambridge, MA: MIT Press, 1998)
8. Jenny Andersson and Eglė Rindzevičiūtė, "The Political Life of Prediction: The Future as a Space of Scientific World Governance in the Cold War Era," *Les Cahiers européens de Sciences Po* 4 (2012): 1–25.
9. Donella H Meadows, Dennis L Meadows, Jorgen Randers, William W Behrens III, *The Limits to Growth: A Report to the Club of Rome* (New York: Universe Books, 1972)
10. Paul R Ehrlich, *The Population Bomb* (New York: Sierra Club/Ballantine Books, 1968); Garrett Hardin, "The Tragedy of the Commons," *Science* 162 (1968): 1243–1248.
11. Ossip K Flechtheim, *Futurologie—Der Kampf Um die Zukunft* (Köln: Wissenschaft und Politik, 1970).
12. Jenny Andersson, this volume.
13. International Institute for Applied Systems Analysis [IIASA], *Energy in a Finite World: Paths to a Sustainable Future* (Cambridge, MA: Ballinger Publishing Company, 1981).
14. See Eglė Rindzevičiūtė, this volume.
15. Don Price, *The Scientific Estate* (Cambridge, MA: Harvard University Press, 1965).
16. Early on, the trend to a multiplication of expert committees was criticized as a "scientizaton of politics" undermining democracy. See Jürgen Habermas, "Verwissenschaftlichte Politik in Demokratischer Gesellschaft," in H. Krauch, W. Kunz, and H. Rittle, eds., *Forschungsplanung* (München: Wien, 1966), 130–144.
17. Theodore M Porter, *Trust in Numbers: The Pursuit of Objectivity in Science and Public Life* (Princeton: Princeton University Press, 1995).
18. Alain Desrosières, *Pour une Sociologie Historique de la Quantification. L'argument Sociologique I* (Paris: Presses de l'Ecole des Mines, 2008).
19. Porter, *Trust in Numbers*, 43.
20. Robert K. Merton, "The Self-Fulfilling Prophecy," *Antioch Review* 8, no. 2 (1948): 198–210.
21. Donald Mackenzie, "Making Things the Same: Gases, Emission Rights and the Politics of Carbon Markets," *Accounting, Organizations and Society* 34, no. 3–4 (2009): 440–455.
22. Thomas Baumgartner and Atle Midttun, *The Politics of Energy Forecasting: A Comparative Study of Energy Forecasting in Western Europe and North America* (Oxford: Clarendon Press, 1987).
23. Jean-Charles Hourcade, "Les modèles dans les débats de politiques climatiques : entre le Capitole et la roche Tarpéienne?," in Amy Dahan, ed., *Les modèles du futur* (Paris: La Découverte, 2007), 140–164.
24. Brian Wynne, "The Institutional Context of Science, Models, and Policy: The IIASA Energy Study," *Policy Sciences* 17, no. 3 (1984): 277–320.
25. Michael Thompson, "Among the Energy Tribes: A Cultural Framework for the Analysis and Design of Energy Policy," *Policy Sciences* 17, no. 3 (1984): 321–339.
26. For this notion, see Sheila Jasanoff, "The Idiom of Co-Production," in Sheila Jansoff, ed., *States of Knowledge: The Co-Production of Science and Social Order* (London, New York: Routledge, 2004), 1–12.
27. German corporatism was characterized by a network of personal and institutional interconnections between major companies, banks and energy utilities. See Jürgen Beyer, "Deutschland Ag A.D.: Deutsche Bank, Allianz und das Verflechtungszentrum Großer Deutscher Unternehmen," *MPIfG Working Paper* 02/4 (2002).

28. Kai Kleinwächter, "Das „Eiserne Pentagramm"—Strommarktregulierung in Deutschland," in Lutz Kleinwächter, ed., *Deutsche Energiepolitik* (Potsdam: Brandenburgische Landeszentrale für politische Bildung, 2007), 65–79.
29. Coal played an important role in German post-war history, as the recovery of the coal extraction industry was synonymous with national reconstruction as it gave both miners work and families coal for heating. This symbolic importance is backed by a strong actor-coalition composed of trade unions, energy utilities, state governments and the Social Democratic Party.
30. Klaus Michael Meyer-Abich and R. A. Dickler, "Energy Issues and Policies in the Federal Republic of Germany," *Annual Review of Energy* 7 (1982): 221–259.
31. Christoph Nonn, *Die Ruhrbergbaukrise. Entindustrialisierung und Politik, 1958–1969* (Göttingen: Vandenhoeck & Ruprecht, 2001), 173–175.
32. The biggest unit for electricity production in 1964 had an output of 300MW. Ten years later, in 1974, this number had risen to 700MW (coal) and 1300MW (nuclear).
33. For an overview over forecasts in this period, see Michael Kraus, *Energieprognosen in der Retrospektive. Analyse von Fehlerursachen der Prognose/Ist-Abweichungen von Energiebedarfsschätzungen in der Bundesrepublik Deutschland von 1950 Bis 1980*, (PhD diss., Universität Karlsruhe, 1988), 18.
34. Despite its denomination as "Energie-Enquete," this commission did not have the same features as the two Enquete Commissions discussed later. In particular, it did not include the participation of experts as regular members of the commission (see next sections for more details).
35. The *Arbeitsgemeinschaft deutscher wissenschaftlicher Institute* was led by *Energiewirtschaftliches Institut* (EWI). See next section for more details on EWI.
36. On the methodology, see Theodor Wessels, *Die Energie-Enquete : Ergebnisse und wirtschaftspolitische Konsequenzen, Tagungsberichte des Energiewirtschaftlichen Instituts an der Universität Köln, Heft 12* (München: Verlag R. Oldenbourg, 1962).
37. These included direct regulations like the *Verstromungsgesetz* and the *Kohlezollkontingentgesetz*, as well as a corporatist agreement between electricity companies and industry to use a minimum amount of domestic coal (*Jahrhundertvertrag*, 1977), a measure financed through the introduction of a levy on electricity prices. See Annika Krisp, *Der Deutsche Strommarkt in Europa—Zwischen Wettbewerb und Klimaschutz* (PhD diss., Fachbereich Sozial- und Kulturwissenschaften der Justus-Liebig-Universität Gießen, 2007), 26, 27.
38. Joachim Radkau, *Aufstieg und Krise der Deutschen Atomwirtschaft 1945–1968. Verdrängte Alternativen in der Kerntechnik und der Ursprung der Nuklearen Kontroverse* (Reinbek bei Hamburg: Rowohlt, 1983), 113.
39. This is particularly true for forecasts after 1965. Based on extrapolations of evolutions in the 1950s and 1960s, they prognosticated a doubling of energy demand per decade. See Hans Diefenbacher and Jeffrey P. Johnson, "Energy Forecasting in West Germany: Confrontation and Convergence," in Thomas Baumgartner and Atle Midttun, eds., *The Politics of Energy Forecasting. A Comparative Study of Energy Forecasting in Western Europe and North America* (Oxford: Clarendon Press, 1987), 61–84.
40. Meyer-Abich and Dickler, "Energy Issues," 223.
41. Joly, "On the Economics."
42. Uwe Fraunholz, Thomas Hänseroth, and Anke Woschech, "Hochmoderne Visionen und Utopien. Zur Transzendenz Technisierter Fortschrittserwartungen," in Uwe Fraunholz and Anke Woschech, eds., *Technology Fiction.*

Technische Visionen und Utopien in der Hochmoderne (Bielefeld: transcript, 2012), 24.

43. Kraus, *Energieprognosen*, 60.
44. Elke Seefried, "Experten Für die Planung? „Zukunftsforscher" Als Berater der Bundesregierung 1966–1972/73," *Archiv für Sozialgeschichte* 50 (2010): 109–152.
45. Other economic research institutes important in the energy field include *Deutsches Institut für Wirtschaftsforschung* (DIW, Berlin, 1925), financed by the federal state and the *Land* Berlin, and *Prognos AG*, a Suisse institute founded in 1959. *Wirtschafts-Wissenschaftliches Institut* (WWI, 1946), the major economic institute of the trade unions, did not have a separate energy division. Technical universities and research institutes with atomic energy departments also played an important role (e.g., Technische Hochschule Karlsruhe, Kernforschungszentrum Jülich).
46. IIASA's 1981 energy forecast innovated by developing a system's approach. See Bill Keepin, "A Technical Appraisal of the IIASA Energy Scenarios," *Policy Sciences* 17, no. 3 (1984): 199–276.
47. For further discussion of the links of economic research institutes to actors in the energy field see Kraus, *Energieprognosen*, 25.
48. On the crisis of forecasting in the 1970s, see Elke Seefried, "Prognostik Zwischen Boom und Krise. Die Prognos AG und Ihre Zukunftsprognosen Für die Entwicklung der Bundesrepublik in den Sechziger und Siebziger Jahren," in Heinrich Hartmann and Jakob Vogel, eds., *Zukunftswissen. Prognosen in Wirtschaft, Politik und Gesellschaft Seit 1900* (Frankfurt, New York: Campus, 2010), 76–106.
49. Berndt Lehbert, "Untersuchung der Kurz- und Langfristigen Elastizitäten der Energienachfrage in Bezug auf die Energiepreise in der Bundesrepublik Deutschland," in *Kieler Arbeitspapiere Vol. 59* (Kiel: Institut f. Weltwirtschaft, 1977).
50. Published forecasts for West Germany went from zero to seven per year in the 1950s and 1960s to seven to thirty per year in the 1970s. See Kraus, *Energieprognosen*, 18.
51. The Reagan administration (1981–1989) phased out the incentive programs in the mid-1980s.
52. Over 16,000 wind turbines of a total of 1.7 GW were installed in California between 1981 and 1990; see John K. Kaldellis and Dimitris Zafirakis, "The Wind Energy (R) Evolution: A Short Review of a Long History," *Renewable Energy* 36, no. 7 (2011): 1887–1901.
53. In the late 1970s, private companies like ARCO Solar constructed in California the world's largest photovoltaic (PV) manufacturing facilities, and federal and regional administrations created Solar One, the first large-scale thermal solar tower power plant. These developments were widely discussed in Germany, in particular by politicians like Herman Scheer (SPD) and researchers like Carl Friedrich von Weizsäcker and his son Ernst Ulrich von Weizsäcker (both physicists).
54. Howard T. Odum, *Environment, Power, and Society* (New York: Wiley-Interscience, 1971).
55. E.g. Alan Kneese, Robert U. Ayres, and Ralph C. D'Arge, *Economics and the Environment: A Materials Balance Approach* (Baltimore: John Hopkins Press, 1970).
56. Michael Common and Sigrid Stagl, *Ecological Economics: An Introduction* (Cambridge, UK: Cambridge University Press, 2005).
57. Robert A. Frosch and Nicholas E. Gallopoulos, "Strategies for Manufacturing," *Scientific American* 261, no. 3 (1989): 144–152.

58. Amory B. Lovins, "Energy Strategy: The Road Not Taken," *Foreign Affairs* 55 (Fall 1976): 65–96; Amory B. Lovins, *Soft Energy Paths: Toward a Durable Peace* (New York: Penguin Books), 1977.
59. Joel B. Hagen, *An Entangled Bank: The Origins of Ecosystem Ecology* (New Brunswick, NJ: Rutgers University Press, 1992).
60. Klaus Michael Meyer-Abich, *Energieeinsparung Als Neue Energiequelle. Wirtschaftspolitische Möglichkeiten und Alternative Technologien. Mit Einem Vorwort von Carl Friedrich von Weizsäcker* (München: C. Hanser, 1979).
61. Klaus Michael Meyer-Abich and Bertram Schefold, eds., *Wie Möchten Wir in Zukunft Leben? Die Sozialverträglichkeit von Energiesystemen. Band 1* (München: C. H. Beck, 1981).
62. Florentin Krause, Hartmut Bossel, and Karl-Friedrich Müller-Reissmann, *Energie-Wende: Wachstum und Wohlstand Ohne Erdöl und Uran. Ein Alternativ-Bericht des Öko-Instituts* (Frankfurt am Main: Fischer, 1980).
63. While bottom-up models do a better job in representing the effects of specific policies or technological developments on energy demand, they aren't designed to calculate the macroeconomic effects of these policies. "Coupled" models combine both modeling traditions.
64. Paul Hockenos, "The Energiewende," *Zeit Online* (November 15, 2012). Amory B. Lovins is the first name the authors cite in their acknowledgment section. See Krause, Bossel, and Müller-Reissmann, *Energie-Wende*, 10.
65. Although probably not directly attributable to him, the proverbial quote was referring to Mao Tse-Tung's writings on tactics of guerilla warfare. See Mao Tse-Tung, *On Guerrilla Warfare* (Champaign, IL: University of Illinois Press, 1961).
66. Such transformations were predicted at that time by some sociologists.
67. Krause, Bossel, and Müller-Reissmann, *Energie-Wende*, 10.
68. Ibid., 27.
69. Ibid., 31. This was later popularized by Amory B. Lovins through the notion of "Negawatt Power Plant". See Amory B. Lovins, "The Negawatt revolution." *Across the Board* XXVII, no. 9 (1990): 18–23.
70. Ibid., 156–166.
71. In energy forecasts, business-as-usual or reference scenarios sketch futures where no policy intervention is assumed and major trends continue unchanged. They are often used as a backdrop against which different policy options are discussed. We use the terms "reference scenario" and "business as usual" as synonyms.
72. Against the IIASA World scenario that the authors call "suicide scenario," see Krause, Bossel, and Müller-Reissmann, *Energie-Wende*, 167.
73. Ibid., 155.
74. Kenneth Ewart Boulding, "The Economics of the Coming Spaceship Earth," in Henry E. Jarrett, ed., *Environmental Quality in a Growing Economy* (Baltimore: Resources for the Future/Johns Hopkins University Press, 1966), 3–14.
75. Herman E. Daly, *Toward a Steady-State Economy* (San Francisco, CA: Freeman, 1973)
76. *Die Zeit* and *Der Spiegel* published (very) brief discussions in which they gave more space to its detractors than to the actual study. See Günter Haaf, "In den Kloaken der Zivilisation." *Die Zeit* 42 (10 October 1980): 101; Der Spiegel, "Tüftler im Grünen." *Der Spiegel* 36 (1980), 225–229.
77. Kurt Schmitz and Alfred Voß, "Energiewende? Analysen, Fragen und Anmerkungen zu dem vom Öko-Institut vorgelegten 'Alternativ-Bericht'." *Aktuelle Beiträge zur Energiediskussion Nr. 2, Jül-Spez-73* (Jülich: Kernforschungsanlage Jülich, 1980).
78. Der Spiegel "Jülicher Eigentor." *Der Spiegel* 48 (1980): 140. See also Karl-Friedrich Müller-Reissmann and Hartmut Bossel, *Ein Bärendienst Für*

die Kernenergie. Stellungnahme zur Kritik der Kfa Jülich A.D. 'Energiewende' (Freiburg i. Br.: Inst. f. angewandte Ökologie, 1982).
79. Krause, Bossel, and Müller-Reissmann, *Energie-Wende*, 10.
80. Francis Chateauraynaud, "Des prises sur le futur. Regard analytique sur l'activité visionnaire." In *Du risque à la menace. Penser la catastrophe*, edited by D. Bourg, P.-B. Joly, and A. Kaufmann (Paris: PUF, 2014), 287–309.
81. Hans Christoph Binswanger, Holger Bonus, and Manfred Timmermann, *Wirtschaft und Umwelt. Möglichkeiten einer ökologieverträglichen Wirtschaftspolitik* (Stuttgart: Kohlhammer, 1981).
82. Huber, a sociologist and economist inspired by Nikolai Kondratjew's and Joseph Schumpeter's work on long economic cycles, posited that a new economic cycle based on microelectronics and new media and information technology would replace resource intensive industrialization.
83. Jänicke, a political scientist, concentrated on "state failure" in environmental and social policy and possible reforms that would enable society to redirect the economy in an environmentally friendly way.
84. For a history of this concept, see Arthur P. J. Mol and Martin Jänicke, "The Origins and Theoretical Foundations of Ecological Modernisation Theory," in Arthur P. J. Mol, David A. Sonnerfeld, and Gert Spaargaren, eds., *The Ecological Modernisation Reader: Environmental Reform in Theory and Practice* (London and New York: Routledge, 2009), 17–27.
85. For an analysis and an overview, see Wilhelm Knelangen, "Enquete-Kommissionen," in Uwe Andersen and Wichard Woyke, eds., *Handwörterbuch des Politischen Systems der Bundesrepublik Deutschland. (4te Auflage)* (Opladen: Leske+Budrich, 2000), 157–159.
86. Article 56 of the internal rules of procedure (Geschäftsordnung) of the *Bundestag*. URL: http://www.bundestag.de/bundestag/aufgaben/rechtsgrundlagen/go_btg/go07/245166 [retrieved 26.01.2015]
87. Knelangen, "Enquete-Kommissionen."
88. Sheila Jasanoff, *Designs on Nature: Science and Democracy in Europe and the United States* (Princeton: Princeton University Press, 2005).
89. Jasanoff compares this to the American tradition of publicly orchestrated contradictory expertise, or to the British tradition of civil servant experts.
90. Silke Beck, "Localizing Global Change in Germany," in Sheila Jasanoff and Marybeth Long-Martello, eds., *Earthly Politics: Local and Global in Environmental Governance* (Cambridge, MA and London: MIT Press, 2004), 173–194.
91. For an analysis of the working of the commission and its political impact, see Cornelia Altenburg, *Kernenergie und Politikberatung. Die Vermessung einer Kontroverse* (Wiesbaden: VS Verlag für Sozialwissenschaften, 2010); Cornelia Altenburg. "Wandel und Persistenz in der Energiepolitik: Die 1970er Jahre und die Enquete-Kommission 'Zukünftige Kernenergie-Politik'," in Hendrik Ehrhardt and Thomas Kroll, *Energie in der modernen Gesellschaft. Zeithistorische Perspektiven*, (Göttingen: Vandenhoeck & Ruprecht, 2012) 245–264.
92. PEK, "Bericht der Enquete-Kommission 'Zukünftige Kernenergie-Politik'." in *Drucksache 8/2628* (Bonn: Deutscher Bundestag, 1980).
93. After a strong pro-nuclear position in the Godesberg program (1959), the SPD had become more and more divided internally on the nuclear question. The Party Congress in Nürnberg on the 27th of August 1986 decided that the new party line would be to abandon nuclear energy within ten years, a position confirmed in the Berlin program (1989).
94. Der Spiegel, "Ozon-Loch, Pol-Schmelze, Treibhaus-Effekt, Forscher Warnen: Die Klima-Katastrophe," *Der Spiegel* 33 (1986): 1.
95. Peter Weingart, Anita Engels, and Petra Pansegrau, "Risks of Communication: Discources on Climate Change in Science, Politics and the Mass Media," *Public Understanding of Science* 9, no. 3 (2000): 261–283.

96. A fourth, more radical *Öko-Institut* scenario was not included in the official energy paths of the commission.
97. PEK, "Schutz der Erde," *Dritter Bericht der Enquete-Kommission "Vorsorge zum Schutz der Erdatmosphäre" des 11. Deutschen Bundestages (Drucksache 11/8020)* (Bonn: Deutscher Bundestag, 1990), 453. Translated by the author.
98. Francis Chateauraynaud, "Sociologie argumentative et dynamique des controverses: l'exemple de l'argument climatique dans la relance de l'énergie nucléaire en Europe," *A contrario* 16, no. 2011/2 (2011): 131–150.
99. Armin Grunwald, "Der Lebensweg von Energieszenarien—Umrisse eines Forschungsprogramms," in Christian Dieckhoff, Wolf Fichtner, Armin Grunwald, Sarah Meyer, Michael Nast, Linda Nierling, Ortwin Renn, Alfred Voß, Martin Wietschel, eds., *Energieszenarien. Konstruktion, Bewertung und Wirkung—"Anbieter" und "Nachfrager" im Dialog* (Karlsruhe: KIT Scientific Publishing, 2011).
100. Nick Brown, Brian Rappert, and Andrew Webster, eds., *Contested Futures. A Sociology of Prospective Techno-Science* (Aldershot: Ashgate, 2000).
101. Charles Tilly, "Contentious Repertoires in Great Britain, 1758–1834," *Social Science History* 17, no. 2 (1993): 253–280.
102. Harro Van Lente and Arie Rip, "Expectations in Technological Developments: An Example of Prospective Structures to Be Filled in by Agency," in Cornelis Disco and Barend van der Meulen, eds., *Getting New Technologies Together: Studies in Making Sociotechnical Order* (Berlin: Walter de Gruyter, 1998).

REFERENCES

Altenburg, Cornelia *Kernenergie und Politikberatung. Die Vermessung einer Kontroverse*. Wiesbaden: VS Verlag für Sozialwissenschaften, 2010.

Altenburg, Cornelia "Wandel und Persistenz in der Energiepolitik: Die 1970er Jahre und die Enquete-Kommission 'Zukünftige Kernenergie-Politik'." In *Energie in der modernen Gesellschaft. Zeithistorische Perspektiven*, edited by Hendrik Ehrhardt and Thomas Kroll, 245–264. Göttingen: Vandenhoeck & Ruprecht, 2012.

Andersson, Jenny "The Great Future Debate and the Struggle for the World." *American Historical Review* 117, no. 5 (2012): 1411–1431.

Andersson, Jenny, and Eglė Rindzevičiūtė. "The Political Life of Prediction: The Future as a Space of Scientific World Governance in the Cold War Era." *Les Cahiers européens de Sciences Po* 4 (2012): 1–25.

Baumgartner, Thomas and Adle Midttun. *The Politics of Energy Forecasting: A Comparative Study of Energy Forecasting in Western Europe and North America*. Oxford: Clarendon Press, 1987.

Beck, Silke "Localizing Global Change in Germany." In *Earthly Politics. Local and Global in Environmental Governance*, edited by Sheila Jasanoff and Marybeth Long-Martello, 173–194. Cambridge, MA and London: MIT Press, 2004.

Beyer, Jürgen "Deutschland AG a.D.: Deutsche Bank, Allianz und das Verflechtungszentrum großer deutscher Unternehmen." *MPIfG Working Paper* 02/4 (2002).

Binswanger, Hans Christian, Holger Bonus, and Manfred Timmermann. *Wirtschaft und Umwelt. Möglichkeiten einer ökologieverträglichen Wirtschaftspolitik*. Stuttgart: Kohlhammer, 1981.

Boulding, Kenneth E. "The Economics of the Coming Spaceship Earth." In *Environmental Quality in a Growing Economy*, edited by Henry E. Jarrett, 3–14. Baltimore: Resources for the Future/Johns Hopkins University Press, 1966.

Brown, Nick, Brian Rappert, and Andrew Webster, eds. *Contested Futures: A Sociology of Prospective Techno-Science*. Aldershot, UK: Ashgate, 2000.

Chateauraynaud, Francis "Des prises sur le futur. Regard analytique sur l'activité visionnaire." In *Du risque à la menace. Penser la catastrophe*, edited by Dominique Bourg, Pierre-Benoît Joly, and Alain Kaufmann, 287–309. Paris: PUF, 2013.

Chateauraynaud, Francis "Sociologie argumentative et dynamique des controverses: l'exemple de l'argument climatique dans la relance de l'énergie nucléaire en Europe." *A contrario* 16, no. 2011/2 (2011): 131–150.

Common, Michael, and Sigrid Stagl. *Ecological Economics: An Introduction*. Cambridge, UK: Cambridge University Press, 2005.

Daly, Herman E. *Toward a Steady-State Economy*. San Francisco, CA: Freeman, 1973.

Der Spiegel. „Tüftler im Grünen." *Der Spiegel* 36 (1980): 225–229.

Der Spiegel. "Jülicher Eigentor." *Der Spiegel* 48 (1980): 140.

Der Spiegel. "Ozon-Loch, Pol-Schmelze, Treibhaus-Effekt, Forscher warnen: Die Klima-Katastrophe." *Der Spiegel* 33 (1986): 1.

Desrosières, Alain *Pour une sociologie historique de la quantification. L'argument sociologique I*. Paris: Presses de l'Ecole des Mines, 2008.

Dieckhoff, Christian *Modellierte Zukunft. Zur Theorie und Praxis von Energieszenarien im Kontext wissenschaftlicher Politikberatung*. PhD diss., Karlsruher Institut für Technologie, 2014.

Diefenbacher, Hans, and Jeffrey P. Johnson. "Energy Forecasting in West Germany: Confrontation and Convergence." In *The Politics of Energy Forecasting: A Comparative Study of Energy Forecasting in Western Europe and North America*, edited by Thomas Baumgartner and Adle Midttun, 61–84. Oxford: Clarendon Press, 1987.

Ehrlich, Paul R. *The Population Bomb*. New York: Sierra Club/Ballantine Books, 1968.

Flechtheim, Ossip K. *Futurologie—Der Kampf um die Zukunft*. Köln: Wissenschaft und Politik, 1970.

Fraunholz, Uwe, Thomas Hänseroth, and Anke Woschech. "Hochmoderne Visionen und Utopien. Zur Transzendenz technisierter Fortschrittserwartungen." In *Technology Fiction. Technische Visionen und Utopien in der Hochmoderne*, edited by Uwe Fraunholz and Anke Woschech, 11–24. Bielefeld: transcript, 2012.

Frosch, Robert A. and Nicholas E. Gallopoulos. "Strategies for Manufacturing." *Scientific American* 261, no. 3 (1989): 144–152.

Grunwald, Armin "Der Lebensweg von Energieszenarien—Umrisse eines Forschungsprogramms." In *Energieszenarien. Konstruktion, Bewertung und Wirkung—"Anbieter" und "Nachfrager" im Dialog*, edited by Christian Dieckhoff, Wolf Fichtner, Armin Grunwald, Sarah Meyer, Michael Nast, Linda Nierling, Ortwin Renn, Alfred Voß, Martin Wietschel, 11–24. Karlsruhe: KIT Scientific Publishing, 2011.

Haaf, Günter "In den Kloaken der Zivilisation." *Die Zeit* 42 (10 October 1980): 101.

Habermas, Jürgen "Verwissenschaftlichte Politik in demokratischer Gesellschaft." In *Forschungsplanung. Eine Studie über Ziele und Strukturen amerikanischer Forschungsinstitute*, edited by Helmut Krauch, Werner Kunz, and Horst Rittel, 130–144. Wien, München, 1966.

Hagen, Joel B. *An Entangled Bank: The Origins of Ecosystem Ecology*. New Brunswick, NJ: Rutgers University Press, 1992.

Hardin, Garrett "The Tragedy of the Commons." *Science* 162 (1968): 1243–1248.

Hecht, Gabriele *The Radiance of France: Nuclear Power and National Identity after World War II*. Cambridge, MA: MIT Press, 1998.

Hockenos, Paul "The Energiewende." *Zeit Online* (November 15, 2012). URL: http://www.zeit.de/2012/47/Energiewende-Deutsche-Begriffe-Englisch [retrieved 27.01.2015]

Hourcade, Jean-Charles "Les modèles dans les débats de politiques climatiques : entre le capitole et la roche tarpéienne ?" In *Les modèles du futur*, edited by Amy Dahan, 140–164. Paris: La Découverte, 2007.

Hulme, Mike "After Climategate . . . Never the Same." In *Exploring Climate Change through Science and in Society: An Anthology of Mike Hulme's Essays, Interviews and Speeches*, 252–264. Abingdon, UK: Routledge, 2013.

International Institute for Applied Systems Analysis [IIASA]. *Energy in a Finite World: Paths to a Sustainable Future*. Cambridge, MA: Ballinger Publishing Company, 1981.

Jasanoff, Sheila *Designs on Nature. Science and Democracy in Europe and the United States*. Princeton: Princeton University Press, 2005.

Jasanoff, Sheila "The Idiom of Co-Production." In *States of Knowledge: The Co-Production of Science and Social Order*, edited by Sheila Jasanoff, 1–12. London and New York: Routledge, 2004.

Joly, Pierre-Benoît "On the Economics of Techno-Scientific Promises." In *Débordements. Mélanges offerts à Michel Callon*, edited by Madeleine Akrich, Yannick Barthe, Fabian Muniesa, and PhilippeMustar, 203–222. Paris: Presse des Mines, 2010.

Kaldellis, John K. and Dimitris Zafirakis. "The Wind Energy (R) Evolution: A Short Review of a Long History." *Renewable Energy* 36, no. 7 (2011): 1887–1901.

Keepin, Bill "A Technical Appraisal of the IIASA Energy Scenarios." *Policy Sciences* 17, no. 3 (1984): 199–276.

Kleinwächter, Kai "Das 'Eiserne Pentagramm'—Strommarktregulierung in Deutschland." In *Deutsche Energiepolitik*, edited by Lutz Kleinwächter, 65–79. Potsdam: Brandenburgische Landeszentrale für politische Bildung, 2007.

Kneese, Alan, Robert U. Ayres, and Ralph C. D'Arge. *Economics and the Environment: A Materials Balance Approach*. Baltimore: John Hopkins Press, 1970.

Knelangen, Wilhelm "Enquete-Kommissionen." In *Handwörterbuch des politischen Systems der Bundesrepublik Deutschland. (4te Auflage)*, edited by Uwe Andersen and Wichard Woyke, 157-159. Opladen: Leske+Budrich, 2000.

Kraus, Michael *Energieprognosen in der Retrospektive. Analyse von Fehlerursachen der Prognose/Ist-Abweichungen von Energiebedarfsschätzungen in der Bundesrepublik Deutschland von 1950 bis 1980*. PhD diss., Universität Karlsruhe, 1988.

Krause, Florentin, Hartmut Bossel, and Karl-Friedrich Müller-Reissmann. *Energie-Wende: Wachstum und Wohlstand ohne Erdöl und Uran. Ein Alternativ-Bericht des Öko-Instituts*. Frankfurt am Main: Fischer, 1980.

Krisp, Annika *Der deutsche Strommarkt in Europa—zwischen Wettbewerb und Klimaschutz*. PhD diss., Fachbereich Sozial- und Kulturwissenschaften der Justus-Liebig-Universität Gießen, 2007.

Lascoumes, Pierre and Patrick Le Galès. "From the Nature of Instruments to the Sociology of Public Policy Instrumentation." *Governance* 20, no. 1 (2007): 1–21.

Lehbert, Berndt "Untersuchung der kurz- und langfristigen Elastizitäten der Energienachfrage in bezug auf die Energiepreise in der Bundesrepublik Deutschland." *Kieler Arbeitspapiere Vol. 59*. Kiel: Institut f. Weltwirtschaft, 1977.

Lovins, Amory B. "Energy Strategy: The Road Not Taken." *Foreign Affairs* 55 (Fall 1976): 65–96.

Lovins, Amory B. *Soft Energy Paths: Toward a Durable Peace*. New York: Penguin Books, 1977.

Lovins, Amory B. "The Negawatt revolution." *Across the Board* 27, no. 9 (1990): 18–23.

Mackenzie, Donald "Making Things the Same: Gases, Emission Rights and the Politics of Carbon Markets." *Accounting, Organizations and Society* 34, no. 3–4 (2009): 440–455.

Meadows, Donatella H., Denis L. Meadows, Jorgen Randers, and William W. Behrens III. *The Limits to Growth: A Report to the Club of Rome*. New York: Universe Books, 1972.

Merton, Robert K. "The Self-Fulfilling Prophecy." *Antioch Review* 8, no. 2 (1948): 198–210.

Meyer-Abich, Klaus Michael *Energieeinsparung als neue Energiequelle. Wirtschaftspolitische Möglichkeiten und alternative Technologien. Mit einem Vorwort von Carl Friedrich von Weizsäcker*. München: C. Hanser, 1979.

Meyer-Abich, Klaus Michael and Robert A. Dickler. "Energy Issues and Policies in the Federal Republic of Germany." *Annual Review of Energy* 7 (1982): 221–259.

Meyer-Abich, Klaus Michael, and Bertram Schefold, eds. *Wie möchten wir in Zukunft leben? Die Sozialverträglichkeit von Energiesystemen. Band 1*. München: C.H.Beck, 1981.

Mol, Arthur P.J., and Martin Jänicke. "The Origins and Theoretical Foundations of Ecological Modernisation Theory." In *The Ecological Modernisation Reader: Environmental Reform in Theory and Practice*, edited by Arthur P.J. Mol, David A. Sonnenfeld, and Gert Spaargaren, 17–27. London and New York: Routledge, 2009.

Müller-Reissmann, Karl-Friedrich, and Hartmut Bossel. *Ein Bärendienst für die Kernenergie. Stellungnahme zur Kritik der KFA Jülich a. d. 'Energiewende'*. Freiburg i. Br.: Inst. f. angewandte Ökologie, 1982.

Nonn, Christoph *Die Ruhrbergbaukrise. Entindustrialisierung und Politik, 1958–1969*. Göttingen: Vandenhoeck & Ruprecht, 2001.

Odum, Howard T. *Environment, Power, and Society*. New York: Wiley-Interscience, 1971.

Parlamentarische Enquete-Kommission [PEK]. "Bericht der Enquete-Kommission 'Zukünftige Kernenergie-Politik'." *Drucksache 8/2628*. Bonn: Deutscher Bundestag, 1980.

Parlamentarische Enquete-Kommission [PEK]. "Schutz der Erde." *Dritter Bericht der Enquete-Kommission "Vorsorge zum Schutz der Erdatmosphäre" des 11. Deutschen Bundestages (Drucksache 11/8020)*. Bonn: Deutscher Bundestag, 1990.

Porter, Theodore *Trust in Numbers: The Pursuit of Objectivity in Science and Public Life*. Princeton: Princeton University Press, 1995.

Price, Don *The Scientific Estate*. Cambridge, MA: Harvard University Press, 1965.

Radkau, Joachim *Aufstieg und Krise der deutschen Atomwirtschaft 1945–1968. Verdrängte Alternativen in der Kerntechnik und der Ursprung der nuklearen Kontroverse*. Reinbek bei Hamburg: Rowohlt, 1983.

Schmitz, Kurt and Voß, Alfred "Energiewende? Analysen, Fragen und Anmerkungen zu dem vom Öko-Institut vorgelegten 'Alternativ-Bericht'." *Aktuelle Beiträge zur Energiediskussion Nr. 2, Jül-Spez-73*. Jülich: Kernforschungsanlage Jülich, 1980.

Seefried, Elke "Experten für die Planung? „Zukunftsforscher“ als Berater der Bundesregierung 1966–1972/73." *Archiv für Sozialgeschichte* 50 (2010): 109–152.

Seefried, Elke "Prognostik zwischen Boom und Krise. Die Prognos AG und ihre Zukunftsprognosen für die Entwicklung der Bundesrepublik in den sechziger und siebziger Jahren." In *Zukunftswissen. Prognosen in Wirtschaft, Politik und Gesellschaft seit 1900*, edited by Heinrich Hartmann and Jakob Vogel, 76–106. Frankfurt and New York: Campus, 2010.

Thompson, Michael "Among the Energy Tribes: A Cultural Framework for the Analysis and Design of Energy Policy." *Policy Sciences* 17, no. 3 (1984): 321–339.

Tilly, Charles "Contentious Repertoires in Great Britain, 1758–1834." *Social Science History* 17, no. 2 (1993): 253–280.

Tse-Tung, Mao *On Guerrilla Warfare*. Champaign, IL: University of Illinois Press, 1961.

Van Lente, Harro and Arie Rip. "Expectations in Technological Developments: An Example of Prospective Structures to be Filled in by Agency." In *Getting New Technologies Together: Studies in Making Sociotechnical Order*, edited by Cornelis Disco and Barend van der Meulen, 202–230. Berlin: Walter de Gruyter, 1998.

Weingart, Peter, Anita Engels, and Petra Pansegrau. "Risks of Communication: Discources on Climate Change in Science, Politics and the Mass Media." *Public Understanding of Science* 9, no. 3 (2000): 261–283.

Wessels, Theodor, *Die Energie-Enquete : Ergebnisse und wirtschaftspolitische Konsequenzen, Tagungsberichte des Energiewirtschaftlichen Instituts an der Universität Köln, Heft 12*. München: Verlag R. Oldenbourg, 1962.

Wynne, Brian "The Institutional Context of Science, Models, and Policy: The IIASA Energy Study." *Policy Sciences* 17, no. 3 (1984): 277–320.

4 Technoscientific Cornucopian Futures versus Doomsday Futures

The World Models and *The Limits to Growth*

Elodie Vieille Blanchard

The publication of the *Limits to Growth*[1] report in 1972 was the starting point of a heated debate. *Limits to Growth* asserted that the growth of world population had to be stabilized and industrial production halted in order to avoid a global collapse. Retrospectively, the report is remembered for its Malthusianism, because it put forward arguments for the fundamental incompatibility between the exponential growth of several variables (population, production) and the limits of others (natural resources, capacity of pollution absorption in the atmosphere). As such, *Limits to Growth* has become an icon of a radical environmentalist position, remembered as a justified warning to take the environment into consideration.[2] Yet its conclusions are also remembered as excessive, as contemporary positions follow, rather, in the continuity of the Brundtland report, and stress that protecting ecosystems and achieving high economic growth are compatible aims.

This, however, is based on a selective reading of the report and the circumstances of its production. Malthusianism does not sum up the view, fails to identify the true complexity of the positions taken in the *Limits to Growth* and by the Club of Rome itself. Indeed, a closer examination of the Club of Rome and of the ideas expressed in the report shows the complexity of 1960s and 1970s discourses of the future and brings out how these borrowed from several competing and coexisting visions of the relationship between technology and the future. In the report, these include an apocalyptic or catastrophic view centered on the destruction of the environment, and a much more optimistic technoscientific outlook promising salvation. Whereas studies tended for a long time to dismiss the *Limits to Growth* publication as a singular and somehow unexplainable event, insisting that the report's conclusions were rapidly discarded, I suggest that on the contrary, this ambiguity in the Club of Rome's representation of the future allows us to understand the long-term significance of the report.

This chapter will do two things. First, it will offer a complex and sympathetic account of the *Limits to Growth* report. Most of the commentary on the report was in fact produced very quickly upon its publication and was motivated much more by the wish to disqualify its conclusions than to understand how and why it was written.[3] By focusing on the context of its production, and by bringing out the complexity of its conception of the future, the chapter will show the extent to which the *Limits to Growth* report (henceforth *Limits*) was rooted in its time. It was a striking expression of the contradictions of the "*Trente Glorieuses*"; in other words, of the coexistence, in the post-war era, of unprecedented faith in technology and affluence with sharp anxiety of the future. Applying this sympathetic perspective to the analysis of the debate provoked by *Limits* will also enable us to understand the development of the sustainable development paradigm, which progressively took precedence over the zero growth approach that was originally set out in *Limits*. Second, the chapter will use the *Limits* report as a prism for investigating post-war visions of the future, and for analyzing how in fact very different statements pertaining to the role of technology as apocalypse or salvation were put forward, by whom, and how the balance of power between catastrophic narratives and technoscientific ones evolved over time, as the first shock effect of the report gave way to a gradual disqualification of its results. By doing this, we can posit the *Limits to Growth* debate in a much larger framework of twentieth century history, as the history both of material affluence and of the reactions and criticisms that the production of this affluence provoked.[4]

The chapter is therefore organized in three parts. The first part accounts for the context in which the *Limits to Growth* report was produced and brings out some of the ambiguities that surrounded this production, between the radical environmental critique offered by the report and the fact that the members of the Club of Rome, which commissioned the report, were not radical environmentalists but OECD planners, industrialists and business leaders. The second part shows that the Club of Rome was informed by four dominant representations about the relationship between technology and the future, which were expressed in the final report, and the third part examines how these different future visions resurfaced in the debate that followed its publication. In particular, I seek to explain why a position of technological utopianism progressively took precedence over the environmentalist–catastrophist position.

PARADOXES OF THE *LIMITS TO GROWTH*

The *Limits to Growth* report was the final result of a complex story that involved actors from many different backgrounds. Before examining the ideological background of the report, the key elements influencing the production of the report need to be laid out. The report was commissioned by an organization created in 1968 under the name of the Club of Rome. Its founder was the Italian industrialist Aurelio Peccei. The purpose of the organization was to study the great problems of "advanced societies," and to offer concrete advice to policymakers in order to solve what the Club referred to as the world "*Problématique*." The *Problématique* referred to an inextricable cluster of problems of different kinds, including limited natural resources, the uneven development of technology between the industrialized and the developing world, and the inadequacy of political institutions in handling problems on a planetary scale. To Peccei, the very future of mankind was threatened by problems whose scale and complexity were unprecedented. The problems of pollution, of natural resources or of population growth could no longer be treated by individual nations but rather had to be tackled at the global level. To his Club, Peccei recruited an eclectic group consisting mostly of industrialists, OECD executives, futurologists, and policymakers. Once the Club was created, Peccei started looking for a tool that would allow for nothing less than the thorough examination of the "*Problématique*," make the interrelationships between its different problem areas clear, and promote the development of concrete measures that might help push the world toward a "willed future." He was very impressed by the computerized management methods in use on the other side of the Atlantic, and with the assistance of the first Club of Rome members he drafted the project of constructing a mathematical model capable of representing the world. At first, this model was envisioned as a comprehensive and flexible tool whose structure would not be fixed but would evolve over time, as political measures developed that would help guide humanity along a more rational and desirable path. This project had been drafted by the Turkish American system scientist Hasan Ozbekhan. It was a very ambitious project, both costly and time consuming.[5] But Peccei was in a hurry, as he understood the threat facing humanity as urgent. He therefore decided to subcontract the construction of the model to Jay Forrester, from the Massachusetts Institute of Technology (MIT). The later suggested a turn-key solution: a System Dynamics model of the world, which was much simpler than what had originally been planned.

The System Dynamics model had been developed by Forrester in the 1950s and 1960s. Its origins can be traced back to the Operations Research and military systems management during World War II. During the war, Forrester, who was a trained electrical engineer, was working on the development of automatic radar systems. His experience of feedback effects in electrical systems helped him shape the methodology that went into the System Dynamics model, which was first designed to solve managerial problems in firms. The first versions of Industrial Dynamics models represented stocks and flows of goods in firms.[6] They were designed to help firms answer customers' demands in real time, and they allowed for the simulation of effects of different management methods by varying specific parameters in the model input. Forrester saw these models as having the potential to revolutionize management, by allowing for "virtual experiments" that would have been impossible without the help of computers.

At a later stage, System Dynamics was applied to urban problems, similar to how other predictive models were used as tools of planning in the 1960s.[7] From 1970 on, Jay Forrester cast his net even larger when he agreed to

Figure 4.1 The *Limits to Growth* team: Jorgen Randers, Jay Forrester, Donella Meadows, Dennis Meadows, William Behrens; the early 1980s. Courtesy of Dennis Meadows.

create a mathematical model of the world for the Club of Rome. He himself drafted the first two versions of the world model, called World 1 and World 2. While there are no records of World 1, we know that this very first version of the world model laid the structural bases for the subsequent models, World 2 and World 3.[8] The structure of these models and the results of their simulation was described in *World Dynamics* and the *Limits to Growth*, respectively. Whereas World 2 was elaborated by Forrester alone, World 3 was built by a multidisciplinary team of young researchers, headed by the 28-year-old Dennis Meadows from the MIT Sloan School of Management (see Figure 4.1). The development of World 3 took much more time than the elaboration of World 2 (a year and a half for the former as opposed to one month for the latter), it had eight times as many equations than World 2, and it was based on more precise data. Another difference was the style of the reports: whereas *World Dynamics* was a short report written in a purely technical style, which was confined to the presentation of the model runs, the *Limits to Growth* benefited from the contribution of Meadows's wife, biophysicist Donella Meadows, who brought to it her own reflections about the evolution of mankind, and made it a best seller. Nevertheless, the models World 2 and World 3 shared essentially the same structure and projected the same results. These models had five main variables: population, industrial production, agricultural production, natural resources and pollution. These variables depicted change at the global level. They were interrelated by a set of equations that described how the evolution of each variable led to change in the others.

The World models were calibrated to fit the historical trends of the period 1900–1970, and allowed the simulation of the world system over the period 1900–2100. Their standard run, which assumed an exponential growth of population and production at a global level, exhibited a characteristic overshoot and collapse behavior: population and production would grow for some time, before collapsing due to the exhaustion of natural resources. Allowing for technological progress in the simulation did not change this output, it only delayed the point of collapse. Meanwhile, the system did stabilize over the long term if, and only if, population and production growth could somehow be expected to come to a halt. Such a halt, the model presumed, could only be actively induced, in other words deliberate and depending on human action.

The *Limits to Growth* report was launched publicly in March 1972 at the prestigious Smithsonian Institute in Washington, DC. The ceremony was sponsored by Xerox Company. For several months, Peccei had sent reports about the work being done at MIT to eminent scientists, such as Nobel laureate Jan Tinbergen and biologist Barry Commoner, but also to policymakers, such as European commissioner Sicco Mansholt. Peccei also hired a communication company, Calvin Kytle Associates, which sent press releases to all major newspapers on the East Coast, including the *New York Times*, the *Boston Globe* and the *Washington Post*. A few days before the ceremony, these newspapers published the main conclusions of the report, which led to a major public interest in it. Two hundred fifty people attended the ceremony, including Secretary of Health, Education and Welfare Elliot Richardson, UN Deputy Secretary-General Philippe de Seynes and former Secretary of the Interior Stewart Udall.

Most of the many reactions to the report were critical. They came from scholars, policymakers, and environmentalists; from representatives of industrialized and developing countries; and from members of the political left and right. The interesting point is that the *Limits of Growth* report's conclusions about the necessary end to material growth were almost unanimously rejected. These rejections expressed clear bias in the report's reception.[9] For instance, US economist William Nordhaus was one of the first to refute the hypothesis that natural resources were limited. Nordhaus defended his position with the affirmation that in any case "the American way of life is hard to visualize without commuters, television, overheated houses, aluminum cans, and jet setters."[10] Meanwhile, for the Norwegian political scientist Johan Galtung, *Limits* did not represent a threat to the richest nations of the world, but was rather evidence of their will to maintain dominance over the whole world, as the conditions of development of both population and growth in developing nations were circumscribed.[11] We need to remember that Mankind was conceived as a question of the totality of humanity in *Limits*, and that there was no geographical differentiation in the model. Its variables were global in essence: it dealt with global population, global pollution, global natural resources, and so on. At the UN conference in Stockholm in 1972, this globalizing perspective of the report was seen as a convenient way of hiding the responsibility of industrialized countries in the destruction of the environment.[12] In France, the debate about *Limits* erupted when the European commissioner for agriculture, Sicco Mansholt, publicly asserted the necessity for the European community to get away from productivity-driven objectives and to completely redirect European policies toward ecological stability. This proposition was vigorously dismissed by the main trade-union organization *Confédération Générale du Travail*, as well as by the main employers' lobby, the *Conseil National du Patronat Français*.[13] Partly these violent reactions were due to the fact that the Club of Rome was an elitist organization whose members belonged to the highest spheres of the business world. The report was publicly defended by Mansholt, who had until then been a strong proponent of agricultural industrialization and now changed his position. The report relied on a mathematical model of the world developed at MIT. MIT was a beacon of technological progress, but at the same time, *Limits* called for the halting of industrial growth on a global scale. This sharp paradox between the actors behind the report and its message led to perplexity. In addition, the report actually put forward several contradictory visions of the future, including visions where technology precipitated a doomsday future and scenarios where it helped build a sustainable one.

THREATS AND PROMISES OF TECHNOLOGY

From the moment on that Peccei began to consider the ominous state of the future of the world, he would describe technology as the very factor that could both precipitate mankind's ruin and lead it to a glorious future.[14] In fact, in Peccei's writings, in the drafts of the Club of Rome project and in *Limits* report itself, technology is the most recurrent theme, and appears as the fundamental factor determining the future of mankind. Four different representations of the relationship between technology and the future can be identified throughout this documentation, and several contradictory understandings of the very concept of technology. Analyzing these representations allows us to shed some light on the complex set of ideas behind the Club of Rome and to elucidate further what stands out as the great paradox of the *Limits* report, namely the fact that it was defended by representatives from the highest spheres of the Western business world even though it recommended putting a halt to industrial growth in the interest of all Mankind.

The paradoxes of *Limits* were in fact personified by Peccei himself. Peccei was a very singular actor in the development of a global consciousness, in that he borrowed from very contrasted visions about the future of the world. This was due to his experience of doing business in Western, Eastern and

Third World countries. For the main part of his life, Peccei, born in 1908, was a very enthusiastic promoter of industrial development. After receiving his PhD (in economics), he was hired by the Fiat company, who sent him to China, and at the end of the 1940s to Argentina. There, he established a Fiat plant, for which he prepared a five-year plan coordinated with the national five-year plan of the Perón government. Juan Perón led a program of forced industrialization aimed at the development of the national gas network as well as rail, road and air transportation infrastructures. To do this, his government developed contracts with several European firms and with Fiat in particular. This partnership, for which Peccei was responsible, led to the building of a network of Fiat plants in Argentina, resulting in the building of the first Argentinean tractor as well as in the extension of the Argentinian electrical network and in the development of its railways. Peccei took a personal interest in these projects. He firmly believed that each industrial enterprise had to be planned precisely, and that every detail had to be studied, in order for industrialization to be efficient and benefit the greatest number of people.[15]

Peccei was thus an important actor in the Western-led development of Latin America. He was also involved in the founding of several research and development organizations: the Oficina de Estudios para la Colaboración Económica Internacional (OECEI, founded in 1956), Italconsult (founded in 1957), and the Atlantic Development Group for Latin America (ADELA), founded in 1964. The OECEI was dedicated to the study of the Argentinean economy and became one of the largest private research institutes in South America. Italconsult aimed to stimulate technological transfer from industrialized countries to developing countries, with companies from industrialized countries providing their expertise freely to projects in developing countries, in return for selling equipment for this project. ADELA was a cooperative investment and management organization that aimed to help technology transfer from industrialized countries to South America.

In other words, Peccei's first interest in technology was in the question of technology transfer from industrialized to developing countries. When he began to formulate his preoccupations about the future of the world, the idea of spreading technological development was at the core of his analysis. In 1965, he gave a conference at the National Military College in Buenos Aires entitled "The Challenge of the 1970s to the World of Today." He held the same talk again in 1966 on the East Coast of the US, where he addressed businessmen and policymakers, as well as students from Harvard, MIT and Cornell. His speech was supposedly devoted to "the trouble and dangers" of his era, but he actually evoked only very rapidly the well-known future threats that scared everyone at that time—overpopulation and nuclear escalation—in order to deal instead with his favorite theme: technological development. To Peccei, the world was threatened by fragmentation, because the various regions of the world had achieved different levels of technological development. He saw the US as "the real leaders" in "the effervescence" of his period, and considered that other countries had "lesser roles" or were "no more than spectators."[16] He thus called for a global policy that would aim to "enlarg[e] and consolidat[e] the area of prosperity [. . .] exist[ing] in the world"[17] based on the setting up of economic markets between industrialized and developing countries, and between the US and Europe. To him, such a policy was likely to homogenize technological development on a global scale. This was necessary in order to allow "the technological revolution" to "attack [. . .] the real problems of the next decade," including "survival in the nuclear age," "overpopulation" and "hunger in large parts of the world."[18] It is important to note that the very expression "technological gap"[19] (referring to the gap between Europe and the US), which was cited as such in Peccei's talk, had been coined a few months earlier by the OECD Committee on Science and Technology.[20] The purpose of this committee was to boost the industrial development of Europe, and it was chaired by Alexander King, who would soon become the co-president of the Club of Rome.

At the end of the 1960s, Peccei wrote a book entitled *The Chasm Ahead*.[21] The book was designed to lay the foundations for a new organization that would help drive mankind toward a desirable future (the organization that would materialize as the Club of Rome). There were many themes in Peccei's book, and some of them were in contradiction with others. But most of the book's content was devoted to "the chasm between the two sides of the Atlantic Ocean." Peccei expressed a great fascination for American technological development. He praised the civilian nuclear industry and the aeronautical industry, but most of all the computer revolution, which to him constituted a vector of deep social transformation. To Peccei, there was a great danger that Europe would fall behind the US-driven technological revolution: "the direct and second-order consequences of a permanent Atlantic fracture would be incalculable."[22]

The first vision of "technology and the future" expressed by Peccei was thus coherent with the main policy approaches developed at his time, at both global and European levels. According to these approaches, industrial and technological development was a valuable goal, which had to be pursued for its own sake, by the poorest as well as by the richest countries. The second representation of technology given shape in Peccei's writings and in the Club of Rome documents was practically the opposite of this positive idea of the spread of technologies to the world. It depicted technological development as an uncontrollable process, threatening even the very survival of mankind.

At the time that Peccei founded the Club of Rome, in 1968, the environmental movement was reaching its apex, especially in the US. This movement brought together several specialized citizen mobilizations, which came into being after 1945, in an unprecedented period of affluence and in reaction to unprecedented attempts to radically control the environment.[23] The most influential of these were directed against the massive use of pesticides in the US,[24] and against the military and civilian use of nuclear power. At that period, when population growth was at the highest level in the whole of human history, there were also numerous voices that described global population growth as the greatest danger to the future of mankind. This discourse, which articulated ecological concepts with geopolitical considerations, was successfully disseminated by the greatest American industrial foundations (Rockefeller and Ford), which lobbied for several decades and finally gained support from the US government and from the United Nations for population control programs.[25] Environmental discourses and alarm calls about global population growth added up to a prevailing narrative of doom, and ideas of an impending global crisis were fed by the scholarly arguments of scientists like Barry Commoner and Paul Ehrlich. In their books (carrying eloquent and apocalyptic titles such as *The Doomsday Book, The Population Bomb* and *The Closing Circle*),[26] they described how the radical technological and industrial changes of the period were threatening the ecosystems by depleting natural resources and creating unprecedented levels of pollution.[27]

The theme of "uncontrollable" character of technological progress stands side by side with his ideas of technology transfer in Peccei's talks from 1965 on. In his 1965 talk, he evoked "the exponential growth" of knowledge, the danger that humans could become "slaves of [their] machines," and the risk that the progress in machines could create "new problems, which other machines may have to be employed to resolve."[28] In *The Chasm Ahead*, he defines "the uncontrolled growth" of the human system (an interdependent man-technology-environment world) as a central threat to the future, along with "the technological gap" mentioned earlier. Peccei describes the "phenomenal increases, rapidly approaching critical maxima along exponential curves, [which were] happening in population, pollution, energy release, speed, automation and other key areas revolutionized by technology."[29] "[M]ankind is confronted with [. . .] an unprecedented complex of explosive problems,"[30] posing the risk of "*economic, social, political, demographic or ecological debacle*" if man is not able to control the development of technology and "*guide his destiny.*"[31]

To Peccei, technological progress was now the most decisive determinant of human history. He referred to the French planner Louis Armand and to the heterodox economist Kenneth Boulding when he described human history as the succession of thresholds putting a limit to human existence. Among these thresholds were thresholds of speed (of supersonic transport, of news, of machines) and thresholds of scale (of energy produced and materials extracted). Peccei was clearly also inspired by the two men in the Club of Rome responsible for methodology, the futurologists Erich Jantsch and Hasan Ozbekhan. Both were born in Europe, and had been influenced by the American approaches to future studies, as well as by environmentalist thinking. Jantsch and Ozbekhan brought to the Club a representation of world evolution inspired by the American catastrophism, and a normative vision of futurology inherited from the European prospectivists.[32] Jantsch had coined the concept of "uncontrolled growth" of the human system, which he described in cybernetic terms. To him, man had to become the self-regulator automatism in that system in order to design his own future. Ozbekhan had drafted a project of world-scale planning with the purpose of restoring ecological balance. This project described "two major disruptive events" in the relationship between man and his natural environment: "population increase" and "the imposition upon the earth of our technological creations." To him these two factors had caused humanity "to destroy forests, flood deserts, change climates, divert rivers" in a possibly irreversible way: to Ozbekhan, humans had "use[d] the natural resources in such quantities that their normal replenishment [could] no longer subvene to [their] needs": they had "subjugate[d] animals and plants to serve [their] techno-economic purposes." He described this process as "spoilage," "ugliness," and "discomfort."[33]

Ozbekhan's ideas resonated with Peccei's own fears about the future. These fears, which derived more from intuition than from rational reasoning, were quite a central aspect of Peccei's vision of the world. Peccei thus hired Ozbekhan to develop a methodology for the Club of Rome. The latter drafted a project that was not retained, because it was considered too expensive and too complicated, hence not competitive with Forrester's mathematical world model. Among the various visions of the future that were expressed by Peccei and his coworkers, Forrester selected Jantsch and Ozbekhan's perspective, which described technological progress as a disruptive force, and completely omitted Peccei's preoccupations about the "technological gap". This was probably not a conscious selection process, but we may infer that the aspects chosen by Forrester perfectly fitted with the concepts of System Dynamics, which was conceived initially in order to represent stocks and flows in a firm. The World 2 and World 3 models described the world as a global entity. As all variables were global averages (which attracted many criticisms), the model was unable to represent a phenomenon like a "technological gap" between developed and developing nations. Moreover, the core structure of the model represented the contradiction between exponentially growing variables—population and industrial production—and the steady-state character of other factors—the supply of natural resources and the capacity for pollution absorption in the atmosphere. This conflict, between projected exponential growth on the one hand and limits in planetary capacity on the other, was responsible for the "overshoot and collapse" behavior of the model, which was eloquently illustrated by the dramatic charts in the report, reproduced in Figure 4.2.

A third vision of technology can be seen as emerging out of Peccei's very first stances about the future of mankind. In his 1965 speech "The Challenge of the 1970s," Peccei expressed his fascination for the revolutionary management methods in use in the US, which relied on computer programming. These approaches, which had been developed for military purposes

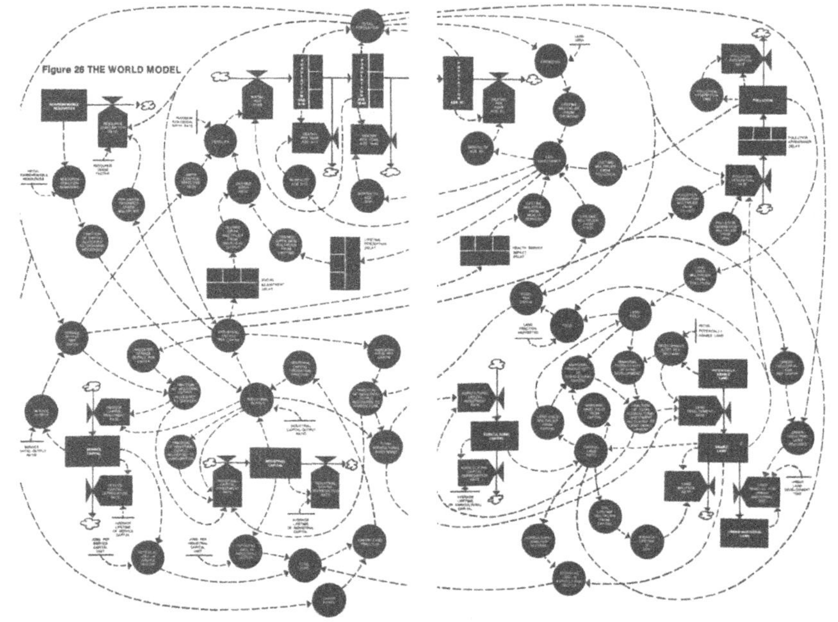

Figure 4.2 The World 3 model flow diagram (Donella H. Meadows, et al., *The Limits to Growth: A Report on the Club of Rome's Project on the Predicament of Mankind* (New York: Universe Books, 1972), Figure 26, p. 102–103. Courtesy of Dennis Meadows.

during World War II and the Cold War,[34] had been successfully applied to civil matters. In the 1960s, "technological forecasting" was booming in industry and was increasingly being applied also to areas such as urban planning, transportation systems, natural resources and education. Several organizations were pioneering in these domains, among them the RAND, the System Development Corporation, and the Hudson Institute.[35] This process was supported by the US Government itself, which had generalized the planning, programming and budgeting system (PPBS), initially developed in the Defense Department, to the whole of the federal administration. This planning system relied on methods such as systems analysis, operational research and cost-efficiency studies, and aimed to help rationalize management toward national objectives.

In many ways, the Club of Rome project perfectly fit with this perspective. It was also nourished by developments in European futures studies, which had an explicitly normative purpose.[36] At the same time, the global scale of the Club of Rome's projections and its apocalyptic views contrasted greatly with planning rationalities developed until then. This paradoxical combination of a radical ecological discourse and a technocratic planning ambition was at the core of what Peter Moll has called "the cross-fertilization between environmentalism and future studies,"[37] and ran right through the Club of Rome. In the first project he drafted,[38] Ozbekhan suggested developing a world-scale plan that would aim to restore ecological balance on Earth. To him, it was necessary to completely redefine the relationships between Man and his environment, to stop viewing nature as an object of conquest, and consider, rather, the human system as a whole. Later on, Ozbekhan refined this perspective. In the presentation he gave at the Bern meeting of the Club of Rome in 1970, he described a very ambitious world modeling project. This project was built on a truly revolutionary method, in which the whole model could be manipulated, rather than just specific parameters, as it would be the case with models World 2 and World 3. In Ozbekhan's interpretation, system models could not have fixed structures, but a model of the world would need to evolve continuously. At first, the model would represent the problems associated with the ecological imbalance of the world. But as governments worldwide were alarmed by these projections and applied the recommendations for achieving ecological balance of the systems model, then the situation would evolve and the structure of the model would change in consequence.

Modeling, however, was a both time-consuming and costly business. Time and cost were decisive factors on the final choice of modelling methodology, and because Ozbekhan's proposition appeared complicated, the Club of Rome chose Jay Forrester's World model, which projected overshoot and collapse, but did not allow for systemic change as a result of political action. Jay Forrester was very confident in his model. He believed that virtually any system, including the whole planetary system, could be translated into a System Dynamics model. To Forrester, who stated this conception very explicitly, each real system had its own characteristic behavior, and undertaking virtual experiments with the corresponding System Dynamics model revealed this behavior and determined how to best manage that system.[39] In short, modeling the planetary system was no different from modeling flows of stock in a firm, and the planetary system could, with the right model, be managed as efficiently as a corporation. Because the *Limits to Growth* depended on the World 3 model for its dramatic results, it also reproduced this belief in the superiority of modeling. The report was presented as an indisputable account of world tendencies in the areas of population growth, natural resources exhaustion and pollution, and these tendencies were portrayed as authoritative because of the way that they had been produced through a pioneering mathematical model of the world. The charts of the report, which showed the evolution of the five main variables of the world model, reinforced the message of the model. They were the perfect graphic illustration of doom. In fact, the doom predicted by the World model was not so different from the doom predicted by environmentalist discourse in the late 1960s and early 1970s, but whereas environmentalists warnings of environmental apocalypse could be ignored, the model, with its scientific authority, could not.

The final representation of the complex interplay between technology and the future that emerges from the Club of Rome was a total reiteration of the catastrophism of the environmentalist movement. This catastrophism reconciled two rather different ideas of the role of technology. First, environmentalist thinking emphasized technology as a global force that transformed the material basis of human societies and threatened the future by generating its own growth. The process of technological development was thus understood, in the thinking of intellectuals such as American biologist Barry Commoner or Frenchman Alexandre Grothendieck, as being largely governed by technology itself in a process outside of human control. Second, other strands of environmentalist thinking dismissed the idea that technological solutions could solve environmental problems, and saw technology, instead, as a quick fix, susceptible of temporarily hiding certain problems while creating new and more difficult ones.

We find the roots of this representation of technology highly representative of doomsday writings of the 1960s and 1970s, as found in the writings of Ivan Illich and Jacques Ellul.[40] Illich was famous for coining the concept of counterproductivity of technological artefacts. Ellul spent his whole life studying and interpreting technological development and its meaning to global civilization. Amongst their followers, Commoner, who had been influential in the movement against the civilian use of nuclear power, called for a re-foundation of science and technology, in order to transform the whole economic and political system toward ecological balance.[41] British author and journalist Gordon Rattray Taylor, author of *The Doomsday Book*, denounced the false solutions of "technological optimists" who relied on abstract reasoning and deliberately ignored the ecological unbalances that their agro-industrial planned had provoked.[42] The *Blueprint for Survival*, which was the emanation of the British review *The Ecologist*, described a reconversion plan to be applied by the whole of human society for a period of one hundred years. Its authors considered that there can be no technological solutions to the disruption of ecological balance. Consequently, the organization of human society had to be completely rethought. Until societies were composed of self-sufficient and self-regulating communities, which they saw as possible sometime between 2050 and 2075, this process of reorganization would have to involve new systems of planning for the purpose of ecological balance.

Even if the approach taken to technology in the Club of Rome was, as we have seen, much more complex and ambiguous, these radical environmentalist positions were also relayed in *Limits*. The report reiterated the rejection of technology as the solution to the impending environmental crisis. The World 3 model, described in *Limits*, simulated an overshoot and collapse pattern caused by the depletion of natural resources. The report examined closely the possibilities of technological solutions that might push these natural limits back. A range of hypotheses concerning possible technological solutions were considered; this led to the modification of the settings in the model in order to consider, for instance, the possible doubling of the stock on initial natural resources. It followed from this assumption that a collapse of the system due to a depletion of resources could be avoided. But the simulations of the World 3 model then showed another type of overshoot, this one due to the increase in pollution following from continued industrial production.

Other "technological hypotheses" were examined, but none of them modified the system's behavior; they only postponed the point of collapse. This led the authors of the report to emphasize that both population growth and industrial production imperatively had to be stabilized. With these two hypotheses (the stabilization of population and industrial production at a global level), the global system would grow and then quickly stabilize. The latter scenario, integrating these hypotheses, was more valued by the report's authors because they advocated structural political breaks in order to move toward stability.

TECHNOLOGY AS SALVATION: THE CONTROVERSY OVER THE WORLD MODELS

In the previous pages, I have shown that the Club of Rome project was structured around a number of potentially contradictory representations of the role of technology. Meanwhile, the Club of Rome would be informed by one dominant notion of technology, namely the idea of modeling as the necessary tool with which to project the long term. To the *Limits* report's authors, forecasts about the future of mankind had authority because they were produced through mathematical simulations of the future of the world. In the introduction to the report, the authors opposed "mathematical modelling" to "mental modelling." Whereas they described mental modeling as based on partial and confused perceptions of reality, they stated that mathematical modeling, based on accurate representations of "real systems," was able to provide undisputable information about the characteristic behavior of these systems. They presented the World model as a model that "aimed to improve [the] mental models of long-term, global problems by combining the large amount of information that is already in human minds and in written records with the new information-processing tools that mankind's increasing knowledge has produced—the scientific methods, systems analysis, and the modern computer."[43]

At the symposium set up for the publication of the report on the 2nd of March 1972 in Washington, DC, the very use of a mathematical model, which was an unprecedented approach for political matters, was the object of heated discussion. To Stewart Udall, the model was a legitimizing tool for pre-existing environmentalist discourse, as he expressed it to *The New York Times*, "You have made us respectable."[44] To environmentalist Lester Brown and member of the executive committee of the Club of Rome Eduard

Pestel, the model was what would allow for the taking of concrete political decisions. To other commentators, however, the forecasts presented in the report were perceived as dangerously simplistic portrayals of complex problems. The model, they argued, could be contested. Peccei, who was more interested in creating a shock about the environmental crisis than about the scientific coherence of the enterprise he initiated, expressed his strong confidence in the World model, while admitting to its preliminary character:

> What we needed was a stronger tool of communication to move men of the planet out of their ingrained habits. This is the reason for the MIT study and the book. Its conclusions are preliminary, but it is a key which permits every layman to enter the labyrinth of the fantastic problems towering over mankind.[45]

The model thus appeared as a tool of communication and of persuasion, rather than as an investigation tool about the impacts of several great policies on the future of Mankind.

In the months that followed the publication of the report, these criticisms were further developed. The Club of Rome members and MIT Scientists were accused of technocracy. For instance, Harvey Simmons, from the Science Political Research Unit (SPRU) of the University of Sussex, drew parallels between the approach of the MIT team and the technocracy movement of the 1930s.[46] Simmons showed how both groups shared a faith in engineering techniques (in designating the cause of social problems, and in creating solutions to these problems), skepticism about citizens' capacity to understand their own situation, and the call to an elite to solve future problems. The use of a mathematical model was also denounced as giving a false impression of validity to political prescriptions, which were in reality disputable. Several commentators pointed out the fetishism of the computer among MIT scientists and dismissed computers as just calculators that draw conclusions from the assumptions submitted to them by human beings. To the economist Carl Kaysen, the argument of the *Limits* report was "presented with a sufficient panoply of graphs, flow diagrams [. . .] and invocations of the computer to produce an aura of scientific authority to the conclusions."[47]

But computer modeling would become the widely accepted approach for studying global problems in the wake of the Club of Rome report. Specifically, *Limits* spawned the genre of global modeling. In the ten years following its publication, a dozen global models were developed, each one bringing its own framework to the debate on growth and development, and each one commenting on *Limits*. The International Institute for Applied Systems Analysis (IIASA) would, as demonstrated by other chapters in this volume, play a central role in the coordination of this emerging modeling field.[48] IIASA would become the meeting point of several groups of global modellers who included not only the American economist William Nordhaus of the Cowles Commission (the same Nordhaus who was outspokenly dismissive of the *Limits to Growth* report at the time of its publication), but also the futurologist Herman Kahn. Most of these "global models" explicitly aimed to either complement or contradict the initial report, and they often led to the publication of a specific "Club of Rome report," such as the report issued by the Bariloche Foundation (see below). Several authors were themselves members of the Club. In other words, IIASA and the Club were privileged spaces where modelers could experiment, exchange ideas and create a new consensus of what could be reasonably expected from global modelling.

Nordhaus's critique of *Limits* contributed directly to the development of the first "integrated assessment model" of climate change. Nordhaus blamed *Limits* for failing to take proper account of economic, political and technological mechanisms that might change the overshoot and collapse scenario by transcending the physical limits to growth, if these in fact existed. Whereas the World 3 model was based on assumptions of technology as a fundamentally uncontrollable process and as impotent in providing real solutions to systemic collapse, most critics of the model defended the idea that a more positive understanding of technology could lead to more optimistic conclusions about the sustainability of industrial growth.[49] According to these critics, technology could help slow down population growth, increase food production, control pollution, and develop the mining of low-grade ores, which until then were considered too expensive. Provided that enough energy was available, the problem of the scarcity of natural resources could under such premises completely disappear: it would be increasingly easy to extract and recycle those resources, and there would be plenty of energy available. It was noticed that in the World models, population and industrial production grew exponentially, while technology did not follow this pattern. If technology would, in contrast, be allowed to grow, assumptions of the limits of natural resources could be fundamentally challenged. This was in essence the critique of Nordhaus, whose rival model introduced the assumption of unlimited natural resources as well as the possibility of industrial production without pollution. Nordhaus also assumed that production inputs (resources, capital and labor) were perfectly substitutable, in accordance with neo-classical theory, but in complete contradiction to the message in *Limits*. Not surprisingly, Nordhaus's model stated that per capita income could continue to grow, fast, and multiply by six by the year 2060.

The critique of World 2 and World 3 emanating from the Sussex team at SPRU was the other fundamental challenge to the report.[50] In a series of articles, SPRU stated that the very structure of the World models was responsible for the behavior of the global system. Indeed, they showed how the positive and negative feedbacks in the "industrial production" sub-model were such that the collapse of the world economy was inevitable if industrial growth was to continue. To the Sussex team, this was an overly fatalist vision of the relationships between humans and their environment, and they suggested various modifications of the world models.[51] These modifications included the stock of natural resources growing exponentially; a much weaker link between industrial growth and the stock of natural resources, assuming that substitution between energy sources, the extraction of unconventional hydrocarbons, and the development of nuclear fusion render energy shortages virtually impossible; decreasing future costs for cultivating new land, and the production of synthetic food; and changes in the relationship between production and pollution, as new technologies might be able to control pollution. As with Nordhaus's propositions, these hypotheses drastically altered the behavior of the World models. In the modified versions, growth could go on indefinitely without system collapse.

Important sets of objections also came from groups that intended to produce their own rival world models in response to the *Limits* report. For instance, researchers from the Bariloche foundation in Argentina developed a world model focused on the perspective of developing countries, which aimed to investigate the possibility for developing countries to reach an acceptable level of life after a few generations. They did not worry about environmental problems, which to them were typical of a Western point of view. On the contrary, they assumed technological improvements, which would not only prevent the exhaustion of natural resources over time, but allow for their stock to increase. They also believed, like the SPRU team, that technical improvements would allow for efficient pollution control. Consequently, the Bariloche model considered variables like "resources" to be irrelevant, and only took them into account indirectly with respect to the effect they had on production costs in certain sectors. Importantly, then, the causality implied between the World models, the SPRU models and the Bariloche models were in complete contradiction. Whereas the World model postulated that more growth meant more pollution and fewer natural resources, later models postulated that more growth meant more capacity to control pollution and extract natural resources.[52]

Every modeling project developed in the aftermath of *Limits* had its own ideological framing of world problems. In fact each one, except the World models presented in *Limits*, assumed the pursuit of growth and development as something natural. As a consequence, only the very first model reached the conclusion that industrial growth had to be halted. The Bariloche model was focused on the needs of developing countries and examined the perspectives of development for these countries. Its main variable was life expectancy, which the model aimed to maximize. Another model, built by Wassily Leontief in collaboration with the United Nations was focused on the distribution of wealth among regions of the world.[53] As to William Nordhaus, he asserted, as early as 1973, that the greatness of the US was to him inseparable from "television, overheated houses, aluminum cans, and jet setters," a statement that informed his subsequent search for alternative models.[54] Herman Kahn and his colleagues from the Hudson Institute, which authored a report entitled *The Next 200 Years*, objected that the authors of the *Limits* report had underestimated the positive role of technology.[55] Only this misconception had led them to emphasize serious issues, issues that were in any case not so serious according to the Hudson Institute.[56] Kahn and his coworkers believed that global society would naturally enter a more stable "postindustrial" state. For instance, the "population explosion" would cease to be a problem once all countries had entered an industrial era and their inhabitants changed their reproductive habits. Energy would be increasingly supplied from "eternal sources" like "solar, geothermal, fission and fusion,"[57] and the transition would be quasi-complete around the year 2050. Raw materials would never be exhausted. Food would not be in shortage as conventional agriculture would increase its productivity with the help of chemical products and as unconventional forms of agriculture (the production of food requiring no soil) developed. The maintenance of a satisfactory environment would be "both economically and technically feasible during the next 200 years [. . .] With improvement in technology and practice, eventually the costs [would] go down."[58]

In other words, subsequent generations of global models projected, against the backdrop of a series of different motivations the fundamental sustainability of industrial growth. Even the second Club of Rome report, published in 1974 asserted that industrial growth was not condemnable per se, but that a specific model of growth, which would be shared more equally and would be less detrimental to the environment, had to be pursued.[59]

POSTSCRIPT: FROM LIMITS TO GROWTH TO THE SUSTAINABLE DEVELOPMENT PARADIGM

William Nordhaus's critique of *Limits* led him to build several models of his own. Progressively, he would start examining the question of climate change, as public awareness was raised about this problem at the end of 1970s.[60] This work was conducted for about fifteen years, and led to the construction of the so-called Dynamic Integrated Climate-Economy (DICE) model in 1992. In two areas, energy and pollution, Nordhaus's conclusions differed radically from the *Limits* report's work two decades earlier. Nordhaus determined a "time path of optimal US technological processes in solving the energy problem, by demand categories, 1970–2120 and beyond,"[61] and asserted that there may be some difficulties in implementing new energy production technologies, but that "the current 'energy crisis' [would] blow over eventually" and that "[man] should not be haunted by the specter of the affluent society grinding to a halt for lack of energy resources."[62] The model developed by Nordhaus does not seek to stabilize emissions, but frames environmental problems as purely economic problems. As concerns climate change, he first asserts the necessity to avoid outstripping the 2°C threshold. Then his model determines the optimal economic path leading to various levels of greenhouse gases concentration in the atmosphere. In

1991, he continued to assert the necessity to stabilize the climate. But in 1992, in a *Science* article that summarized the main conclusions of his work, he freed himself from physical objectives and claimed an optimal trajectory leading to a warming of +3.2°C in 1975. This freeing from the physical reality of climate change lies in abandoning a cost-efficiency approach and adopting a cost-benefit one.

From the technological point of view, several assumptions reflect the great optimism of the approach. In 1974, in his model intended to optimize energy production and distribution, Nordhaus assumed that around year 2010, a new technology, called "backstop technology" would allow the production of an infinite amount of energy without producing any pollution. Consequently, his model allowed the optimization of the production of energy before the generalization of this technology, and the crisis announced by the Club of Rome was to him absolute nonsense.[63] This "backstop technology" was also implicitly present in subsequent works. In 1980, it was seen as a technology allowing to reduce the greenhouse gases emissions at fixed cost, even if this reduction represented a great part of the emissions. In 1992, the geoengineering assumption allows avoiding climate change at a virtually null cost. Apart from that, Nordhaus appears very optimistic concerning the potential effects of climate change. For instance he assumes that the effects of marginal emissions are decreasing, which contradicts the thesis that beyond a given warming, the ecosystems could suffer irreversible dislocations.

This chapter has shown that the question of the relationship between technological development and the future of humanity was a central theme in the Club of Rome project and remained so in the debates triggered by *Limits* and the World models. But whereas the Club of Rome incorporated several different representations of the role of technology for the future, including a highly dystopian vision of doom borrowing from catastrophist environmentalist writings of the late 1960s and early 1970s, the subsequent critiques of the Club of Rome mostly displayed a very strong technological optimism. We have seen how this reevaluation of the role of technology influenced the building of alternative world models, which in the end asserted, to the world, that there was no reason for concern about the future of industrial growth. These various critiques and models thus paved the way for the emergence of the "eco-development" paradigm, which was officially adopted at the 1972 Stockholm Conference, and later on the emergence of the sustainable development paradigm. This model of conciliation between environment and development would become dominant and shape political discourses and arrangements including around the issue of climate change.

The Club of Rome, despite its technocratic approach, thus contributed to the opening up of a space of debate in academic and political arenas, one that global modeling and environmental economics seem to have closed in the following decades. In retrospect, the *Limits* debate can appear as an ephemeral transgression or a parenthesis which was quickly closed.

Nevertheless, that debate gave visibility to an array of subjects (pollution, the scarcity of industrial and food resources) which have gained social and political acceptance since.

NOTES

1. Donella H. Meadows, John Richardson, and Gehrart Bruckmann, *Groping in the Dark: The First Decade of Global Modelling* (New York: Wiley, 1982).
2. Donald Worster, *Nature's Economy: A History of Ecological Ideas* (San Francisco: Sierra Club Books, 1977).
3. Robert Golub and Joe Townsend, "Malthus, Multinationals and the Club of Rome," Social Studies of Science 7, no. 2 (1977): 201–222; Francis Sandbach, "The Rise and Fall of the Limits to Growth Debate," *Social Studies of Science* 8, no. 4 (1978): 495–520; Brian P. Bloomfield, "Anomalies and Social Experience: Backcasting with Simulation Models," *Social Studies of Science* 15, no. 4 (1985): 631–675.
4. Céline Pessis, Sezin Topçu, and Christophe Bonneuil, eds., *Une autre histoire des "Trente Glorieuses"* (Paris: La Découverte, 2013).
5. Hasan Ozbekhan, *The Predicament of Mankind: Quest for Structured Responses to Growing World-Wide Complexities and Uncertainties*. Club of Rome (1970). http://quergeist.net/Christakis/predicament.pdf.
6. This is what the methodology was called initially.
7. Andersson, this volume.
8. Peter Moll, *From Scarcity to Sustainability* (Frankfurt am Main: Peter Lang, 1991).
9. There was virtually no reaction to the recommendation to stabilize the global population.
10. William D. Nordhaus, "The Allocation of Energy Resources," *Brookings Papers on Economic Activity* 3 (1973), 529.
11. Johan Galtung, "'The Limits to Growth' and Class Politics," *Journal of Peace Research* 10, no. 1–2 (1973): 101–114.
12. Claude Vadrot, *Déclaration des Droits de la Nature* (Paris: Stock, 1973).
13. Jean-Claude Thill, ed., *La Lettre Mansholt—réactions et commentaires* (Paris: Jean-Jacques Pauvert, 1972).
14. Aurelio Peccei, *The Challenge of the 1970s to the World of Today* (Buenos Aires: National Military College, 1965).
15. Aurelio Peccei, *The Human Quality* (Oxford: Pergamon Press, 1977).
16. Ibid., 7.
17. Ibid., 11.
18. Ibid.
19. Peccei, *The Challenge of the 1970s*, p. 114.
20. Pierre Cognard, *Recherche scientifique et indépendance* (Lausanne: DGRST, Centre de recherches européennes, 1965).
21. Peccei, *The Chasm Ahead* (Toronto: MacMillan, 1969).
22. Ibid., 70.
23. John McCormick, *Reclaiming Paradise: The Global Environmental Movement* (Bloomington, IN: Indiana University Press, 1989).
24. Rachel Carson, *Silent Spring* (Cambridge, MA: Riverside Press, 1962).
25. Matthew Connelly, *Fatal Misconception: The Struggle to Control World Population* (Cambridge, MA: Belknap, 2008); Laura Landolt, "Constructing Population Control: Social and Material Factors in Norm Emergence and Diffusion," *Global Society* 21, no. 3 (2007): 393–414; Élodie Vieille Blanchard, *Les Limites*

à la croissance dans un monde global—Modélisations, prospectives, réfutations (PhD diss., École des Hautes Études en Sciences Sociales, 2011).

26. Barry Commoner, *The Closing Circle: Nature, Man, and Technology* (New York: Knopf, 1971); Paul R. Ehrlich, *The Population Bomb* (New York: Ballantine Books, 1968); Gordon Rattray Taylor, *The Doomsday Book* (London: Thames and Hudson, 1970a).
27. Vieille Blanchard, *Les Limites à la croissance dans un monde global*, 218–242.
28. Peccei, *The Challenge of the 1970s to the World of Today*, 115.
29. Peccei, *The Chasm Ahead*, xiv.
30. Ibid., xiv (bolding in the original text).
31. Ibid., 136 (italics in the original text).
32. Moll, *From Scarcity to Sustainability*.
33. Hasan Ozbekhan, *Towards a General Theory of Planning* (Paris: OCDE, 1969), 102.
34. Amy Dahan Dalmedico and Dominique Pestre, eds., *Les sciences pour la guerre 1940–1960* (Paris: Presses de l'EHESS, 2004).
35. Erich Jantsch, *La Prévision technologique* (Paris: OCDE, 1967).
36. Moll, *From Scarcity to Sustainability*.
37. Ibid.
38. Ozbekhan, *Towards a General Theory of Planning*.
39. William Thomas and Lambert Williams, "The Epistemologies of Non-Forecasting Simulations, Part I: Industrial Dynamics and Management Pedagogy at MIT," *Science in Context* 22, no. 2 (2009): 245–270.
40. Ivan Illich, *La Convivialité* (Paris: Seuil, 1973); Jacques Ellul, *La Technique ou l'enjeu du siècle* (Paris: Armand Colin, 1954).
41. Commoner, *The Closing Circle*.
42. Gordon Rattray Taylor, *The Doomsday Book* (London: Thames and Hudson, 1970).
43. Donella H. Meadows et al., *The Limits to Growth*, 21.
44. Robert Gillette, "The Limits to Growth: Hard Sell for a Computer View of Doomsday," *Science* 175, no. 4026 (1972): 1092.
45. Ibid.
46. Harvey Simmons, "Systems Dynamics and Technocracy," in Hugh Cole, ed., *Thinking About the Future: A Critique of the Limits to Growth* (London: Chatto & Windus, 1973).
47. Carl Kaysen, "The Computer that Printed Out W*O*L*F*," *Foreign Affairs* 50, no. 4 (1972): 660.
48. Meadows, Richardson, and Bruckmann, *Groping in the Dark*. See also Rindzevičiūtė, this volume.
49. Philip H. Abelson, "Limits to Growth," *Science* 175, no. 4027 (1972): 1197; Robert Boyd, "World Dynamics: A Note," *Science* 177, no. 4048 (1972): 516–519 ; Hugh Cole, ed. *Thinking About the Future: A Critique of the Limits to Growth* (London: Chatto & Windus, 1973); Kaysen, "The Computer"; Peter Passell, Marc Roberts, and Leonard Ross, "The Limits to Growth, World Dynamics and Urban Dynamics," *The New York Times Book Review* (2 April 1972): 1, 10–13; Martin Shubik, "Review: Modeling on a Grand Scale," *Science* 174, no. 4013 (1971): 1014–1015.
50. Cole, *Thinking About the Future*.
51. Christopher Freeman, "Introduction," in Hugh Cole, ed., *Thinking About the Future: A Critique of the Limits to Growth* (London: Chatto & Windus, 1973).
52. Amilcar Herrera, *Catastrophe or New Society? A Latin American World Model* (Ottawa: International Development Research Center, 1976).
53. Wassily A. Leontief, A. Carter, and P. Petri, *The Future of the World Economy* (New York: Oxford University Press, 1977).
54. Nordhaus, "The Allocation of Energy Resources," 529.

55. Herman Kahn, William Brown, and Leon Martel, *The Next 200 Years: A Scenario for America and the World* (New York: William Morrow and Company, 1976).
56. Ibid., 4.
57. Ibid., 67.
58. Ibid., 139.
59. Mihajlo Mesarovic and Eduard Pestel, *Mankind at the Turning Point : The Second Report of the Club of Rome* (New York: Dutton, 1974)
60. William D. Nordhaus, "Strategies about Carbon Dioxide," *Cowles Foundation Discussion Paper* 443 (1977); William D. Nordhaus, "Thinking about Carbon Dioxide: Theoretical and Empirical Aspects of Optimal Control Strategies," *Cowles Foundation Discussion Paper* 565 (1980).
61. Nordhaus, "The Allocation of Energy Resources," 552.
62. Ibid., 570.
63. Nordhaus, "Resources as a Constraint on Growth." *American Economic Review* 64, no. 2 (1974): 22–26.

REFERENCES

Abelson, Philip H. "Limits to Growth." *Science* 175, no. 4027 (1972): 1197.

Bloomfield, Brian P. "Anomalies and Social Experience: Backcasting with Simulation Models." *Social Studies of Science* 15, no. 4 (1985): 631–675.

Boyd, Robert. "World Dynamics: A Note." *Science* 177, no. 4048 (1972): 516–519.

Carson, Rachel. *Silent Spring*. Cambridge, MA: Riverside Press, 1962.

Cole, Hugh, ed. *Thinking About the Future: A Critique of the Limits to Growth*. London: Chatto & Windus, 1973.

Commoner, Barry. *The Closing Circle: Nature, Man, and Technology*. New York: Knopf, 1971.

Connelly, Matthew. *Fatal Misconception: The Struggle to Control World Population*. Cambridge, MA: Belknap, 2008.

Cognard, Pierre. *Recherche scientifique et indépendance*. Lausanne: DGRST, Centre de recherches européennes, 1965.

Dahan Dalmedico, Amy and Dominique Pestre, eds. *Les sciences pour la guerre 1940–1960*. Paris: Presses de l'EHESS, 2004.

Ehrlich, Paul R. *The Population Bomb*. New York: Ballantine Books, 1968.

Ellul, Jacques. *La Technique ou l'enjeu du siècle*. Paris: Armand Colin, 1954.

Freeman, Christopher. "Introduction." In *Thinking About the Future: A Critique of the Limits to Growth*, edited by Hugh Cole, pp. 5–13. London: Chatto & Windus, 1973.

Galtung, Johan. " 'The Limits to Growth' and Class Politics." *Journal of Peace Research* 10, no. 1–2 (1973): 101–114.

Gillette, Robert. "The Limits to Growth: Hard Sell for a Computer View of Doomsday." *Science* 175, no. 4026 (1972): 1088–1092.

Golub, Robert and Joe Townsend. "Malthus, Multinationals and the Club of Rome." *Social Studies of Science* 7, no. 2 (1977): 201–222.

Herrera, Amilcar. *Catastrophe or New Society? A Latin American World Model*. Ottawa: International Development Research Center, 1976.

Illich, Ivan. *La Convivialité*. Paris: Seuil, 1973.

Jantsch, Erich. *La Prévision technologique*. Paris: OCDE, 1967.

Kahn, Herman, William Brown, and Leon Martel. *The Next 200 Years: A Scenario for America and the World*. New York: William Morrow and Company, 1976.

Kaysen, Carl. "The Computer that Printed Out W*O*L*F*." *Foreign Affairs* 50, no. 4 (1972): 660–668.

Landolt, Laura. "Constructing Population Control: Social and Material Factors in Norm Emergence and Diffusion." *Global Society* 21, no. 3 (2007): 393–414.
Leontief, Wassily, A. Carter, and P. Petri. *The Future of the World Economy*. New York: Oxford University Press, 1977.
McCormick, John. *Reclaiming Paradise: The Global Environmental Movement*. Bloomington, IN: Indiana University Press, 1989.
Meadows, Donella H., Dennis L. Meadows, Jørgen Randers, and William W. Behrens III. *The Limits to Growth: A Report for the Club of Rome's Project on the Predicament of Mankind*. New York: Universe Books, 1972.
Meadows, Donella H., John Richardson, and Gehrart Bruckmann. *Groping in the Dark: The First Decade of Global Modelling*. New York: Wiley, 1982.
Mesarovic, Mihajlo and Eduard Pestel. *Mankind at the Turning Point : The Second Report of the Club of Rome*. New York: Dutton, 1974.
Moll, Peter. *From Scarcity to Sustainability*. Frankfurt am Main: Peter Lang, 1991.
Nordhaus, William D. "The Allocation of Energy Resources." *Brookings Papers on Economic Activity* 3 (1973): 529–576.
Nordhaus, William D. "Resources as a Constraint on Growth." *American Economic Review* 64, no. 2 (1974): 22–26.
Nordhaus, William D. "Strategies about Carbon Dioxide." *Cowles Foundation Discussion Paper* 443 (1977).
Nordhaus, William D. "Thinking about Carbon Dioxide: Theoretical and Empirical Aspects of Optimal Control Strategies." *Cowles Foundation Discussion Paper* 565 (1980).
Ozbekhan, Hasan *The Predicament of Mankind: Quest for Structured Responses to Growing World-Wide Complexities and Uncertainties*. Club of Rome (1970). http://quergeist.net/Christakis/predicament.pdf.
Ozbekhan, Hasan. *Towards a General Theory of Planning*. Paris: OCDE, 1969.
Passell, Peter, Marc Roberts, and Leonard Ross. "The Limits to Growth, World Dynamics and Urban Dynamics." *New York Times Book Review* (April 2, 1972): 1, 10–13.
Peccei, Aurelio. "The Challenge of the 1970s to the World of Today." In *Crusader for the Future*, edited by Gunter A. Pauli, pp. 105–124. Oxford: Pergamon Press, 1987.
Peccei, Aurelio. *The Chasm Ahead*. Toronto: MacMillan Press, 1969.
Peccei, Aurelio. *The Human Quality*. Oxford: Pergamon Press, 1977.
Pessis, Céline, Sezin Topçu, and Christophe Bonneuil, eds. *Une autre histoire des "Trente Glorieuses."* Paris: La Découverte, 2013.
Sandbach, Francis. "The Rise and Fall of the Limits to Growth Debate." *Social Studies of Science* 8, no. 4 (1978): 495–520.
Simmons, Harvey. "System Dynamics and Technocracy." In *Thinking About the Future: A Critique of the Limits to Growth*, edited by Hugh Cole, pp. 192–208. London: Chatto & Windus, 1973.
Shubik, Martin. "Review: Modeling on a Grand Scale." *Science* 174, no. 4013 (1971): 1014–1015.
Taylor, Gordon Rattray. *The Doomsday Book*. London: Thames and Hudson, 1970.
Thill, Jean-Claude, ed. *La Lettre Mansholt—réactions et commentaires*. Paris: Jean-Jacques Pauvert, 1972.
Thomas, William and Lambert Williams. "The Epistemologies of Non-Forecasting Simulations, Part I: Industrial Dynamics and Management Pedagogy at MIT." *Science in Context* 22, no. 2 (2009): 245–270.
Vadrot, Claude. *Déclaration des Droits de la Nature*. Paris: Stock, 1973.
Vieille Blanchard, Élodie. *Les Limites à la croissance dans un monde global—Modélisations, prospectives, réfutations*. PhD diss., École des Hautes Études en Sciences Sociales, 2011.
Worster, Donald. *Nature's Economy: A History of Ecological Ideas*. San Francisco: Sierra Club Books, 1977.

5 Toward a Joint Future beyond the Iron Curtain

East–West Politics of Global Modelling

Eglė Rindzevičiūtė

This chapter argues that computer-based global modelling produced particular long-term horizons which played an important, transformative role in Soviet governance by opening it up to East–West cooperation. Global modellers conceptualized the planet as a complex, interconnected system, the understanding of which required transnational scientific cooperation, enabling both scientists and data to cross national boundaries and Cold War divisions.[1] In turn, Soviet scientists forged and used the idea of the long-term future of the world to reveal and criticize problems being experienced, but not always acknowledged, in the Soviet Union. A history of computer-based global modelling is, therefore, a history of intertwining globalization, the transformation of the Soviet regime, and East–West transfer.

The first computer-based global models of social and economic development were produced under the auspices of international organizations, which brought together individuals from the Eastern and Western blocs: the Club of Rome, the United Nations agencies and, most importantly, the first international think tank, the International Institute for Applied Systems Analysis (IIASA) in Austria.[2] Although historians habitually use references to these international organizations as examples of the emergence of global governance, we still lack empirical knowledge about the professional networks that formed the framework of these organizations and the concrete projects that were pursued there.[3] This chapter fills this gap in knowledge by examining several cases of East–West cooperation in computer-based global modelling carried out through IIASA and the United Nations (UN).

But what is global modelling? Indeed, global modelling refers to a great many different concepts and techniques, which could be digital or analogue, purely conceptual or calibrated to run on particular computers. This chapter focuses on computer-based global modelling, a phenomenon that so far has been overlooked in histories of computing, although the impact of computer-based global modelling on modern governmentality cannot be overestimated.[4] First, global models encouraged policy makers to look farther ahead and evaluate present-day policies in the light of their long-term consequences. Even when computer power was quite limited, the idea of computer-assisted long-term planning commanded authority. Thus in 1961 the United Nations adopted the resolution *Planning for Economic Development*, which called for long-term projection and planning of the world economy.[5] Second, to be able to plan for the long-term became synonymous with being an advanced, post-industrial state, and the foremost tool for this kind of planning was a computer. In line with Peter Galison and Bruce Hevly, I suggest that being an expensive undertaking, requiring huge investment in computer technologies and transnational cooperation in collecting and sharing data, global modelling was part of Big Science and, as such, a symbol of state power.[6]

Another important aspect of global modelling is its connection to a very particular social world. Global computer models were traditionally associated with the small and closely-knit teams that created them. As a result, this technology could not be easily decoupled from its producers: the majority of models could not be easily reproduced or circulated through anonymous channels. Unlike computer hardware, the blueprints of which could be stolen through espionage, transferred and reproduced in another context, computer software for global modelling often had to be *co-produced* through *face-to-face* collaboration in order to be transferred. This is because the ability to use global models depended on almost tacit knowledge of particular systems. As a result, computer modelling platforms were disseminated through personal connections among the modellers.[7] Hence the history of global modelling is also a story of the emergence and spread of particular informal groups of both scientists and policy-makers. These often informal groups of global modellers were probably too loose and ad hoc to be described as transnational communities, but they certainly could be understood as distinct thought collectives, mobilized by their aim to produce a new type of science, global modelling.[8]

This chapter discusses several such thought collectives, based at the Computer Centre and the All-Union Institute for Systems Research (VNIISI) of the Soviet Academy of Sciences in Moscow. Both the Computer Centre and VNIISI were strongly anchored in international networks through the United Nations and the Institute of Applied Systems Analysis in Laxenburg, Austria. In what follows, I briefly review the origins of global modelling in the Soviet Union and the West. Then I proceed to describe the development of several international nodes, by which I mean *ad hoc*, temporary constellations of technology, scientists and political rationales, which enabled Soviet and Western scientists to co-produce a new, long-term and global future.

HOW TO JOIN CAPITALIST AND COMMUNIST FUTURES

How could it be possible to accommodate capitalist and communist futures in one world model? Did not the communist future exclude capitalism by default? The case of global modelling clearly shows that technology has its own politics that are able to transform existing ideological systems. Global modelling was based on mathematical methods and computer technology and as such belongs to the exact sciences. But it also drew on universalism and global thinking, which have a long cultural and political history and offered a particularly powerful form to the idea of global interconnectivity, which became increasingly influential from the 1960s.

Computer-based global modelling derived its influence from a wider fascination with new technologies as tools of transformation of economies and societies. Such post-war innovations as computer technologies, cybernetics, and systems theory were described by Western commentators as drivers of deep changes in industry, society and the economy. In the late 1950s, such thinkers as Donald Michael in his *Cybernation: The Silent Conquest* (1962), and also Marshall McLuhan, conceptualized these changes as the third technological or cybernetic revolution. Daniel Bell, who was sceptical about many of the postulates about cybernation, later developed the theory of a post-industrial society, leading to the convergence of capitalist and communist regimes, as political ideology would be replaced with technical systems of control.[9] Bell's ideas had some basis in reality: struggling to develop their economy, the Soviets embraced Western science and technology, but the Soviets also fiercely rejected any possibility of convergence with capitalism. A discursive resolution to the problem of how to combine the communist project with Western technoscience was found in 1954, when Soviet philosophers adopted John Desmond Bernal's notion of a scientific-technical revolution (STR).[10] The British thinker Bernal not only espoused leftist principles, but also postulated that progressive science, including the scientific-technical revolution, was by its nature "incompatible with capitalism."[11] Drawing on Bernal's formula, the Soviets posited that although the scientific-technical revolution was identical in East and West, its effects were different: dogma said that the scientific-technical revolution alone could not solve class antagonism and ameliorate the negative effects of private capital ownership; therefore the communist system had nothing to fear.[12] In this way, the scientific-technical revolution was recognized on both sides of the Iron Curtain as a driver of universal change that produced a new future which could not be strictly divided into capitalist and communist camps.[13]

Indeed, what brought the communist and capitalist regimes together was not even the scientific-technical revolution *per se*, but the insight that economic growth, driven by the scientific-technical revolution, had complex global consequences. In a somewhat round-about way the idea of a worldwide scientific-technical revolution significantly changed the meaning of "global" in Soviet scientific and policy thinking. If, as archival documents show, Soviet economists used the notion of "global models" to name models of the national economy in the 1960s, a decade later in the 1970s the notion of "global" economic models was used to refer to the world economy.[14] At the same time, Soviet international relations theorists used the word "globalism" to designate US ambition for the world hegemony. Accordingly, in Soviet international relations discourse, the word "global" was charged with negative undertones.[15] A completely different use of the term "global" emerged in Soviet geophysical sciences, where scholars used "global" to describe planetary processes as early as the 1950s. I suggest that it was through computer modelling that this geophysical notion of the "global" eventually migrated into Soviet economic and, at a later stage, political discourses. The culmination of Soviet global thinking was reached in 1985 when the notion of "global problems" was used for the first time to describe world issues in the official documents of the Congress of the Communist Party.[16]

The contention that the global system referred to a phenomenon that should be understood as both natural, that is, geophysical, and human-made, that is, induced by the scientific-technical revolution, was actively promoted by a new type of actor on the stage of world politics: international organizations. The first impetus to computerize global, that is, planetary processes involving nature, economy and population, came from the Club of Rome, an international organization established by the Italian industrialist Aurelio Peccei. The Club of Rome linked members of state governments, industries and academia hailing from both East and West. In the early 1970s the Club of Rome commissioned the creation of a world model from the American engineer Jay Forrester and a group of researchers directed by the young Dennis Meadows at the Massachusetts Institute of Technology.[17] Consisting of five interacting blocks of agriculture, natural resources, pollution, population and capital, this model was used to demonstrate the power of relations between these different sectors. The key goal was to demonstrate that such relations existed and were strong, rather than to produce a reliable forecast; in fact, the latter was impossible because of the absence of robust and detailed empirical data.[18] It is important to note that in Meadows's model, the long-term dimension emerged as an unintended side effect: this model extrapolated the possible development of world economic growth until it obtained an interesting result, namely, a dramatic decline of the world economy, population and living standards in 2050. Publicized in the report *The Limits to Growth* (1972), the message was clear: the long-term effects of current developments had to be considered in order to avoid a disaster. If humanity wished to maintain its living standards in the future, the leading Western nations had to revise their consumption habits and accept the idea of no-growth.[19]

How did the Soviet Union, struggling to "catch up with and overtake" the West, react to *The Limits to Growth*? First of all, this report did not take the Soviets by surprise, because the Soviet government was already part of the network in which this study originated. The vice-chairman of the state committee of science and technology (GKNT) and son-in-law of Prime Minister Aleksei Kosygin, Dzhermen Gvishiani, was a member of the Club of Rome. In the early 1960s, Gvishiani first met Aurelio Peccei, then the head of Olivetti, during the latter's business trip to Moscow, and since then Gvishiani interacted regularly with Peccei, Alexander King of the OECD and other members of the Club.[20] This network was used to bring some innovative ideas to the Soviet Union before they were made public in West. Thus Gvishiani invited Forrester and Meadows to Moscow to present their World model in the winter of 1970. East-West scholars also met to discuss the theses and methodology that would be used in the study underlying *The Limits to Growth* in a seminar, organized in Italy in 1971.[21]

The very organization of the visit of Forrester's team to Moscow testifies to the fact that the Soviet research administrators, such as Gvishiani, were not only seriously interested in global modelling, but that they also sought to communicate the importance of this approach to the political elite: the American scientists were taken straight to the villa of the mayor of Moscow, where, in an informal environment, they briefed high-ranking Soviet officials, including Gvishiani and his protégé, the future head of global modelling at VNIISI, Viktor Gelovani.[22] Moreover it is clear that the Soviets were able to differentiate between the fiercely Malthusian implication of *The Limits to Growth* and global modelling as a new type of technique for generating policy-relevant knowledge. This dual approach was evident in the Russian translation of this report: the thesis of no-growth was censored, whereas its author, Dennis Meadows, was warmly welcomed to the Soviet Union. Indeed, Meadows visited the Soviet Union more than twenty times, lecturing on computer-based modelling in Moscow and a dozen other cities. Before the publication of *Limits*, Gvishiani initiated the translation of Forrester's *Industrial Dynamics* (1961), which was published in Russian under the title *The Foundations of Cybernetics of Firms* in 1971.[23] Indeed, Gvishiani also supported the translation of *The Limits to Growth*, which was done at the Institute of Scientific Information on the Social Sciences of the Soviet Union (INION), Moscow. However, the Russian translation of *Limits* was distributed only in limited circles within the Soviet Academy of Sciences and held in the tightly regulated special collection at the Lenin Library.[24] Although some entrepreneurial individuals secretly copied the INION's translation of *The Limits to Growth* and sold these copies for 300 USD on the black market, the wider Soviet public only had access to the ideological commentaries on this report.[25]

In this way, in the Soviet Union, just like in the West, *The Limits to Growth* was received with both scepticism (it was described as the most criticized model ever) and fascination. In any case, the path-breaking role of *The Limits to Growth* in the opening up Soviet interest to a new understanding of the basic parameters necessary for scientific governance cannot be denied. The long-term perspective and an understanding of complexity were beginning to be integrated into the Soviet governmentality: both retrospective accounts and published sources reveal the strong interest of Soviet scientists in developing the technique of global modelling, applying it to different policy areas and interacting with their Western colleagues

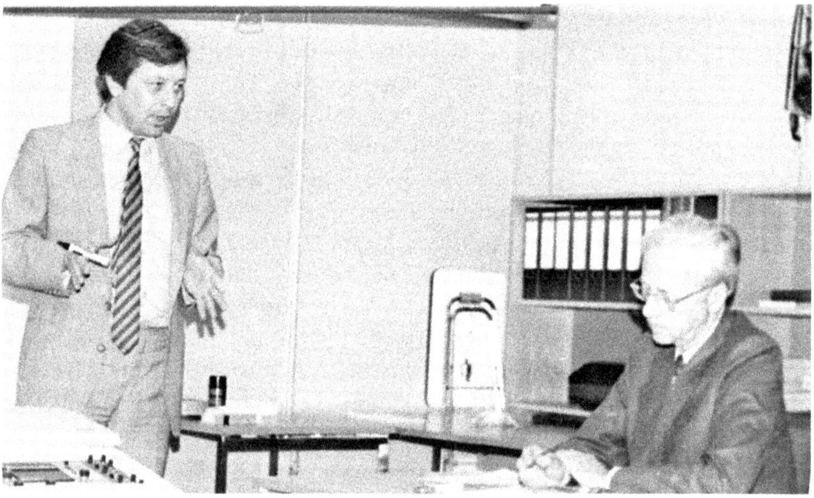

Figure 5.1 Seminar at the laboratory of Viktor Gelovani, presented by Jay Forrester, at the All-Union Scientific Institute for Systems Research (VNIISI), Moscow; the 1970s. Courtesy of the Institute for Systems Analysis, Moscow.

(see Figure 5.1). For instance Mihajlo Mesarovic, a prominent US systems theorist and computer scientist of Serbian origin, the author of another global model also sponsored by the Club of Rome, presented his work at the House of Friendship in Moscow, the public forum from which many prominent Western scientists addressed Soviet audiences.[26] In the next section, I show how global modelling was developed at two international nodes of East–West interactions: IIASA and the UN. Then I return to the developments inside the Soviet Union to discuss the consequences of these international interactions within the authoritarian regime.

INTERNATIONAL NODE 0.1: IIASA

IIASA played a fundamental role in the development of global modelling thanks to its unique institutional design and scientific agenda for developing cutting-edge policy sciences. The idea of an East–West institute was adopted by Lyndon Johnson's administration as part of their bridge-building policy and announced officially in October 1966.[27] Many different actors were behind this proposal: similar ideas for an international think-tank were articulated by econometricians who required access to data on the world economy. For instance, Wassily Leontief called for an East–West centre as early as the late 1950s. Also, the US president's economic advisor, Francis Bator, who contributed to the actual writing of Johnson's bridge-building speech, suggested the idea of an East–West centre for agriculture and management in the summer of 1966.[28] Concrete steps to organize an institute that would address the common problems were taken by former US presidential national security advisor George McBundy and Soviet Prime Minister Aleksei Kosygin. It took more than four years of negotiations among high-level policy makers, such as Gvishiani, McBundy, British government science advisor Solly Zuckerman and President of the US National Academy of Sciences Philip Handler, to agree on the institutional design, research agenda and location of the East–West institute.

In these negotiations the trajectories of the Club of Rome and the US–Soviet negotiators often intersected: for instance, Peccei not only facilitated the meeting of Gvishiani, Bundy and Zuckerman in 1968, but was involved, although not always directly, in the negotiations. This intertwining of the networks of the Club of Rome and the East–West institute turned out to be both an asset and a problem.[29] Purely coincidentally, the IIASA charter was signed several months after the publication of *The Limits to Growth* in 1972. The signatories worried that public opinion might confuse the IIASA and the Club of Rome, especially because some of the IIASA's founding members were fiercely critical of the Forrester/Meadows model.[30] One such was Zuckerman, who in his address to the UN conference on the Human Environment in Stockholm in 1972 argued that global problems should be faced "in a hopeful and scientific spirit and not in one of hysterical computerized gloom."[31] MIT scholar Carl Kaysen, who was McBundy's right hand man in the negotiations over the IIASA, was similarly sceptical about Meadows's model.[32]

In this context, it was not self-evident for IIASA to include global modelling in its research agenda. On the one hand, the newly born IIASA still had to earn its scientific reputation and, consequently, avoid risky projects. On the other hand, global modelling was a genuine innovation and therefore an opportunity to appear at the forefront of science. The dilemma of whether to model or not to model at IIASA was resolved by its first director, American mathematician and decision-scientist of Harvard, Howard Raiffa. IIASA, Raiffa proposed, would not develop any original models, but instead would become a clearing house for global modelling experiments undertaken in different countries.[33] Hence IIASA's research strategy of 1973 included methodological studies of "long-run global simulation" and a series of conferences on this topic.[34] Having hosted six symposia on global modelling from 1974. IIASA became the first platform for sustained international exchange in the area of global modelling.[35]

The IIASA's global modelling conferences played an important role in socializing scientists from East and West into a shared understanding of the possibilities, but also, importantly, the limitations of global modelling. First of all, global modelling was institutionalized as a "normal," albeit post-positivist science. In their internal discussions, as well as published papers, scientists acknowledged that many of the projections generated by global models could not be verified by empirical experiments. Furthermore, the modellers recognized that modelling results were often messy and inconclusive; many modellers, although not all of them, never attempted to hide the inconclusive character of their studies. In this way, it was neither accuracy nor proof, but uncertainty which loomed large behind global modelling.[36] Scientists, for example, agreed that precision was at best something to be aspired to, but that could hardly ever be reached: the data produced by computer models was subject to random errors. Another peculiar feature of global modelling was the discrepancy between the shortage of input data, which was often severely limited and imperfect, and the overflow of output data. Indeed, computers would churn out such volumes of alternative calculations that further software filters had to be designed to figure out which results made sense and which did not.[37]

Special social skills were needed to be able to navigate this complex world of global modelling. For instance, one interviewee, a Russian mathematician, emphasized that a particularly high "mathematical culture" was needed to use a global computer model properly. According to this informant, such a mathematical culture could not be learned from books, but could only be acquired from close and lengthy interaction in modelling teams.[38] It is doubtful that IIASA, where most scientists were visiting and the directors were appointed on temporary contracts, could ever become such a high-brow milieu of mathematical modelling, where sustained face-to-face contacts were paramount. However, IIASA could and did provide mathematicians from East and West with a unique place for meeting.

Discretion was another important quality that IIASA conferences could offer the emerging world community of computer modellers. Being an international, non-governmental organization, the IIASA could not easily be accused of being biased toward particular national or industrial interests. Both printed sources and oral interviews underscore the importance of IIASA's organizational culture of discretion, which enabled the modellers to discuss the most politically unorthodox versions of development. For example, in 1980 an IIASA global econometric modelling conference discussed the implications of Poland and Hungary joining the European Economic Community.[39] Discretion was highly important not only for political, but also for commercial reasons. Scientists were anxious about having their models in progress secretly copied, because then a modeller risked losing his or her potential income from commissions. However, complete discretion also posed a serious problem: without access to the model's architecture, an outsider could never tell if a particular model really worked, that is, if a model had an inner dynamic and if inputs did not straightforwardly determine outputs. The history of modelling shows that refusal to disclose the inner architecture of models ultimately jeopardized their authority.[40]

INTERNATIONAL NODE 0.2: THE UN

Whereas the IIASA offered a place for scientists to develop their global models in a discrete environment, where informal exchange of ideas and mutual scrutiny was made possible behind closed doors, the UN operated on different principles. Based on governmental membership, UN agencies could not offer the same level of discretion and informality (IIASA's members were not governments, but academic organizations). Nevertheless, global modelling pursued at UN agencies was significant, because the UN had a particularly important mandate to collect and share data from the West with the East. The importance of personal contacts and, to a more limited extent, face-to-face cooperation, is revealed in the models of the world economy, which were developed under the aegis of the UN.

It has to be recalled that the institutionalization of mathematical modelling in economics dates back to 1930, when the US-based Econometric Society was established by Ragnar Frisch. However, these early models were mainly theoretical exercises; econometricians began to fill their models with data only after the Second World War. As mentioned earlier, in 1961 the UN began promoting long-term economic planning, equipped with what were then new computer technologies. In 1965, the UN acquired its first mainframe computer and from about the same time began organizing a series of econometric conferences. To meet its needs for international data calculation, the UN established its International Computing Centre in 1971.[41]

The UN supported econometric research as part of its worldwide development program, the rationale for which was shaped in line with modernisation theory. According to modernization theory, Third World countries should imitate Western standards and economic structures. At a later stage, the UN development agenda was widened to include environmental issues, the importance of which were indicated in *The Limits to Growth*. In 1973 the UN initiated a study of the interrelationships between growth, resources, pollution and abatement policies.[42] It was this coupling of the economy and environment that justified the inclusion of communist and capitalist regimes into a single modelling system: the geophysics of the Earth did not respect national boundaries and ideological divides.

In this context, the key Soviet organization to liaise with the UN world development planning program was the Central Institute for Economic Mathematics (TsEMI) at the Soviet Academy of Sciences in Moscow. Established in 1963 and directed by Nikolai Fedorenko (also a member of the Club of Rome), TsEMI enjoyed limited scientific autonomy in the Soviet empire of science and actively sought to join the most prominent international activities.[43] Hence in 1965, Fedorenko attended the first econometrics congress in Rome on the invitation of Wassily Leontief,[44] and TsEMI was also invited to the Copenhagen conference on long-term economic planning that was organized by the UN Economic Commission for Europe in 1966.[45] During the 1960s, TsEMI corresponded and exchanged publications with such pioneering modellers of the long-term as Ragnar Frisch, Jan Tinbergen, and Richard Stone.[46]

Yet the key actor in my narrative is Wassily Leontief, a tireless mediator between Western and Soviet econometricians, as well as a recipient of the Nobel Prize for his method of calculating inter-branch balance in 1973. Leontief was born into a well-off family of Russian industrialists and academics in 1909 and grew up in Saint Petersburg, where he witnessed the Russian revolution unfold.[47] Leontief left Russia in 1925 and returned for the first time in 1959. At the beginning of his exile, he worked at Kiel University in Germany, one of the first institutions to study the world economy in Europe. In the 1930s, Leontief was invited to advise the Chinese government on the development of railway infrastructure. It was during his long trip to China and back that he first encountered the Third World. In 1931, he was invited to join the National Bureau of Economic Research and soon thereafter became a professor at Harvard. Leontief first presented his theory of systems dynamics to the military in Washington, DC; in 1948–1949 his empirical input-output studies were funded by both the Rockefeller and Ford Foundations under the Harvard Economic Research Project.[48] Leontief's mathematical skills, his life experience and proximity to government agencies made him a rather unusual non-academic economist, who was keenly interested in the development of large scale and long-term models.

It was during de-Stalinizaton that Leontief's work entered the Soviet space to later become a standard reference in Soviet global thinking.[49] Thanks to the efforts of the mathematician Vasilii Nemchinov, the Soviet Union legitimized mathematical methods in economics in the mid-1950s. Then Leontief's pupil, Polish economist Oskar Lange, inspired the Soviets to use input-output methods for planning purposes. In 1959, Leontief was officially invited to Moscow, a visit which he described in his memoir as unsatisfactory, having discovered insufficient competence amongst the Soviet economists. However, following this visit Leontief established and chaired the US-Soviet Statistics Bureau in Cambridge, Massachusetts, where many young Soviet administrators were subsequently trained. In this context, it is not surprising that it was Leontief, so well personally integrated in East–West networks, who was commissioned to direct the first study of the world economy for the UN.

At the requested of the UN Centre for Development Planning, Forecasting and Policies, Leontief created the first world trade balance model, the results of which were reported in *The Future of the World Economy* (1976). One of his co-authors was Stanislav Men'shikov, a Russian economist who would later feed the data gathered for Leontief's report to the information-starved economists in Moscow. The data, typically, did not flow in the opposite direction: Leontief's report did not list any Soviet sources.

Outlining scenarios for world development for the next twenty-five years, *The Future of the World Economy* treaded carefully in the terrain of Cold War political divisions. First, the rationale for making such a model was derived from environmental concerns, deemed to be globally relevant and universal irrespective of political ideology. Although the logic of the model was primarily economic (investment and trade flows), it was precisely the environmental effects of economic growth, argued Leontief, which required the introduction of a long-term perspective into the study of development.[50] Second, the political implications of Leontief's analysis were carefully managed. One finding was that without foreign investment, it was impossible to narrow the income gap between developed and developing countries by the year 2000. But such a statement would have placed a direct responsibility on Western governments and was therefore understood as politically controversial at the UN. Accordingly, this finding was left implicit in the report.[51] Third, Leontief depoliticized the very conceptual structure of the world model in many ways. For instance, the model elaborated on possible changes in internal economic structures in developing countries, but no change at all was modelled for the communist regions. Then, world regions were defined according to their economic-administrative system and geographical features. Hence the Soviet Union and Eastern Europe were called "developed centrally planned regions"; meanwhile Western Europe was split into high and medium income regions.[52] As a result, Leontief's model, on the one hand, erased the communist and capitalist divide from the future of the world economy. Yet, on the other hand, Leontief's model conserved the political status quo by avoiding the modelling of changes within the communist system.

A FUTURE LESS THAN BRIGHT? GLOBAL MODELS OF SOVIET DECLINE

Both Meadows's and Leontief's models grew out of attempts to clarify the possibilities of economic development and its consequences for the environment from a long-term perspective. Although the computer technologies that were used were new, the concern with the environment was not. In Soviet Russia, a particularly important role was played by the Russian intellectual tradition which was conducive to the emerging global environmentalist thinking. Indeed, the Soviet intellectual interest in the modelling of global processes predated both Meadows's and Leontief's studies because it stemmed from pre-war thinking, in particular from Vladimir Vernadskii's idea of the noosphere, which was formulated in the 1930s.[53] From the 1960s Vernadskii's thought was promoted by the prominent Soviet biologist Nikolai Timofeev-Resovskii and the prominent mathematician and research director of the Computer Centre, Nikita Moiseev.[54]

As in Western scholarship, Soviet efforts in global modelling oscillated between the economic and geophysical poles. Under Moiseev, the Computer Centre became the heart of geophysical modelling with a particular focus on climate and ecological systems. The focus on interaction between the economy and the environment, as well as the interest in systemic breakdown, was inspired by *The Limits to Growth*, which was discussed at Moiseev's seminar at the Computer Centre.[55] Indeed, Moiseev was first introduced to the *global problematique* of the Club of Rome by the prominent Canadian economist of Russian origin Paul Medow, who lectured on Forrester's model at the Centre in the early 1970s. The networks intertwined: Medow, in turn, invited Moiseev to take part in a meeting organised by the Club of Rome and the RAND Corporation.[56]

Just like his Western colleagues, Moiseev found Forrester's and Meadows's World models mathematically imperfect and limited in their conceptual structure. According to Moiseev, the World models were not good as policy tools for decision-making; indeed, Moiseev was generally sceptical about the use of modelling in economic planning. This scepticism was rooted in his hands-on experience of defining statistical variables for social and economic indicators at the State Planning Committee (Gosplan). Moiseev did, however, compromise, contending that global economic models could be created in principle, but only on the basis of "proper" geophysical modelling. For Moiseev, if the natural processes were not properly understood and represented, it was pointless to model the economy, dependent on natural resources.[57]

A different view was held at VNIISI, which listed global modelling as one of its research priorities. Established under the patronage of Gvishiani, who personally promoted global modelling in the Soviet Union and at IIASA, VNIISI was a very important institute, which, however, so far has been overlooked by historians of Soviet science.[58] VNIISI's global modelling program was directed by another Georgian, Gelovani, who kept direct and close connections with Meadows. If the Computer Centre made a major contribution to the field of geophysical modelling, VNIISI innovated in global economic modelling in the Soviet context.[59]

Both the Computer Centre and VNIISI were closely connected with IIASA and the UN.[60] Gvishiani was the director of VNIISI from 1976 to 1992 and vice director of IIASA from 1972 to 1987. Moiseev was involved with water projects at IIASA and was also behind the Computer Centre's participation in the UNESCO program *Man and Biosphere*, which launched an ambitious international study of the intertwining of human-made and natural systems at the planetary level. Although Moiseev held different views from Gvishiani, their networks closely intertwined. For example, the Balaton Group that modelled the pollution of Lake Balaton in Hungary at IIASA included junior scientists from the Computer Centre and was established jointly by Meadows and Gelovani in 1982.[61]

Importantly, global modelling also entailed horizontal, transnational relations between strong researcher milieus. The first computer-based world model, simulating the interaction between the ocean and atmosphere, was developed by American scientists Syukuro Manabe and Kirk Bryan in 1972. Somewhat later the Soviets also began developing their own geophysical global models. From 1977, the Computer Centre launched a research program to build a world ocean-atmosphere model suitable for environmental analysis; this model was completed in 1982.[62] It should be noted that the development of world models was more creative bricolage than creation *ex nihilo*. For example, in an interview, a Russian mathematician involved in the development of the first global models stressed that his team did not compete for originality; on the contrary, they found it acceptable to borrow already existing models, created by Western scientists. These Western models were adjusted to both local research goals and computer equipment, namely, the Centre's BESM-6.[63] The Computer Centre borrowed a global circulation model, created by Yale Mintz and Akio Arakawa at the University of California Los Angeles, and then improved by Lawrence Gates, first at RAND (1971) and later at Oregon University (1978).[64] Well anchored in Soviet networks, Gates did not mind giving his model to the Soviets and even proposed sending two American scientists to Moscow to help adjust the model to the Russian computer BESM-6.[65] Indeed, not only models, but also data was shared transnationally: the Computer Centre received atmospheric data from the Norwegian Meteorological Centre.[66]

The conceptual rationale of Soviet global models addressed concerns similar to those of *The Limits to Growth*, but also extended them further in reconceptualizing the role of humanity on Earth. Thus, Moiseev envisioned an integrated model of the biosphere, putting the natural, socio-economic and cognitive environment into one model in an attempt to study the large scale effects of anthropogenic activities.[67] This model simulated interconnections among global climate, ecology and economic systems, aiming to identify the conditions under which environmental change would set boundaries for economic development.[68] Yet the economy was of secondary interest for Moiseev's group: this model, involving land, ocean and atmosphere blocks, was first used to simulate CO_2 emissions and climate change in the early 1980s. Another aspect illustrating the difference and possibly some rift between Moiseev's and Gelovani's teams is that the Computer Centre's global model was created independently of IIASA: the Computer Centre's scientists did not participate in IIASA's global modelling conferences.[69] In the late 1970s, Moiseev's group directly cooperated with American and French modelling teams, and their cooperation with IIASA would intensify only in the 1980s.[70]

It soon became clear that although global models were meant to study apolitical, geophysical processes, they led to the formulation of innovative political and policy ideas. Here the most prominent study done on the Computer Centre's model was a simulation of the long-term environmental consequences of nuclear war, conducted in cooperation with Western scientists within the framework of the Scientific Committee on Problems of the Environment (SCOPE) of the International Council of Science Unions (ICSU). The result was a Soviet-American report about nuclear winter, released in the autumn of 1983.[71] The models simulated the effect of nuclear explosions on world climate, ecological systems and population. The American team first traced the consequences of nuclear war over the period of one month, whereas Moiseev's team traced the consequences over one year, discovering that nuclear war would cause darkening of the atmosphere, significant cooling and extinction of the majority of plants and animals in the Northern, and parts of the Southern hemisphere. At a later stage, scientists examined even longer term consequences of nuclear war, tracing changes to the global environmental system for thirty years. These findings showed that even a limited nuclear war would cause irreversible environmental change and affect the whole of humanity.[72] The nuclear winter study was used by disarmament activists, deeply influenced public opinion in both West and East, and contributed to incremental change in nuclear strategy thinking.[73]

The nuclear winter simulation introduced a long-term element which previously had not been considered in nuclear strategies concerned with short-term damage calculations. Unlike *The Limits to Growth*, the nuclear winter study was made public in the Soviet Union. The modelling exercise transformed the understanding of nuclear war from a potential conflict between the US and the Soviet Union into a global disaster, which would affect both North and South. Just like Leontief's *Future of the World Economy*, albeit for different reasons, the nuclear winter study made redundant the idea of a systemic divide between capitalism and communism by showing that the ash of both systems would be indistinguishable.[74] The Soviet Union, in other words, could not win the nuclear war in any meaningful sense.

The impact of global modelling efforts at VNIISI was quite different but no less significant. Global modelling was a prominent part of VNIISI's research agenda from its establishment in June 1976: the first report of annual activities included the development of a conceptual framework for global modelling.[75] VNIISI was exceptionally well positioned to tap into international science because it was created to be the Soviet counterpart of IIASA. Claiming that the Eastern bloc lagged behind the West in global modelling, VNIISI's purpose was to "catch up" with the West by developing interdisciplinary research on large-scale, complex and global problems.[76] Thanks to Gvishiani, VNIISI had a direct link to the heart of Soviet power. For instance, in 1977 a high-level meeting of global modellers, including members of the Club of Rome, was organised in Moscow, where five members of Politburo, the de-facto highest decision-making body in the Soviet government, attended.[77] If global modellers at the Computer Centre first and foremost developed their models as tools for gaining new scientific knowledge about geophysical systems, VNIISI sought to generate policy-relevant knowledge.

In the context of Soviet academia, VNIISI was an important, large and well-funded organization. The global modelling program at VNIISI was co-chaired Gvishiani and Gelovani.[78] This global modelling group stemmed from a GKNT team for operations research that was involved in creating complex models of world development.[79] This team also included a prominent scientist, Sergei Dubovskii, with experience of modelling from the highly esteemed Institute for Control Sciences. These and other scholars who later shaped the core of VNIISI were closely involved in the formation of IIASA's research agenda from 1972.[80] Unlike IIASA, which never hosted more than 100 scholars at a time, and in the true spirit of a Soviet organization, VNIISI employed more than 300 staff and grew to almost 700 in the late 1980s.[81] The Institute was well provided with a large building and its technical equipment was more than adequate: VNIISI modellers used PDP-11/70, an American computer.[82]

The principal task of VNIISI was to forecast the development of countries and regions for a twenty to thirty year period.[83] The idea to forecast social and economic development for the year 2000 stemmed from the work of the US Commission for the Year 2000.[84] Such forecasts were made in the Soviet Union from the 1960s, although most of them were kept secret. A glimpse at the archives and memoirs reveals a much more complex and diverse landscape of Soviet scientific expertise than previously thought: at VNIISI, scientists looked farther ahead to test the impact of globally significant changes on the Soviet Union. For instance, the first global dynamics model developed at VNIISI forecasted the impact of armament on the Chinese economy. The model showed that increased investment in defence would devastate the Chinese economy; it was therefore concluded that China was not likely to consider military expansion. Accordingly, the Soviet government did not have to invest to counteract Chinese military growth. Ironically, this model used the existing intelligence data on China, but could not model fine-grained impacts on the Soviet Union, because Gosplan refused to give them Soviet data.[85]

Nevertheless, other studies explored the development of the Soviet economy as part of global dynamics. If in 1981 VNIISI had a model which consisted of three blocks, representing the US, Japan and China, in 1983 this model was expanded to include the communist bloc. Unlike Leontief's model for the UN, which did not divide the world according to nations or political regimes, VNIISI's model divided the world along political allegiances into nine blocks: the Soviet Union, the Eastern European bloc, the European community, the US, China (the power of which was of growing concern to the communist leaders), Japan, "other capitalist countries," OPEC countries, and developing countries. Sector-wise, this model included demography, trade, energy resources, the environment and climate.

However, just as before, Gosplan was not forthcoming with the Soviet data; only highly aggregated statistics were available. What did VNIISI modellers do? They turned to their personal, transnational contacts to solve this data gap. This was the colleague of Leontief mentioned earlier, the Russian economist Stanislav Men'shikov, the vice-director and then director of the UN Department of Prognosis, Planning and Development, 1974–1980. Men'shikov boasted good personal relations with some of the most prominent US intellectuals, such as the economist John Kenneth Galbraith; the two met at a lunch with David Rockefeller at the US Embassy in Moscow in the mid-1960s. In 1988, Men'shikov and Galbraith co-authored the book *Socialism, Capitalism and Co-existence*.[86] Indeed, the cooperation between VNIISI and the UN was built on strikingly intertwined socio-technical networks, which joined machines, organisations and individuals: Men'shikov worked with Leontief on the UN world economy model. Furthermore, Leontief's world economy model was computed on a PDP-10 at the Feldberg

Computer Centre, the same type of machine as VNIISI had.[87] IIASA provided the data about global markets.[88] Scholars recalled that they could easily obtain CIA reports on the Soviet economy, industry and society, but not the data from the State Committee for Statistics (Goskomstat).[89] The result was a gargantuan modelling system joining 47 models, 4,700 averaged points and 5,000 variables, and based on quantification of 370,000 empirical observations. On this basis, the world system and Soviet development was projected for the next twenty years.[90]

It should be clear by now just how painstaking of a process it was for Soviet scientists to make such long-term projections. It is quite significant that Soviet scientists used long-term projections to reveal the current problems that the Soviet Union faced, but which could not be explicitly identified for ideological reasons. Long-term projections into the future, meanwhile, constituted an important rhetorical device to show the roots of the problems. Thus VNIISI scientists reported to Prime Minister Kosygin and, later, Nikolai Tikhonov, in 1979, 1982 and 1984, each time demonstrating that the growth of the Soviet economy would sharply decline in the future unless the Soviet government sharply upped investment in research and development.[91]

This was not a trivial warning. Indeed, very few communist scientists dared to model the slow-down or, worse, non-growth of Soviet economy. For instance, TsEMI's director retrospectively wrote that he "just could not accept" even as a hypothesis the zero-growth option proposed by Meadows's report.[92] In turn, the hypothesis of zero growth was censored out of the Russian translation of *The Limits to Growth*.[93] Yet there was some, albeit limited space for Soviet scientists to offer negative feedback to the government. A well-known example is that of the Russian economist Gregory Khanin, who repeatedly wrote letters to the Central Committee reporting his own estimates of the future Soviet economy, which were much lower than the official figures.[94] Whereas Khanin was tolerated and, probably, ignored, other scientists were less fortunate: for instance, the East German scientist W. Harich calculated a version of non-growth communism, for which he was seriously repressed.[95] Another example of a reaction to economic forecasts showing the decline of Soviet economic power involved IIASA's project on modelling economic growth, directed by the German economist Wilhelm Krelle. Dissatisfied with Krelle's results, several Russian scholars complained that it was "a big mistake" to show that the impact of the Soviet Union on world economic development was minor. They wrote that if Krelle used the official forecast for the year 2000, the global role of Soviet trade would have been much more significant.[96] Indeed, according to my interlocutors, who were involved in these discussions, VNIISI scientists did fear repression and this is why the scenario of the collapse of the Soviet Union was not tested at all.[97]

Furthermore, the VNIISI model revealed deep divisions among the scientists involved, who disagreed about the actual purpose of long-term analysis. One scientist told me that several VNIISI economists involved in the development of this model simply refused to believe that the model was expected to produce unanticipated results. Well drilled in the communist planning system, these economists assumed that the modelling exercise was merely a ritual, an attempt to create "a mechanic proof" for plan targets, specified in the Party directives. Others were anxious that their results might be understood as a criticism of the standard of Soviet life, so demographers simply refused to take into account the influence of the quality of life on birth rates.[98] Finally, it is very likely that the curve pointing out the decline of Soviet growth from 4.5 percent in 1980 to 2.1 percent in 2000 was also a cautiously selected one: this curve was diplomatically accompanied by another curve, which showed that US growth would slow down even more.

Nevertheless, it is clear that some Soviet modellers regarded their task as a serious and genuine contribution to policy processes by "speaking truth to power," to use Aaron Wildawsky's terms, and not just as a mere ritual. This is why they sought to make their studies public. In 1984, this VNIISI modelling exercise was described in a report *On the Threshold of the Millennium: The Global Problems and Development Processes in the USSR*; the following year, some of the results were published in VNIISI proceedings. However, the Main Directorate for Literary and Publishing Affais (Glavlit), which was also the central censorship organ, requested that most of the information concerning the Soviet Union was removed in order to be suitable for a wide audience.[99] Whereas Soviet censorship found it acceptable to publish studies on the complete extinction of Soviet citizens during a nuclear winter, it would not release a forecast of the slowing down of Soviet economic growth from the optimistic five percent to what was considered a meagre two percent.

These examples show that although there were pretty clear boundaries to the criticism of the Soviet regime, some Soviet global modellers persistently tried to push them. Soviet scientists used a sophisticated tool, computer-based global modelling, as a vehicle to criticize the Soviet economy by showing its imminent failure to the Politburo. In this way, the long-term projections, I suggest, enabled new kinds of criticism before the new policy of openness, or *glasnost*. launched by Mikhail Gorbachev in 1987.

For the Soviets, the struggle for the long-term was inevitably a struggle for access to models, data and computers. It is difficult to overestimate the role that the scientific methodology of global modelling played in international cooperation. No global model could run without empirical data. No national model of natural or economic systems could be realistic if global processes were not plugged into it. Nothing clashed harder with the Soviet bureaucracy, pervaded with secrecy and compartmentalisation, than the idea of unrestricted international circulation of data. Here, the modelling of geophysical processes, particularly the studies of environmental change, enabled Soviet scientists to experiment with formulating alternative visions of the Soviet future. If, in the 1960s to 1980s, many of the Soviet demographic statistics were not available, as Gosplan would not disclose the population mortality rates from the 1930s to 1940s,[100] the data on the atmosphere and the ocean could be circulated more easily, which explains the Computer Centre's focus on geophysical global models. But then, models and data were co-produced: for instance, the global models required new kinds of data drawn from specially conducted experiments, because, for example, nitrogen reactions were different in Siberia and Latin America. The modelling software itself was not easy to replicate; without direct face-to-face communication, wrote Moiseev, sophisticated mathematical models could never become "real."[101]

CONCLUSION

For the Soviet government global modelling was both an instrument for policy making and a symbol, part of the struggle for super power status. Soviet scientists aspired to use big computers to project big data over the long-term and long-range world future and do this just as well as the US scientists. Brimming with political prestige, global modelling served as an important source of authority for Soviet scientists, who wished to innovate not only in science, but also in policy-making. And they were innovative: global modelling posed deep challenges to the secrecy and compartmentalization of Soviet scientific expertise. This chapter shows that global modelling demanded international, face-to-face cooperation to co-produce both the models and data. As a result, the Soviet government eventually began to release control over small communities of modellers, which remained close to, albeit at arm's length from, central power. Second, global models made visible through graphs, maps and statistical curves different, unexpected and negative consequences of long-term developments. In some areas, such as the natural environment or global trends, this long-term future was actively portrayed as global and politically neutral. It is highly significant that Soviet scientists used references to such a neutralized global future to criticize contemporary Soviet realities.

Global models were constitutive to the emerging understanding of the global future as a truly interdependent affair. The discourse of interdependence became a new diplomatic language of non-zero-sum games. For instance, the GKNT's head of foreign relations would assure his Japanese visitors that the Soviets understood the world "as a system of partners," where "when the system as a whole wins, each partner wins."[102] Deeds, unsurprisingly, did not always follow from the words: Soviet statistical agencies refused to provide the data either to Soviet or Western scientists.[103] In spite of these difficulties, Soviet global models were, to use Brian Wynne's words, "more than its final results." As showed in this chapter, global modelling transformed from being a mere channel for feeding data to the policy maker's office into a large enterprise of "policy argumentation."[104]

Furthermore, global modelling drove a deep epistemological transformation of the computer-based Soviet governmentality. The governmental role of computers was not limited to what Donald MacKenzie called the mechanization of proof.[105] Rather, Soviet discussions about the methodology of global computer-based modelling articulated and disseminated a non-determinist worldview, in which a great many areas of nature and human activity were understood as probabilistic or even purely uncertain. The message of the global models to the communist government was that total control was impossible in the long-term. In this way, instead of producing certainty, global computer models time and again reminded officials of the boundaries of the human knowledge and scientific methods of control.[106]

Global modelling, in this way, permitted a different way of relating to the future of communist society. Although Soviet scientists cautiously avoided directly challenging the ideological dogma of the superiority of the communist system, the uniqueness of the communist system was simply made redundant. By the early 1980s, the scientists framed global problems as an issue of a metabolic relation between the man and biosphere, something which was beyond the Cold War struggle for global hegemony, became legitimate and central in the Soviet Union.[107] This globalist environmental discourse slowly but steadily accumulated power as the key framework for economic development strategies and, in doing so, as Dmitrii Efremenko notes, the global environmental framework was parallel to and only rarely intersected with Marxist-Leninist political economy.[108] The focus on long-term global and environmental processes enabled Soviet scholars and policy makers to point out that the Soviet economy and society also had serious problems, which were of universal and global character and which could not be internally resolved.

Moiseev was especially acute on this, claiming as early as the 1970s that there was a need to focus on new problems in order to prepare for the new world of advancing computer technologies.[109] We should not dismiss this call as trivial rhetoric: it was, indeed, a smart way of suggesting that the Soviet system was stuck in solving its old problems. To suggest changing the whole system would be too revolutionary, even for such an independent-minded Soviet scholar as Moiseev. Instead, he suggested turning to new problems, ones of global and long-term character. The attempt to solve these new problems could and did transform the Soviet system.

NOTES

1. UNESCO was a central platform for the articulation of these views. See Perrin Selcer, *Patterns of Science: Developing Knowledge for a World Community at UNESCO* (PhD diss., University of Pennsylvania, 2011); and Sibylle Duhautois, *The Future of World Problems* (PhD diss., Sciences Po, in progress).
2. See, for example, Dzhermen Gvishiani, "Methodological Problems of Global Development Modelling", in E. P. Velikhov, J. M. Gvishiani, and S. R. Mikulinsky, eds., *Science, Technology and the Future: Soviet Scientists' Analysis of the Problems of and Prospects for the Development of Science and Technology and their Role in Society* (Oxford: Pergamon Press, 1980): 21–35.

3. For instance, see Fernando Elichirigoity, *Planet Management: Limits to Growth, Computer Simulation, and the Emergence of Global Spaces* (Evanston, IL: Northwestern University Press, 1999); Giuliana Gemelli, "Building Bridges in Science and Societies During the Cold War: The Origins of the International Institute for Applied Systems Analysis (IIASA)," in Giuliana Gemelli, ed., *American Foundations and Large Scale Research: Construction and Transfer of Knowledge* (Bologna: Clueb, 2001). The lack of empirical research was pointed out in Michael Barnett and Martha Finnemore, *Rules for the World: International Organizations in Global Politics* (Ithaca, NY: Cornell University Press, 2004). The subject is touched upon, but not yet fully explored in the transnational histories of science, especially with regard to the Club of Rome activities and the development of earth sciences. See Nestor Herran, Soraya Boudia, and Simone Turchetti, *Transnational History and the History of Science* (Cambridge, UK: Cambridge University Press, 2012).
4. One possible reason may be, building on Tatarchenko, the excessive focus on hardware in the histories of computing. In contrast, global modelling is principally about software. Ksenia Tatarchenko, *"A House with a Window to the West": The Akademgorodok Computer Centre (1958–1993)* (PhD diss., Princeton University, 2013). For historiography, see Martin Cambell-Kelly and William Aspray, *Computer: A History of the Information Machine* (Boulder, CO: Westview Press, 2004); Paul Ceruzzi, *A History of Modern Computing* (Cambridge, MA: MIT Press, 2003) and Jon Agar, *The Government Machine: A Revolutionary History of the Computer* (Cambridge, MA: MIT Press, 2003).
5. United Nations, 1084th Plenary Meeting (December 19, 1961), www.un.org.
6. Peter Galison and Bruce W. Hevly, eds., *Big Science: The Growth of Large-Scale Research* (Stanford: Stanford University Press, 1992).
7. Brian Wynne, *Models, Muddles and Megapolicies: The IIASA Energy Study as an Example of Science for Public Policy.* IIASA Reports WP-83–127 (December 1983).
8. Ludwik Fleck, *Genesis and Development of a Scientific Fact* (Chicago: Chicago University Press, 1979).
9. Donald Michael, *Cybernation: The Silent Contest* (Santa Barbara, CA: Centre for the Study of Democratic Institutions, 1962); Marshall McLuhan, "Cybernation and Culture," in Charles Dechert, ed., *The Social Impact of Cybernetics* (Notre Dame, IN: University of Notre Dame Press, 1966); Daniel Bell, *The End of Ideology: The Exhaustion of Political Ideas in the 1950s* (Cambridge, MA: Harvard University Press, 1962/2000); Daniel Bell, *The Coming of Post-Industrial Society: A Venture in Social Forecasting* (New York: Basic Books, 1976); Malcolm Waters, *Daniel Bell* (London and New York: Routledge, 1996), 106.
10. Dmitrii Efremenko, *Ekologo-politicheskie diskursy: vozniknovenie i evoliutsiia* (Moscow: INION, 2006).
11. J. D. Bernal, *The Social Function of Science* (London: Lowe & Brydone Printers, 1939/1946), 409.
12. For the US version of technological modernisation as both solution and problem, see Michael Latham, *The Right Kind of Revolution: Modernization, Development and U.S. Foreign Policy from the Cold War to the Present* (Ithaca, NY: Cornell University Press, 2011).
13. Bruce Allyn, "Fact, Value, and Science," in Loren Graham, ed., *Science and the Soviet Social Order* (Cambridge, MA: Harvard University Press, 1990), 238. For more on Bernal, see Andrew Jamison, "Technology's Theorists: Conceptions of Innovation in Relation to Science and Technology Policy," *Technology and Culture* 30, no. 3 (1989): 505–533.
14. I base this statement on the use of the term "global" in the Gosplan documents, kept at the Russian State Economic Archives (RGAE).

15. Vadim Zagladin and Ivan Frolov, *Global'nye problem sovremennosti: nauchnyi i sotsial'nyi aspekty* (Moscow: Mezhdunarodnye otnosheniia, 1981).
16. Ibid, 25, 33.
17. For more, see Vieille Blanchard, this volume.
18. Paul Edwards, "The World in a Machine: Origins and Impacts of Early Computerized Global Systems Models," in Agathe C. Hughes and Thomas P. Hughes, eds., *Systems, Experts and Computers: The Systems Approach in Management and Engineering, World War II and After*, 221–254 (Cambridge, MA: MIT Press, 2010).
19. Immediately criticized as flawed and imperfect, this experiment of projecting world development up to the year 2050 stirred huge interest from the scientific community: in just four years a further nine major world development models were created. Gvishiani, "Methodological Problems," 22–27.
20. Dzhermen Gvishiani, *Mosty v budushchee* (Moscow: URSS, 2004), 77.
21. For instance, Forrester's and Meadows' model was discussed in a symposium on trends in mathematical modelling, organized by the Italian National Research Council and UNESCO in December 1971. In addition to Dennis Meadows, participants included prominent Russian scientists, such as the mathematician Nikita Moiseev and specialist in economic modelling Kirill Bagrinovskii, as well as the leading American futurologists Olaf Helmer and Alvin Toffler. Nigel Hawkes, ed., *International Seminar on Trends in Mathematical Modelling, Venice, 13–18 December 1971* (Berlin: Springer Verlag, 1973).
22. Interview 35, 4 March 2014; Donella Meadows, John Richardson, and Gerhart Bruckman, *Groping in the Dark: The First Decade of Global Modelling* (Chichester, UK: John Wiley and Sons, 1982); Gvishiani, *Mosty*. All informants signed an agreement to be interviewed, and they are fully anonymized in order to protect their identities.
23. Jay Forrester, "System Dynamics and the Lesson of 35 Years," in Kenyon De Greene, ed., *A Systems-Based Approach to Policymaking* (Cambridge, MA: Springer, 1993), 199–240.
24. V. A. Gelovani, V. B. Britkov, and S. V. Dubovskii, *SSSR i Rossiya v globalnoy sisteme (1985–20130): rezultaty globalnogo modelirovaniia* (Moscow: Librokom, 2009), 48; Efremenko, *Ekologo-politicheskie*, 104; Interview 29, 15 April 2013.
25. Interview 22, 13 December 2012.
26. Sergei Dubovskii, "Global'noe modelirovanie: voprosy teorii i praktiki," *Vek globalizatsii* 2 (2010): 57; Sergei Dubovskii and O. A. Eismont, "Long-Range Modelling of the USSR Economy," in Wilhelm Krelle, ed., *The Future of the World Economy*, 311–324 (Berlin: Springer Verlag, 1989).
27. For more, see Leena Riska-Campbell, *Bridging East and West: The Establishment of the International Institute for Applied Systems Analysis in the United States Foreign Policy of Bridge Building, 1964–1972* (Helsinki: Finnish Society on Science and Letters, 2011); and Thomas Schwartz, "Moving beyond the Cold War: The Johnson Administration, Bridge-Building, and Détente," in Francis Gavin and Mark Atwood Lawrence, eds., *Beyond the Cold War: Lyndon Johnson and the New Global Challenges of the 1960s* (Oxford and New York: Oxford University Press, 2014); see also Eglė Rindzevičiūtė, *Riders on the Storm: How Cold War Scientists Co-Operated to Create Global Governance*, in progress.
28. Schwartz, "Moving beyond the Cold War"; Gvishiani, *Mosty*, 87.
29. Gvishiani, *Mosty*, 141.
30. H.S.D. Cole, C. Freeman, M. Jahoda, and K. Pavitt, eds., *Thinking about the Future: A Critique of the Limits to Growth* (Sussex: Sussex University Press, 1973).

31. Barbara Ward, René DuBois, Thor Heyerdahl, Gunnar Myrdal, Carmen Miró, Lord Zuckerman, and Aurelio Peccei. *Who Speaks for Earth?* New York: Norton, 1973.
32. Carl Kaysen, "Computer that Printed Out W*O*L*F," *Foreign Affairs* 50 (July 1972): 660–668.
33. Howard Raiffa, *Analytical Roots of a Decision Scientist: A Memoir* (28 June 2005), an unpublished manuscript, IIASA archives, Laxenburg; Meadows, Richardson, and Bruckman, *Groping in the Dark*.
34. Howard Raiffa, *The International Institute of Applied Systems Analysis: Background Information, Provisional Research Strategy. Preliminary Version for Restricted Circulation* (April 1973), IIASA Archives, Laxenburg, p. 30.
35. IIASA global modelling conferences scrutinized the Mesarovic-Pestel model (1974), the Latin American labour model Bariloche (1974), the Dutch model of international relations in agriculture (MOIRA, 1975), the British Systems Analysis Research Unit model (SARU), which was developed by the Environment Agency (1976) and hence the only one funded by governmental body. Its version SARUM 76 was used by OECD Interfutures scenarios. Other models discussed at IIASA included the MRI (Polish national model) (1976), as well as the UN world model and the Futures of Global Interdependence model (FUGI, 1977) (Meadows, Richardson, and Bruckman, *Groping in the Dark*, 2–4).
36. A. G. Ivakhnenko and V. G. Lapa, *Cybernetics and Forecasting Techniques* (New York: Elseviere, 1967); compare with Reuben Hersh, "Mathematics Has a Front and a Back," *Synthese* 88, no. 2 (1991): 127–133. See also an internal debate in *Modelirovanie protsessov global'nogo razvitiia: sbornik trudov VNIISI*, vol.8, Moscow, 1979.
37. The inability of global models to produce conclusive results led to redefinition of the purpose of modelling, which shifted from policy prescription to less formal insight. For instance, in 1979 at IIASA, Olaf Helmer created Global Economic Model (GEM), which did not "solve the problems directly," but to lead to "a better intuitive understanding of the problem structure." Olaf Helmer and L. Blencke, *GEM: An Interactive Simulation Model of the Global Economy*. (Laxenburg: IIASA, 1979), 2.
38. Interview 31, 10 April 2013.
39. Moiseev to Gvishiani (1980), Archives of the Russian Academy of Sciences (ARAN), f.1918, op.1, d.463, l.4.
40. See Wynne, *Models*, 5; Bill Keepin and Brian Wynne, "Technical Analysis of IIASA Energy Scenarios," *Nature* 319 (20 December 1984): 691–695; Rindzevičiūtė, *Riders on the Storm*.
41. Information from http://unstats.un.org.
42. Wassily Leontief, A. Carter, and P. Petri, *The Future of the World Economy: A United Nations Study* (Oxford: Oxford University Press, 1977), 1.
43. Jakob Mosak to Fedorenko (5 February 1965), ARAN, f.1959, op.1, d.92, l.19–20.
44. Fedorenko to Jakob Mosak, UN HQ Bureau of General Economic Research and Policies (this document is not dated, but it is filed in the folder for 1965), ARAN, f.1959, op.1, d.92, l.18; Fedorenko to Leontief (no date, the folder is dated 1965), ARAN, f.1959, op.1, d.92, l.60.
45. Gvishiani to Fedorenko (17 February 1966), ARAN, f.1959, op.1, d.129, l.15.
46. ARAN, f.1959, op.1, d.129.
47. Leontief's father was a professor who organized strikes in his grandfathers' factories. Bernard Rosier, ed., *Wassily Leontief: textes et itinéraire* (Paris: Éditions la découverte, 1986), 78–80.
48. Wassily Leontief, "The Decline and Rise of Soviet Economic Science," *Foreign Affairs* (January 1960): 261–272; Rosier, *Wassily Leontief*, 80, 90–92; David

Engerman, *Know Your Enemy: The Rise and Fall of America's Soviet Experts* (Oxford: Oxford University Press, 2009), 97.

49. The Soviet philosophers appreciated Leontief's world model as "more realistic" than the models sponsored by the Club of Rome. Zagladin and Frolov, *Global'nye problem sovremennosti*, 11, 182, 189.
50. Leontief's study involved demographic, economic and environmental spheres with bench-mark years of 1980, 1990 and 2000. The world was divided into fifteen regions, each region comprising forty-five sectors of activities. The regions were linked via imports and exports of forty classes of goods and monetary transfers.
51. Leontief, Carter, and Petri, *The Future*, 1–3.
52. Ibid., 2, 34–35.
53. Nikita Moiseev, *Algoritmy razvitiia* (Moscow: Nauka, 1987), 5.
54. Timofeev-Resovski was the pioneering population geneticist. For more, see V. V. Babkov and E. S. Sakanian, *Nikolai Timofeev-Resovskii* (Moscow: Pamiatniki istoricheskoi mysli, 2002); Y. G. Rokityanskij, "N. V. Timofeef Resovski in Germany, July 1925–September 1945," *Journal of Bioscience* 30, no. 5 (2005): 573–580; on Vernadskii, see Jonathan Oldfield and Denis Shaw, "V. I. Vernadskii and the Development of Biogeochemical Understandings of the Biosphere, c.1880s–1968," *The British Journal for the History of Science* 46 (2013): 287–310.
55. Interview 34, 5 April 2013.
56. A. A. Petrov, *Nikita Nikolaevich Moiseev: sud'ba strany v sudbe uchenogo* (Moscow: Ekologiia i zhizn, 2011), 56.
57. Moiseev was skeptical about econometrics in general and even more so about Soviet econometrics, yet for him economic modelling was a lesser evil. Moiseev wrote that in order to succeed Soviet economic forecasting had to be based on "strictly scientific modelling systems" and not "unreliable expert surveys." Nikita Moiseev, *Prosteishie matematicheskie modeli ekonomicheskogo prognozirovaniia* (Moscow: Znanie, 1975), 62; Moiseev to Gvishiani (1980), ARAN, f.1918, op.1, d.463, l.16.
58. For instance, Gvishiani published regularly articles promoting systems analysis and global modelling. See Dzhermen Gvishiani, "Sistemnyi podkhod k issledovaniiu global'nykh problem." In *Nauka i chelovechestvo*, 237–249. Moscow: Znanie, 1983; Dzhermen Gvishiani, "The Search for a Global Strategy." *Unesco Courier* (November 1979): 25–26.
59. Curiously, even at VNIISI, global modelling was split into two directions, the environmental and economical, and these two groups worked in parallel and did not directly collaborate (Interview 29, 15 April 2013).
60. Other milieus where Soviet global models of various kinds were developed included econometric modelling at the Institute of World Economy and International Relations (IMEMO), the Main Geophysical Laboratory in Leningrad, under Gurii Marchuk in the Novosibirsk branch of the Soviet Academy of Sciences, even at the Moscow State University (Dubovskii, "Global'noe," 55). It was Stanislav Men'shikov, the vice director of IMEMO who conducted econometric research as early as 1968 at IMEMO and later in Novosibirsk (interview 34, 5 April 2013).
61. In 1987, Gelovani was also appointed as the head of the Soviet branch of the World Laboratory. The World Laboratory was an interesting Cold War effort: established in 1986, it was an NGO initiated by Dirac, Kapitsa and Zichichi. It was recognized by the UN, dedicated to facilitate East-West and North-South scientific exchange. S. Emel'ianov, ed., *30 let institute sistemnogo analiza Rossiiskoi akademii nauk: istoriia sozdaniia i razvitiia Institutea sistemnogo analiza, 1976–2006 gg* (Moscow: URSS, 2006):129.
62. Otchet (1982), ARAN, f.1918, op.1, d.492, l.4.

63. Interview April 2013 with a modeller from the Computer Centre.
64. Lawrence Gates developed a model called OSU AGCM, which was detailed, but also faster and, unlike other American models, did not require that much computer memory (only 100Kb), which was very important for Soviet scientists who worked on the slow BESM-6. Otchet (1978), ARAN, f.1918, op.1, d.421, l.88.
65. Ibid, l.89.
66. Interview March 2013 with a modeller from the Computer Centre.
67. For an early publication discussing human impact on environmental change, see William Thomas, *Man's Role in Changing the Face of the Earth* (Chicago: University of Chicago Press, 1956).
68. Interview 31, 10 April 2013.
69. Ibid.
70. Moiseev kept close connections with French scholars: taught French by his grandmother, he was less comfortable with English and there were many Russian exile scientists in Paris who were keen to welcome him. Petrov, *Nikita Nikolaevich Moiseev*, 50.
71. Lawrence Badash, *A Nuclear Winter's Tale: Science and Politics in the 1980s* (Cambridge, MA: MIT Press, 2009); Rindzevičiūtė, *Riders on the Storm.*
72. Yurii Svirezhev, *Ecological and Demographic Consequences of Nuclear War* (Moscow: Computer Centre of the USSR Academy of Sciences, 1985), 205.
73. Matthew Evangelista, *Unarmed Forces: The Transnational Movement to End the Cold War* (Ithaca, NY: Cornell University Press, 2002) and Badash, *A Nuclear Winter's Tale*; as well R. E. Munn, "Discussion," in *The Biosphere and Humanity*, IIASA Report CP-94–4, (April 1994).
74. For instance, the division between capitalism and communism is not mentioned at all in Anatolii Dorodnitsyn ed., *Ekologicheski i demograficheskie posledstviia iadernoi voiny* (Moscow: Nauka, 1986), 132.
75. "Otchetnyi balans," RGAE, f.9480, op.12, d.343, l.9.
76. "Poiasnitel'naia zapiska," RGAE, f.9480, op.12, d.343, l.13, 18.
77. Walter Clemens's interview with Alexander King, 4 February 1987 (Walter Clemens, *Can Russia Change? The USSR Confronts Global Interdependence* (New York: Routledge, 1990), 138); see also Gvishiani, "Methodological Problems of Global Development Modelling," 33.
78. VNIISI was created by the initiative of Gvishiani and in alliance with systems theoretician Boris Mil'ner, economist Stanislav Shatalin and OR specialist Stanislav Emel'ianov. A. Porshnev, "Vklad B.Z. Mil'nera v razvitiiu nauki upravleniia," *Rosiiskii zhurnal menedzhmenta* 4 (2004), 156.
79. Emel'ianov, *30 let*, 132–135.
80. For example, Gvishiani invited V. Sadovskii and Emel'ianov to the meeting with Raiffa at GKNT in November, 1972. RGAE, f. 9480, op.9, d.1716, l.110.
81. RGAE, f.9480, op.12, d.343, l.20; RGAE, f.9480, op.12, d.1865, l.4.
82. Dubovskii, "Global'noe," 56. Given that VNIISI's computer was produced by the major American computer producer, Digital Equipment Corporation, and the embargo on exporting computer technology to the Soviet Union, its road to VNIISI must had been an interesting one. Frank Cain, "Computers and the Cold War: United States Restrictions on the Export of Computers to the Soviet Union and Communist China," *Journal of Contemporary History* 40, no.1 (2005): 131–147.
83. Gelovani, Britkov, and Dubovskii, *SSSR i Rossiya*, 16.
84. See Jenny Andersson, "The Invention of the Long Term: RAND, The Commission for the Year 2000 and the Rise of Futurology", in preparation.
85. Gelovani, Britkov, and Dubovskii, *SSSR i Rossiya*, 18.

86. Stanislav Men'shikov, *O vremeni i o sebe* (Moscow: Mezhdunarodnye otnosheniia, 2007).
87. Leontief, Carter, and Petri, *The Future*, iii; Men'shikov, *O vremeni*.
88. VNIISI was informed by Wilhelm Krelle's study, a joint project between IIASA and the University of Bonn in 1985–1987. Wilhelm Krelle, *The Future of the World Economy: Economic Growth and Structural Change* (Berlin: Springer Verlag, 1989). In 1984–1989, by Krelle's invitation, VNIISI took part in this project where Brekke, Gelovani and Kay developed scenarios of global development on the basis of American, Soviet and Japanese models (Dubovskii, "Global'noe," 57; Emel'ianov, *30 let*,141).
89. Interview 28, 15 April 2013.
90. Gelovani, Britkov, and Dubovskii, *SSSR i Rossiya*, 64; Dubovskii and Eismont "Long-Range."
91. Gelovani, Britkov, and Dubovskii, *SSSR i Rossiya*, 7.
92. N. P. Fedorenko, *Vspominaia proshloe, zagliadyvaiu v budushchee* (Moscow: Nauka, 1999), 387.
93. Interview 31, 10 April 2013.
94. Michael Ellman and Vladimir Kontorovich, eds., *The Destruction of the Soviet Economic System: An Insiders' History* (Armonk, NY: M. E. Sharpe, 1998), 76–85.
95. Efremenko, *Ekologo-politicheskie*.
96. D. Chernikov, A. Batizi, N. Volkov and A. Ivanov to Lee (28 November 1986), USSR NMO, IIASA Archives, Laxenburg.
97. Gelovani, Britkov, and Dubovskii, *SSSR i Rossiya*, 48, 80.
98. Dubovskii, "Global'noe," 56–57.
99. Ibid, 58.
100. Interview 28, 15 April 2013.
101. Moiseev to Gvishiani (1980), ARAN, f.1918, op.1, d.463, l.12.
102. "Zapis' besedy", GKNT (13 April 1972), archive document from RGAE, f.9480, op.9, d.1716, l.37.
103. Gvishiani, *Mosty*, 239.
104. Wynne, *Models*, 12–13.
105. Donald MacKenzie, *Mechanizing Proof: Computing, Risk and Trust* (Cambridge, MA: MIT Press, 2001).
106. For more about the history of probabilistic theory, see Stephen Stigler, *Statistics on the Table* (Cambridge, MA: Harvard University Press, 1999).
107. P. N. Fedoseyev, "Topical Problems of Our Time," in E.P. Velikhov, J.M. Gvishiani, and S.R. Mikulinsky, eds., *Science, Technology and the Future: Soviet Scientists' Analysis of the Problems of and Prospects for the Development of Science and Technology and their Role in Society* (Oxford: Pergamon Press, 1980), 18.
108. Efremenko, *Ekologo-politicheskie*.
109. Petrov, *Nikita Nikolaevich Moiseev*.

ARCHIVES

Raiffa, Howard. *Analytical Roots of a Decision Scientist: A Memoir* (June 28, 2005), an unpublished manuscript, IIASA archives, Laxenburg.

Raiffa, Howard. *An Initial Research Strategy for the International Institute of Applied Systems Analysis* (February 1973). IIASA Archives, Laxenburg.

The Archives at the International Institute of Applied Systems Analysis (IIASA)

The Archives of the Russian Academy of Sciences (ARAN)

The Russian State Economic Archives (RGAE)

REFERENCES

Agar, Jon. *The Government Machine: A Revolutionary History of the Computer*. Cambridge, MA: MIT Press, 2003.

Allyn, Bruce J. "Fact, Value, and Science." In *Science and the Soviet Social Order*, edited by Loren Graham, 225–255. Cambridge, MA: Harvard University Press, 1990.

Andersson, Jenny. "The Invention of the Long Term: RAND, The Commission for the Year 2000 and the Rise of Futurology", in preparation.

Babkov, V. V. and E. S. Sakanian. *Nikolai Timofeev-Resovskii*. Moscow: Pamiatniki istoricheskoi mysli, 2002.

Badash, Lawrence. *A Nuclear Winter's Tale: Science and Politics in the 1980s*. Cambridge, MA: MIT Press, 2009.

Barnett, Michael and Martha Finnemore. *Rules for the World: International Organizations in Global Politics*. Ithaca, NY and London: Cornell University Press, 2004.

Bell, Daniel. *The Coming of Post-Industrial Society: A Venture in Social Forecasting*. New York: Basic Books, 1976.

Bell, Daniel. *The End of Ideology: The Exhaustion of Political Ideas in the 1950s*. Cambridge, MA: Harvard University Press, 1962/2000.

Bernal, J. D. *The Social Function of Science*. London: Lowe & Brydone Printers, 1939/1946.

Cain, Frank. "Computers and the Cold War: United States Restrictions on the Export of Computers to the Soviet Union and Communist China." *Journal of Contemporary History* 40 (2005): 131–147.

Cambell-Kelly, Martin and William Aspray. *Computer: A History of the Information Machine*. Boulder, CO: Westview Press, 2004.

Ceruzzi, Paul E. *A History of Modern Computing*. Cambridge, MA: MIT Press, 2003.

Clemens, Walter. *Can Russia Change? The USSR Confronts Global Interdependence*. New York: Routledge, 1990.

Cole, H.S.D., C. Freeman, M. Jahoda, and K. Pavitt, eds. *Thinking about the Future: A Critique of the Limits to Growth*. Sussex: Sussex University Press, 1973.

Dorodnitsyn, A. A., ed. *Ekologicheski i demograficheskie posledstviia iadernoi voiny*. Moscow: Nauka, 1986.

Dubovskii, Sergei. "Global'noe modelirovanie: voprosy teorii i praktiki." *Vek globalizatsii* 2 (2010): 47–67.

Dubovskii, Sergei and O. A. Eismont. "Long-Range Modelling of the USSR Economy." In *The Future of the World Economy*, edited by Wilhelm Krelle, 311–324. Berlin and Heidelberg: Springer Verlag, 1989.

Duhautois, Sibylle, *The Future of World Problems* (PhD diss., Sciences Po, in progress).

Edwards, Paul. "The World in a Machine: Origins and Impacts of Early Computerized Global Systems Models." In *Systems, Experts and Computers: The Systems Approach in Management and Engineering, World War II and After*, edited by Agathe C. Hughes and Thomas P. Hughes, 221–254. Cambridge, MA: MIT Press, 2000.

Efremenko, Dmitrii. *Ekologo-politicheskie diskursy: vozniknovenie i evoliutsiia*. Moscow: INION, 2006.

Elichirigoity, Fernando. *Planet Management: Limits to Growth, Computer Simulation, and the Emergence of Global Spaces*. Evanston, IL: Northwestern University Press, 1999.

Ellman, Michael and Vladimir Kontorovich, eds. *The Destruction of the Soviet Economic System: An Insiders' History*. Armonk, NY: M. E. Sharpe, 1998.

Emel'ianov, S. V., ed. *30 let institute sistemnogo analiza Rossiiskoi akademii nauk: istoriia sozdaniia i razvitiia Institutea sistemnogo analiza, 1976–2006 gg*. Moscow: URSS, 2006.
Engerman, David. *Know Your Enemy: The Rise and Fall of America's Soviet Experts*. Oxford: Oxford University Press, 2009.
Evangelista, Matthew. *Unarmed Forces: The Transnational Movement to End the Cold War*. Ithaca, NY: Cornell University Press, 2002.
Fedorenko, N. P. *Vspominaia proshloe, zagliadyvaiu v budushchee*. Moscow: Nauka, 1999.
Fedoseyev, P. N. "Topical Problems of Our Time." In *Science, Technology and the Future: Soviet Scientists' Analysis of the Problems of and Prospects for the Development of Science and Technology and their Role in Society*, edited by E. P. Velikhov, J. M. Gvishiani, and S. R. Mikulinsky, 3–20. Oxford: Pergamon Press, 1980.
Fleck, Ludwik. *Genesis and Development of a Scientific Fact*. Chicago: Chicago University Press, 1979.
Forrester, Jay. "System Dynamics and the Lesson of 35 Years." In *A Systems-Based Approach to Policymaking*, edited by Kenyon De Greene, 199–240. Berlin: Springer, 1993.
Galison, Peter and Bruce W. Hevly, eds. *Big Science: The Growth of Large-scale Research*. Stanford: Stanford University Press, 1992.
Gemelli, Giuliana. "Building Bridges in Science and Societies During the Cold War: The Origins of the International Institute for Applied Systems Analysis (IIASA)." In *American Foundations and Large Scale Research: Construction and Transfer of Knowledge*, edited by Giuliana Gemelli, 159–198. Bologna: Clueb, 2001.
Gelovani, V. A., V. B. Britkov, and S. V. Dubovskii. *SSSR i Rossiya v globalnoy sisteme (1985–20130): rezultaty globalnogo modelirovaniia*. Moscow: Librokom, 2009.
Gvishiani, Dzhermen. "Methodological Problems of Global Development Modelling." In *Science, Technology and the Future: Soviet Scientists' Analysis of the Problems of and Prospects for the Development of Science and Technology and their Role in Society*, edited by E. P. Velikhov, J. M. Gvishiani, and S. R. Mikulinsky, 21–35. Oxford: Pergamon Press, 1980.
Gvishiani, Dzhermen. *Mosty v budushchee*. Moscow: URSS, 2004.
Gvishiani, Dzhermen. "Sistemnyi podkhod k issledovaniiu global'nykh problem." In *Nauka i chelovechestvo*, 237–249. Moscow: Znanie, 1983.
Gvishiani, Dzhermen. "The Search for a Global Strategy." *Unesco Courier* (November 1979): 25–26.
Hawkes, Nigel, ed. *International Seminar on Trends in Mathematical Modelling, Venice, 13–18 December 1971*. Berlin: Springer Verlag, 1973.
Helmer, Olaf and L. Blencke. *GEM: An Interactive Simulation Model of the Global Economy*. RR-79–4, Laxenburg: IIASA, 1979.
Herran, Nestor, Soraya Boudia, and Simone Turchetti. *Transnational History and the History of Science*. Cambridge, UK: Cambridge University Press, 2012.
Hersh, Reuben. "Mathematics Has a Front and a Back." *Synthese* 88, no. 2 (1991): 127–133.
Ivakhnenko, A. G. and V. G. Lapa. *Cybernetics and Forecasting Techniques*. New York: Elseviere, 1967.
Jamison, Andrew. "Technology's Theorists: Conceptions of Innovation in Relation to Science and Technology Policy." *Technology and Culture* 30, no. 3 (1989): 505–533.
Kaysen, Carl. "Computer that Printed Out W*O*L*F." *Foreign Affairs* 50 (July 1972): 660–668.
Keepin, Bill and Brian Wynne. "Technical Analysis of IIASA Energy Scenarios." *Nature* 319 (December 20, 1984): 691–695.

Krelle, Wilhelm. *The Future of the World Economy: Economic Growth and Structural Change*. Berlin: Springer Verlag, 1989.
Latham, Michael E. *The Right Kind of Revolution: Modernization, Development and U.S. Foreign Policy from the Cold War to the Present*. Ithaca, NY: Cornell University Press, 2011.
Leontief, Wassily. "The Decline and Rise of Soviet Economic Science." *Foreign Affairs* (January 1960): 261–272
Leontief, Wassily, A. Carter, and P. Petri. *The Future of the World Economy: A United Nations Study*. New York: Oxford University Press, 1977.
MacKenzie, Donald. *Mechanizing Proof: Computing, Risk, and Trust*. Cambridge, MA: MIT Press, 2001.
Meadows, Donella, John Richardson, and Gergart Bruckman. *Groping in the Dark: The First Decade of Global Modelling*. Chichester, UK: John Wiley, 1982.
Men'shikov, Stanislav. *O vremeni i o sebe*. Moscow: Mezhdunarodnye otnosheniia, 2007.
Michael, Donald. *Cybernation: The Silent Contest*. Santa Barbara, CA: Centre for the Study of Democratic Institutions, 1962.
McLuhan, Marshall. "Cybernation and Culture." In *The Social Impact of Cybernetics*, edited by Charles Dechert, 95–108. Notre Dame, IN: University of Notre Dame Press, 1966.
Modelirovanie protsessov global'nogo razvitiia. Sbornik trudov VNIISI, vol.8. Moscow, 1979.
Moiseev, Nikita. *Algoritmy razvitiia*. Moscow: Nauka, 1987.
Moiseev, Nikita. *Prosteishie matematicheskie modeli ekonomicheskogo prognozirovaniia*. Moscow: Znanie, 1975.
Munn, R. E. "Discussion." In *The Biosphere and Humanity*, IIASA Report CP-94-4, April 1994.
Oldfield, Jonathan and Denis Shaw. "V. I. Vernadskii and the Development of Biogeochemical Understandings of the Biosphere, c.1880s–1968." *The British Journal for the History of Science* 46 (2013): 287–310.
Petrov, A. A. *Nikita Nikolaevich Moiseev: sud'ba strany v sudbe uchenogo*. Moscow: Ekologiia i zhizn, 2011.
Porshnev, A. "Vklad B.Z. Mil'nera v razvitiiu nauki upravleniia." *Rosiiskii zhurnal menedzhmenta* 4 (2004):153–159.
Rindzevičiūtė, Eglė. *Riders on the Storm: How Cold War Scientists Co-Operated to Create Global Governance*, in preparation.
Riska-Campbell, Leena. *Bridging East and West: The Establishment of the International Institute for Applied Systems Analysis in the United States Foreign Policy of Bridge Building, 1964–1972*. Helsinki: Finnish Society on Science and Letters, 2011.
Rokityanskij, Y. G. "N.V. Timofeef Resovski in Germany, July 1925–September 1945." *Journal of Bioscience* 30, no. 5 (2005): 573–580.
Rosier, Bernard, ed. *Wassily Leontief: textes et itinéraire*. Paris: Éditions la découverte, 1986.
Schwartz, Thomas. "Moving Beyond the Cold War: The Johnson Administration, Bridge-Building and Détente." In *Beyond the Cold War: Lyndon Johnson and the New Global Challenges of the 1960s*, edited by Francis Gavin and Mark Atwood Lawrence, 76–96. Oxford: Oxford University Press, 2014.
Selcer, Perrin. *Patterns of Science: Developing Knowledge for a World Community at UNESCO*. PhD diss., University of Pennsylvania, 2011.
Stigler, M. Stephen. *Statistics on the Table*. Cambridge, MA: Harvard University Press, 1999.
Svirezhev, Yuri. *Ecological and Demographic Consequences of Nuclear War*. Moscow: Computer Centre of the USSR Academy of Sciences, 1985.

Tatarchenko, Ksenia. *"A House with a Window to the West": The Akademgorodok Computer Centre (1958–1993)*. PhD diss., Princeton University, 2013.

Thomas, William. *Man's Role in Changing the Face of the Earth*. Chicago: University of Chicago Press, 1956.

Ward, Barbara, René DuBois, Thor Heyerdahl, Gunnar Myrdal, Carmen Miró, Lord Zuckerman, and Aurelio Peccei. *Who Speaks for Earth?* New York: Norton, 1973.

Waters, Malcolm. *Daniel Bell*. London and New York: Routledge, 1996.

Wynne, Brian. *Models, Muddles and Megapolicies: The IIASA Energy Study as an Example of Science for Public Policy*. IIASA Reports WP-83-127, December 1983.

Zagladin, Vadim and Ivan Frolov. *Global'nye problem sovremennosti: Nauchnyi i sotsial'nyi aspekty*. Moscow: Mezhdunarodnye otnosheniia, 1981.

6 Forecasting the Post-Socialist Future

Prognostika in Late Socialist Czechoslovakia, 1970–1989

Vítězslav Sommer

In the late 1980s, the Czechoslovak public was enthralled by *prognostika*.[1] Scholars and experts, mainly economists who were involved in the forecasting project of the Academy of Sciences (CSAS) Forecasting Institute (*Prognostický ústav*) and in other more informal expert activities, constituted an important part of the newborn critically minded intelligentsia, which was composed of professionals increasingly dissatisfied with growing economic and political malaise; their field of inquiry became a political phenomenon of Czechoslovak *perestroika*.[2] Official opening of reform debates in the Soviet Union by Mikhail Gorbachev had its impact in Czechoslovakia as well. Although the public discussions were more restricted than in the Soviet Union, Poland and Hungary, because every call for fundamental political reform was read as reference to officially still condemned Prague Spring, there were also less restricted fields enabling more open and critical debates.

The most important of these fields was economy. Its serious problems and the failure of all attempts to resolve them forced communist authorities to establish more relaxed rules of discussion about economic issues. It was partially enabled also by the fact that Party leadership recognized the economy as a relatively safe territory, a complex field accessible only to experts and thus prevented from broader involvement of wider public. However, this opened the possibility to critically analyze economic problems and propose various paths for the way out of the economic crisis, which was discussed primarily as a crisis of predictability and efficiency. As a result of urgent necessity to deal with economic decline, discourse of economics was less controlled and provided opportunity to challenge prevailing ideological orthodoxy. Regarding the organization of socialist economy on the basis of central planning, issues like planning, forecasting, management, and organization of production were of utmost importance. It was thus not surprising that prognosticating gained such prominent voice in *perestroika* period and that was possible to gather such diverse group of experts under the banner of *prognostika*. Research undertaken by forecasting scholars, which was closely connected with these debates about economic reform, meant that *prognostika* was perceived as an empirically and theoretically founded critique of stagnating late socialism. In August 1989, the popular scientific journal *Technický magazín* (the Technical Magazine) published an article *Prognostika and Perestroika*, written by forecasting scholar Miloš Zeman. His short text about possible scenarios of the Czechoslovak future caused a real sensation. According to Zeman, "dialogue about *perestroika*" was "dialogue about the future" of Czechoslovakia. Zeman highlighted a "plurality of possible futures" and called for critical research on the contemporary systemic crisis. In his article, which very soon achieved broad notoriety, *prognostika* was characterized as an important source of critical thinking. Zeman emphasized the necessity of publicly discuss social and economic problems as really existing and serious threats. In the center of his argument, apart from the pessimist narrative of Czechoslovak decay and backwardness, was a call for thinking about alternatives, which was read by public as a critique of authoritarian rule in the socialist state. As a consequence of the publication of this article, Zeman lost his job. In a subsequent appearance on Czechoslovak television in late August 1989, Zeman repeated his critique of the current state of the Czechoslovak economy. This act of public criticism was the source of his later reputation as a courageous and controversial public intellectual and political actor. After 1989, Zeman enjoyed a successful political career. In 1998, he became the first social democratic prime minister after the Velvet Revolution, and in 2013 Zeman was elected president of the Czech Republic.[3]

In the autumn of 1988, the Forecasting Institute completed a document called *General Prognosis*. This forecasting report on future development in Czechoslovakia until 2010 was awaited with some excitement and the findings of this most influential forecasting enterprise of the 1980s were posited as a challenge to the helpless official policies of a stagnating regime. After 1989, many scholars, who were experts in *perestroika* future-oriented studies, became influential actors in the process of post-socialist transformation. It was no coincidence that numerous rumors about late socialist *prognostika* emerged during the 1990s. Nevertheless, historiographical accounts of this phenomenon are still very rare.[4] The aim of this chapter is to assess this scholarly field and analyze its development from the late 1960s to 1989. The text is divided into three sections. The initial part examines the emergence of *prognostika* in the aftermath of the 1968 Prague Spring. It deals with connections and discontinuities between a previous wave of futurological studies in the late 1960s and subsequent developments, which led to the establishment of *prognostika* as a centralized system of future-oriented forecasting expertise in the service of the late socialist regime. The emergence of this particular forecasting expertise in direct relationship to political power in the 1970s is described in the second part of the chapter. The final section discusses forecasting in the *perestroika* period and is concerned with the "critical turn" in economic forecasting. This chapter, thus, follows the development of future studies in Czechoslovakia from the collapse of reform communist scholarship after 1968 to the construction of new, future-oriented project in the *perestroika* period that remained critical of communism but was focused on economic liberalization.[5]

FROM FUTUROLOGY TO *PROGNOSTIKA*: FUTURE STUDIES, THE PRAGUE SPRING AND THE COLLAPSE OF REFORM COMMUNISM

Future studies in Czechoslovakia began with the interdisciplinary team of scholars headed by the philosopher Radovan Richta in the 1960s. His research project, which was based in the CSAS Institute for Philosophy, was concerned with the "social and human implications of the scientific and technological revolution (STR)." STR attracted a large group of researchers, mostly philosophers and sociologists, and aimed to develop a coherent account of social change in socialism. Its attempt was to interconnect empirical research on science, technology, economy, and organization of labor with future-oriented theorizing about "productive forces" in post-industrialism and changes of the self in the process of large-scale introduction of advanced technologies in the sphere of production. In 1966, the most important outcome of the STR collective's research, *Civilization at the Crossroads*, was published.[6] Richta analyzed the future paths of socialism in the context of technological development and an emerging, increasingly post-industrial labor force. His research was a consciously Marxist answer to the Western conceptualizations of modernity and social change. Although Richta proposed concrete governmental strategies that fell within the established policies of the STR and projected images of future "scientific civilization" in accordance with official ideas of technologically advanced socialism, his work was however also a sophisticated attempt to critically rethink the Marxist theory of labor in order to reform the communist political project after its Stalinist turmoil.

The Czechoslovak version of STR emerged within the specific conditions posed by post-Stalinist dictatorship. As a theory of social change, it was based on a firm belief in the transformation of "productive forces" and future domination of non-manual labor. Richta characterized science as a "direct productive force" and the most important actor of social change. Labor of the future was, according to Richta, organized around scientific research and application of its outcomes in the sphere of production. Reflecting the Marxist core of STR thinking, the new structure of productive forces was starting point for far-reaching changes in human subjectivity, as well as the praxis of socialist governance. Future post-industrial socialism, based on the large-scale mobilization of science, was dependent on the mastering of increasingly complex scientific knowledge. It follows that a civilization of scientists, engineers and managers had to be governed by different governmental setting reflecting the crucial importance of individual intellectual capacity, creativity and responsibility. If the industrial era, characterized by factory and conveyor belt manual worker, was governed by direct and authoritative control of the social, socialist post-industrialism required synthesis of broader social participation in the governance and employing of the most advanced methods of organization and planning. Richta was convinced that the future of socialism was based not solely on market-oriented economic reform or the introduction of "socialist democracy" but, first of all, on the creation of a new socialist man and new socialist society. This was nevertheless grounded in historical materialism and in his understanding of changing productive forces. Technological and scientific progress was seen as a source of fundamental economic, social and political changes. Future communism was portrayed as a world of scientific objectivity, rational planning, automatized production and perfect organization of all social processes through the development of organizational science. At the same time, Richta predicted a new society composed of educated individuals actively participating in decision-making and consciously striving for personal self-realization and self-cultivation. Tensions between the technocratic appraisal of scientific rationality and humanistic appeals were thus more than evident. These tensions mirrored the inner contradictions of reform communism and the Prague Spring. It was not an easy task to find a middle way between the idea of scientific high-tech governance and the socialist promise of a new society free of alienation and composed of fully emancipated socialist subjects.

The concept of STR reached notoriety during the Prague Spring when it became an important source of reform communist political thought. In the context of reform communism, the STR was important for its explicit critique of various problems caused by Stalinist and post-Stalinist governance. Vision of groundbreaking technological progress and idea for the further development of socialism by the means of science, education and broader political participation were in almost perfect accordance with the growing social demand for prosperous and democratic socialism. Richta himself would become the leading intellectual of the Czechoslovak 1968. He was the co-author of the crucial Communist Party policy document, the *Action Program of the Communist Party of the Czechoslovakia*, which was sanctioned by the Party leadership in late March and early April 1968 and became the official outline of far-reaching reform in all fields of socialist governance from economy to political system.[7] Richta and his collaborators were among the scholars who revealed that the outcomes of the post-war socialist transformation, from the systemic failures to the subtle issues of everyday life under the state socialism, were disputable. Their research on the future was concerned not only with the introduction of automation, scientific governance or broader political participation, but also with the question of how human subjectivities might be transformed by the rapidly changing conditions of modernity. From this perspective, modern socialism was an ongoing project, rather a future perspective than the immediate reality of the 1960s Czechoslovakia, and its realization was fully dependent on the close collaboration between three equally important actors—communist authorities, scientists and society.

Although the idea of STR dominated Czechoslovak discourse on the future, Czechoslovak future studies also reflected influences from Western futurology, and introduced this field into the Czechoslovak intellectual context. This was possible primarily thanks to Ota Šulc, a researcher at the Higher School of Economics in Prague. Šulc published several basic introductions to the different concepts of futurology.[8] Šulc used the term *futurology* as a general label for all activities dealing with the future, including Richta's project, but also the works of officially sanctioned Soviet scholars such as Edvard A. Arab-Ogly and Igor Besthuzev-Lada.[9] Czechoslovak futurology was a field in the making, and its development followed Western influences and the attempts of reform-minded scholars to rethink the philosophical basis of Marxism. Futurology seemed to be new and attractive discipline, scholarship based on transnational cooperation and incorporating different intellectual traditions, and it conformed to reform communist efforts to open socialism to Western influences. Thanks to the relaxing of ideological control, futurological enthusiast were enabled to gather individual supporters of futurology in official organizations and coordinate their activities as a part of flourishing civic life of the Prague Spring period. In 1968, the Czechoslovak Futurological Society was founded as a part of the Socialist Academy.[10] At the same time, students at the Higher School of Economics established the Futurological Club.[11] The Czechoslovak Futurological Society attempted to create a team working on "futurological simulation models" called *Theoretical Foundations of a Complex Social Project* and provide an alternative to official state planning. The most important planned outcome was a "concrete macro-prognosis" for the years 1985–2000.[12] However, the slow but inevitable suppression of all grassroots activities after the Warsaw Pact invasion of Czechoslovakia in August 1968 was unfavorable for this lay expertise rooted in the arising civil society. Ambitious plans of the Czechoslovak Futurological Society were thus in most cases spoiled by political development in the country. The organization was still active in 1970 under the title Central Scientific and Technological Council for Futurology of the Socialist Academy.[13] The bimonthly journal *Trend*, which provided Czechoslovak scholars with information on future studies institutions and literature, was issued from 1969.[14]

In the relatively free intellectual environment of the late 1960s, it was possible to conceptualize Eastern future studies as a part of a wider international field. In 1968, this field went under the label of *futurology*. At the same time, Czechoslovak reform minded economists such as Ota Šik, who was the leading figure of Czechoslovak economic reform and internationally respected theoretician of market socialism, began rethinking the structure of economic planning.[15] Here, the catchword was market socialism, the idea of mixed economy based on less centralized planning, existence of more autonomous enterprises and market exchange among the state-owned companies.

In this context, *prognostika*, which was widely discussed primarily in the Soviet Union and the German Democratic Republic (GDR) as a specific Marxist-Leninist approach to the future, as well as a more advanced technique of economic planning, would be analyzed by reformist Czechoslovak scholars as a particular version of the transnational field of futurology and as sharing important features with Western futures research. The terms *futurology* and *prognostika* were used as more or less synonyms, and original Czechoslovak theoretical accounts of Marxist-Leninist *prognostika* were rather rare.[16] A unique attempt to elaborate a theoretical Marxist-Leninist concept of *prognostika* was made by the philosopher Jindřich Schermer in his 1968 book *Prognostika-Science-Planning*.[17] For Schermer, research on the future meant searching for different paths of development and analyzing developmental alternatives and embryonic phases of crucial social and economic processes. In his book, *prognostika* was interconnected with economic and social planning and its aim was to outline complex alternatives of future development. The result of forecasting research was *prognóza* (prognosis)—the planner's source of information and "set of alternative predictions."[18] Contrary to Schermer's original theory of *prognostika*, Šulc and his young collaborator Miloš Zeman were concerned mainly with the reception of pre-existing forecasting theories and methods.[19] Future studies were thus constructed as an intellectual field open to both Eastern and Western influences. Nevertheless, early Czechoslovak futurology remained in the shadow of the obsession with STR, as the official central research institution, the CSAS, supported by the communist authorities, was charged with research on the future of scientific and technological development. The concept of STR was elaborated in the writings of Richta and his collaborators within the CSAS. In the Czechoslovak context, the impact of Soviet or GDR *prognostika* was limited, because due to the influence of reform communist ideas, market-oriented reform would become more important for Czechoslovak economists than the reception of Marxist-Leninist forecasting methods. Western future studies were widely discussed by Richta and other scholars, but the specific field of research and expertise called futurology also remained a "bottom-up" activity of individual forecasting enthusiasts like Šulc, Zeman or Schermer.

The slow but inevitable collapse of reform communism after the August 1968 Warsaw Pact invasion of Czechoslovakia and the disintegration of the reformist fraction in the Party leadership was a disaster for the politically important STR project and all related scholarship. Radovan Richta followed the path of political pragmatism and renounced all reformist elements in his STR scholarship. This act of intellectual and political self-criticism, as well as close collaboration with several prominent Soviet scholars, enabled the further existence of his research team, although with partially purged personnel and field of inquiry limited to the theory and sociology of science. During the 1970s and 1980s, STR functioned in much more dogmatic ideological framework and suffered serious intellectual decline. Although Richta secured his prominent position in the Czechoslovak academia, STR as a concept of future was transformed to an inanimate and fossilized doctrine describing a non-problematic heading toward communist utopia.

The failure of the Prague Spring was followed by a massive campaign against "revisionism" and "rightist opportunism" in the social sciences. This ideological attack was led from positions emanating from rigid Marxism-Leninism, and it had severe consequences for both the institutional structure and theoretical nature of social scientific scholarship. The clamp down, which was in its broader political context referred to as "consolidation" or "normalization," was characterized by the closing down of institutions, the purging of reform-minded scientists, humiliating self-critiques and publication bans. In addition, the theoretical and conceptual landscape of the social sciences was fundamentally upset. Theories and arguments associated with reform communism were excluded from the official theoretical apparatus. Future studies were by no means exempt from this disastrous development that affected all social scientific disciplines. Both scholarship on the STR and the emerging field of futurology were slowly but inevitably purged. The Czechoslovak Futurological Society as an institution organized from below by scholars themselves was dissolved in 1971.[20] From this moment on, research on the future was possible solely within the boundaries of official institutions. Until the mid-1980s, more or less independent forecasting activities were allowed only on the margins of the social sciences.

The "consolidation" of Czechoslovak future studies in the early 1970s as a question of Party expertise, emptied of its critical philosophical content, was characterized by a unification of terminology. Officially, from the early 1970s onward, forecasting was called *prognostika*, whereas the term *futurology* was used exclusively as a dismissive label for "bourgeois" future studies. This recoding had an important political significance. It served as evidence that Czechoslovak future-oriented scholarship was fully incorporated into the transnational structure of Marxist-Leninist *prognostika*, and all ideological influences of Western futurology were definitively removed. It came with a theoretical clarification of the ideological setting of the field, as part of the early 1970s campaign against so called bourgeois theories.[21] Western futurology was ideologically denounced by its former most prominent Czechoslovak promoter, Ota Šulc, in his article *Marxist-Leninist Prognostika in the Struggle Against Bourgeois Concepts of Future Studies*.[22] This brief text discussed the methods of future studies (scenarios, Delphi) and their alleged ideological background. For the first time, a clear and normative distinction between Marxist-Leninist future studies and Western futurology was made. Future studies were thus divided into two disciplines, Eastern *prognostika* and Western futurology, which shared some methodological tools but differed fundamentally in their broader philosophical background and social functions. Only *prognostika*, or at least the future-oriented research officially called *prognostika*, was allowed to exist in Czechoslovakia after the introduction of this highly ideological critique of Western futurology. This rebranding of Czechoslovak future studies and formal demarcation of the field were widely used by authors from the state socialist countries. The critique of "bourgeois" futurology was an important component of writings on Western future studies already in the 1960s in some countries, such as the Soviet Union and GDR.[23] However, the idea that future studies were a site of ideological struggle between socialism and capitalism was rather new in Czechoslovakia, because in the reform communist period, the critique of Western intellectual concepts had been accompanied by a striving for mutual communication and discussion.[24] The introduction of an ideologically more rigid discourse after 1968 substantially changed the situation, because it imposed an obligatory intellectual framework for research on the future. Although scholars after 1968 would continue to widely reflect theories of Western futurology and attempted to apply its methodological tools in their forecasting praxis, the forced rejection of a "bourgeois" body of knowledge introduced by "consolidation" campaign remained officially valid until 1989.[25]

The ideologically driven critique of futurology was closely connected with the effort to establish a specific discipline of Czechoslovak *prognostika*. Whereas the futurology of the late 1960s had been the product of the quasi-civic society of the Prague Spring, and STR scholarship emerged from the long-lasting intellectual and organizational efforts of Richta and his collaborators, the building of *prognostika* in the early 1970s was a top-down process initiated by the state. The aim was to establish a complex network of forecasting expertise, centralized in the CSAS and consisting of specialized expert institutions. This version of future research would replace reform communist criticism and substitute the calls for broader social participation and thinking about alternative futures that reform communist intellectuals has put forward with a hierarchical and centrally organized technocratic expertise.

THE PRIMACY OF ECONOMICS: FORECASTING IN THE AGE OF "CONSOLIDATION"

In the forecasting field, the first "consolidation" measures were made as early as 1970 and 1971. In February 1970, the Institute for Sociology and the Center for the Theory and Methodology of Science were dissolved and the Institute for Philosophy was transformed into the CSAS Institute for Philosophy and Sociology, which became the hub of social scientific research in Czechoslovakia. Radovan Richta, having pragmatically denounced his reform communist past and reinterpreted his theory, was installed as a director of this academic body. The institute was composed of three main sections: sections for philosophy and sociology and the section for the STR.[26] The specialized department concerned with the theory and methodology of forecasting, headed by Ota Šulc, who was compliant to ideological self-critique and thus was not purged from the academia, became part of the STR section.[27] A specialized forecasting team in the Institute for Economy and Management of the Scientific and Technological Development was also established in 1970, dealing with research on the relationship between the forecasting of scientific and technological development and economic growth.[28] Forecasting units were likewise active at the CSAS Center for the Theory of Architecture and Creation of Environment, the CSAS Institute for Economics, and the Federal Council for the Development of Technologies and Investment.[29]

In the substantial reconstruction of future studies after 1968, economic forecasting would take a leading role. Party demands were for future-oriented expertise in the field of economics, and following this lead, economic forecasting would play a dominant role in the process of "consolidation." Already, in the late 1960s, the first steps to organize economic forecasting as part of economic planning had been carried out. In October 1970, the Ministry of Technological and Investment Development and the National Planning Committee were charged by the Federal government to construct a "proposal for a coordinated program for the preparation of long-term prognoses."[30] This document, which was authorized by the government in March 1971, declared the necessity of long-term perspectives for the Czechoslovak economy.[31] The governmental decree established a "system of forecasting activities," a network of research teams and institutions, which were concerned with social, scientific, technological and economic development. Forecasting expertise was divided into four "sub-systems": prognoses of the "national economy," more general documents composed as so-called summarizing prognoses, prognoses of individual branches and sectors of the economy, and territorial prognoses.[32] The main anticipated outcomes were the prognosis for the Czechoslovak economy until 1985 and the long-term prospect (*výhled*) of economic development. Besides these major documents, the forecasting expertise also produced supportive materials like prognoses of scientific progress, regional development or prospects of individual industrial sectors. At the center of the complex *prognostika* network stood the National Planning Committee.[33]

The first plan of the forecasting activities of the CSAS, the most important scientific institution in Czechoslovakia, was authorized in September 1971. Its forecasting expertise was involved in various topics and played a significant role in the forecasting as such. The initial involvement of the CSAS in *prognostika* was structured around several crucial research topics usable for the general prognosis of the "national economy": "life-style and its connection with the future development of society"; the practical application of physics and microbiology; and the prognosis of Czechoslovak science and its institutional framework.[34] Forecasting in the CSAS was coordinated by a commission composed of influential academic officials. Its aim was to review the outcomes of the research teams and closely cooperate with the Ministry of Technological and Investment Development.[35]

The initial stage of forecasting activities was characterized by serious shortcomings. According to CSAS officials, forecasting reports were rather proposals of possible research topics than documents based on the "prognostic approach" and were thus not usable in policy-making.[36] In July 1973, two important reports by CSAS social scientific forecasting experts were completed. Philosophers and sociologists authored the report on the "making of a socialist life-style."[37] This relatively short document called for long-term social prognoses and characterized forecasting as a blueprint for the "optimization" of governmental interventions in social life. The topics of "socialist life-style" and the relationship between "socialist man" and the STR were further analyzed by Richta's institute in later years.[38]

Also in June 1973 a voluminous document prepared by economists from the CSAS Institute for Economics was discussed. The research team, headed by leading Czechoslovak economists Josef Goldmann and Růžena Vintrová, completed a research report on "macroeconomic prognosis" called *The Problems of the Czechoslovak Economy until 1990*. Here, *prognostika* was used as an applied science, closely connected with economic planning. The report covered several aspects of the "process of reproduction" such as the standard of living, foreign trade, and economic relations with socialist and capitalist countries. "Efficiency" was the catchword of the document, which aimed to detect the "central problems" (*uzlové problémy*) of the Czechoslovak economy in order to prevent it from possible "difficulties and losses." The report argued that the socialist future had to be consciously and complexly constructed with the help of coordinated and interdisciplinary research and the collection of data on social reality. The long-term prognosis was composed from a "complex system of ideas and claims about the future state of the economy." It was a prospective economic analysis with various degrees of certainty. Goldmann and his collaborators argued that "scientific prognosis" possessed the ability to overcome particular social interests and formulate a "superordinate universal interest" (*nadřazený obecný zájem*) based on the "authoritative formulation of the goals of society." According to forecasting experts, particular actors had to tailor their interests and aims to this universal framework in order to fulfill tasks crucial for society as a whole.[39] Economic forecasting in the early 1970s was thus in perfect accordance with the image of the socialist economy as one huge factory optimally managed by sophisticated planning technologies.[40]

Forecasting activities of the 1970s were a typical intellectual product of the more authoritarian post-1968 regime. As an activity of "consolidation," *prognostika* was established and integrated into the networks of Eastern forecasting research on the STR. Scholars from Czechoslovakia became involved in the Eastern Bloc forecasting community as active participants at the international conferences.[41] Some events were even organized by Czechoslovak institutions.[42] During the 1970s, original texts were published on the theory and methodology of forecasting and on the practical application of *prognostika* in different branches of the economy and governance (science, military, cultural policy, management).[43] So-called "social planning," highly technocratic scholarship on the "programming" and "management" (*řízení*) of social development and "social processes," became a specific field of expertise. It was occasionally combined with the theory of management and was obsessed with the idea of centralized administration of social life.[44] Josef Goldmann published his book *Macroeconomic Analysis and Prognosis*, which characterized macroeconomic forecasting as dialectic thought "discovering the laws of the development of society."[45] Goldmann discussed the relationship between macroeconomic analysis and macroeconomic prognosis, and the problem of probability in forecasting. He proposed to combine prospective and retrospective research. Management of the socialist economy required the "integration of analytical and prognostic inquiry"[46] and use of economic forecasting as a prominent policy instrument. In the second half of the 1970s, the former reform communist economist and forecasting enthusiast, Valtr Komárek, published two monographs on *prognostika*.[47] Both texts emphasized the necessity of a well-organized forecasting system. The socialist economy was more and more complex, so economists and managers had to carefully analyze economic and social developments. According to Komárek, future studies were a technique of economic governance, capable of integrating precise economic and social analysis with systematic forecasting expertise. Scientific prediction meant searching for the most appropriate paths of economic development, and *prognostika* seemed to reach a high level of predictability in the management of centrally planned Czechoslovak economy.

FORECASTING AND *PERESTROIKA*: FROM THE CRISIS OF PLANNING TO SYSTEMIC CHANGE

In 1979, the Czechoslovak government issued a decree calling for the further development of "prognostics works." CSAS scholars prepared a long-term prognosis for scientific-technological, social and economic development until 2000, as well as reports on the perspectives of the economy.[48] The Institute of Philosophy and Sociology was concerned with the prognosis of socialist society and "socialist life-style" until 2000.[49] These forecasting documents mirrored the technocratic nature of *prognostika*, which emphasized careful and detailed planning and the application of the most advanced technologies in industry. The future of socialism seemed to be secured by technological progress and the introduction of sophisticated techniques of planning.

The belief that the socialist future could in this way be systematically projected was shaken in the early 1980s, when Czechoslovakia entered into a period of economic crisis. This crisis was interpreted as a crisis of economic planning. In 1983, the Presidium of the Party Central Committee, the most important political body in Czechoslovakia, passed several resolutions on the planning system. In September 1983, the State Commission for Scientific-Technological and Investment Development was established composed of high-ranking apparatchiks, top-managers and representatives of scientific institutions and universities.[50] Its aim was to coordinate and improve "development management." Coordination meant the integration of research with the different sectors of economy in order to establish centralized management of science, industry and agriculture. At the same time, personnel changes in the National Planning Committee, which authored five-year plans, were made.[51] The most important task of politicians, managers and experts was "improving the planned management of the national economy."[52]

Prognostika now became a vehicle for a renewed emphasis on economic planning. From this perspective, *prognostika* was not scientific experimentation or mere speculation about the future, but rather, *prognostika* was a tool for the resuscitation of a dying state socialism. Already in 1982, the CSAS had built its own permanent internal network of forecasting, which aimed to improve the connection between academic research and policy praxis.[53] The next step was the founding of the first specialized forecasting institution in the CSAS. The Center for Forecasting (*Kabinet prognóz*), which was created in January 1984, was the brainchild of Komárek, who was able to promote his idea in the highest ranks of the Party apparatus.[54] With Komárek as director, the center was involved in the writing of *General Prognosis*, the most important forecasting document of the 1980s.[55]

In September 1986, Komárek's center was transformed into the CSAS Forecasting Institute, which was organized into two main research units: the section for macroeconomics and the section for research on "social contexts."[56] Although the main expected outcome of the Institute was the *General Prognosis*, research teams devoted themselves to different sub-projects. *General Prognosis* was thus a synthesis of minor research reports and analyses of various issues from gross domestic product and the structure of prices to a comparison of Czechoslovak and capitalist industry.[57] Komárek proved to be extraordinarily skilled in staff recruitment. The Institute was well known for its open and nonorthodox intellectual environment and decent material background.[58] In the context of Czechoslovak academia, the Forecasting Institute was quite unique establishment. It became a prestigious and well-known institution and, at the same time, was a relatively safe haven for scholars who aimed to discuss controversial and potentially politically sensitive topics. Among the employees were experienced middle-aged economic experts, nonconformist scholars (monetarists as well as reform communists) and ambitious young economists. Several active participants in the market socialism reform of the 1960s were also involved. The granting of positions at the Institute represented these scholars' return to academia after years of publication bans and other forms of persecution. Hiring of former reform communist economists was also carried out in a careful attempt to establish some continuity with the late 1960s reform politics. The Institute gained a controversial reputation after 1989, when several of its researchers entered the political arena. Some of them joined right-wing parties and authored neoliberal economic reforms (Václav Klaus, Vladimír Dlouhý, Karel Dyba, Tomáš Ježek). Others played an important role in the social democratic and communist Left (Valtr Komárek, Miloslav Ransdorf, Jiří Dolejš, Jan Mládek). Even today, the Forecasting Institute is a mysterious presence in Czech political discourse, a place where future influential politicians and experts started their successful post-1989 careers.

In fact, the work on *General Prognosis* held extraordinary political importance. Alongside the specialized Forecasting Institute another important body was formed: the General Prognosis Board. This committee, members of which were mostly scholars and National Planning Committee officials, supervised and coordinated all "forecasting works" on *General Prognosis*.[59] In 1984, the Czechoslovak government emphasized the long-term dimension of the project. Hence, predictions were made not just to 2000, as was initially planned, but for the period up to 2010. Three years later, sub-prognoses dealing with the development of science and the economy were finally finished.[60] The government and the Presidium of the Party Central Committee discussed reports presenting the "basic findings" of the first years of the *General Prognosis* project. In a typically complicated manner, forecasting expertise was described as a "scientific anticipation of the objective developmental tendencies of the economically advanced world in the context of the concrete historical conditions of socialism in the relatively small and open Czechoslovak economy." Authors of the *General Prognosis* suggested that discussion about economic reform had to reflect Czechoslovak involvement in international economic exchange. The report called for an "optimization strategy" based on the "extraordinarily complex and difficult transformation of contemporary socio-economic reality." The proposed outcome was a "comprehensive and coherent strategic project of general prognosis oriented toward optimum perspectives."[61] The critique of economic imperfections was combined with a flood of empirical data. Reform proposals called for the encouragement of "material interests" in the society and for a change in the relationship between plan and market based on the "releasing of market forces" (*uvolnění tržních sil*). Although that the document did not call for radical solutions like large-scale privatization or complete market deregulation, its general narrative was built upon the conviction that introduction of market mechanisms was inevitable and that the only way toward future prosperity was to follow the example of the "advanced" Western capitalist countries. In fact, forecasting experts tried to develop detailed, empirically well-founded and theoretically sophisticated reform programs as a part of Czechoslovak *perestroika*.[62]

General Prognosis was completed in autumn 1988. The activities of the Forecasting Institute received much public attention and were praised as a critical and almost subversive project, quite rare in Czechoslovak public discourse, which was still under the strict control of the communist authorities. *Prognostika* became a form of expertise enabling the revelation of economic defects and shortcomings, and emphasizing the necessity of fundamental reform. However, the crucial prognostic document was not published in its entirety until 1990.[63] Instead, the Czechoslovak public was informed about the content of the *General Prognosis* through unofficial channels or foreign radio broadcasting. Communist party authorities were neither prepared nor willing to publicly discuss this report, which could serve as an impulse for further critique of the Party governance and as a source of alarming data about the state of Czechoslovak economy.[64]

However, this history of the Forecasting Institute would not be complete without emphasizing the controversies and tensions in the research team. Komárek was able to gather a very heterogeneous group of scholars. *General Prognosis* was, first of all, his project, deep-rooted in the official *perestroika* discourse. However, the making of the prognosis required numerous case studies, which were influenced by the theoretical, methodological and political preferences of individual scholars. They all shared the conviction that socialism was in deep crisis and required reform. Nevertheless, opinions about the concrete content of the reform varied from cautious modifications of the existing system to market socialism or even a free-market solution.[65] This was manifested in the May 1989 special issue of the journal *Politická ekonomie* (Political Economy), which published case studies written by the members of the *General Prognosis* research team. Komárek discussed the "restructuring" of the economy, its "effective and dynamic development" as well as the market resuscitation of the economy and the strengthening of its competitive ability. He proposed "planned regulation of the market" and economy based on various socialist forms of property and "totally democratic decision making of the widest strata of society."[66] His monetarist colleagues were concerned rather with the role of the "economic center." Tomáš Ježek and Otakar Turek called for the "renewal of market structures and institutions in order to produce a pricing system." The aim of the "economic center" was not to "dictate and specify particular desirable targets," but to "open space for the application of people's knowledge and creative potential." They proposed a limited or, more precisely, different role for the state, which had to "formulate rules and general conditions for the decision making of economic subjects." The state was an ordinary economic subject subordinated to the common "general rules" and "strict budgetary restrictions" and was not the "creator of structures". Its aim was only to formulate "general rules" and set "normative frameworks for the independent decision making of economic subjects."[67] Economist Václav Klaus even discussed "shock treatment" as a proper reform strategy. In his article, Klaus wrote: "The vast majority of economic arguments are in favor of this, naturally under the assumption that there is a clear political desire to realize this reform, and that there is a clear conception of what it involves."[68] This claim was supported by the example of the West German post-war transition to a social market economy. Instead of one-time reform, Klaus mentioned Hayek's notion of "gradualism that inspires trust."[69]

Members of Komárek's team thus favored fundamentally different reform strategies. The Forecasting Institute was the organizational background for important research activities, which resulted in the formulation of various reform proposals. On the other hand, all were based on a fundamental economic determinism, which later, after 1989, enabled such immediate and broad acceptation of neoliberal framework of economic theorizing and policy-making. The implementation of market structures was seen as an inevitable outcome of reform by all economists involved in the *General Prognosis* project. Regardless of the differences mentioned above, the future was analyzed primarily as economic method, as an attempt to find solutions to the crisis of the late socialist economy. This style of thinking about the future significantly influenced the critical discourse of Czechoslovak *perestroika* and later strategies of post-socialist transformation, in which the future was discussed primarily as a matter of economic reform. Careful macroeconomic calculations and well-organized administration of the national economy were the tools as well as the outcomes of future governance. *Perestroika* reform strategies produced by Forecasting Institute scholars were the result of a serious narrowing of possible ways of seeing the future after 1968 and they provided the intellectual basis for a set of post-socialist discourses on state, economy and society dominated by the neoliberal economism. The first predictions of possible market solutions to the crisis of late socialism paved the way for a discourse on pro-market policies, which could be implemented after the extraordinarily quick collapse of communist rule in Czechoslovakia in 1989.

Even though *prognostika* was developed as part of the "consolidation" attempt, with the ambition to provide a Party expertise on economic futures, perestroika forecasting thus opened up opportunities for a critical evaluation of contemporary social and economic problems and functioned as a space for the initiation of discussions about controversial issues to do with the future of the socialist system. This was carried out in fact by a wider network of forecasting scholars, not only in the Forecasting Institute but also by nonconformist scholars in semi-official research centers. These divergent worlds of official projects like *General Prognosis* and the grassroots activities of critically-minded scholars were by no means isolated from each other. In the course of the 1980s, these structures became increasingly interconnected. The desire for at least partially independent scholarship went hand in hand with the gradual opening of official academic institutions and publication platforms. At the same time, scholars were able to skillfully use official forums and projects for promoting their own research.

An important outcome of semi-official *prognostika* was the collected volume *Methodological Problems of Social Scientific Research of the Future*, published as early as 1983 by the Sportpropag Department of Complex Forecasting Modeling.[70] This small and subaltern research unit, which was headed by Zeman, was concerned with the forecasting of sport and physical education and served as an asylum for several politically unwanted scholars.[71] Its periodical seminars enabled informal discussion about forecasting and related social scientific topics.[72] However, the publication of *Methodological Problems* proved to be the swan-song of the Sportpropag forecasting department. The book contained several unorthodox texts dealing with controversial and ideologically sensitive subjects.[73] As a result of its publication, the department was officially dissolved in early 1984.[74] Other important semi-official networks emerged around periodical seminars organized in Prague by the Czechoslovak Scientific-Technological Society.[75] These seminars engendered relatively free discussion among scholars and interested public. The same institution organized forecasting conferences called *Recent Problems of Prognostics*.[76] Papers were presented mostly by *prognostika* experts, economists and sociologists; among them were reform-minded younger scholars as well as former reform communists who were purged after 1968. In short, a specific critical public emerged in the semi-official structures of 1980s *prognostika*. Discourse on forecasting opened the way for critical evaluations of late socialist economic and social reality and future studies became once again a source of knowledge, which offered alternatives to official policies. In the late 1980s, within the context of *perestroika*, it merged with the technocratic management language, which later became fundamental component of post-socialist neoliberalism.

CONCLUSION: *PROGNOSTIKA* AS A LABORATORY OF POST-SOCIALISM

Future-oriented research was introduced in Czechoslovakia in the 1960s via the STR theory of post-industrial socialism and the enthusiastic reception of Western futurology. After 1968, in the context of personnel purges and ideological campaigns against reform communism, this field was substantially changed. During the 1970s, *prognostika* was built as a complex network of forecasting activities directed from above. The emancipatory visions of the 1960s reform communism were replaced by technocratic expertise dealing primarily with the supervision of the "economic mechanism." Technocratic elements were certainly already present in the 1960s STR and in other Prague Spring social sciences projects, which however sought also to think about the social and philosophical implications of a changing economy. In the 1970s, forecasting left no space for such more ambitious concepts of the future. Thinking about the future was subordinated to the general framework of late socialist economism. Moreover, after 1968, future studies lost to a great extent their transnational perspective and their Western orientation. Whereas Richta in the 1960s discussed "civilizational" change in tandem with the reflections of Western scholars such as Daniel Bell, and offered the concept of a global socialist future based on transition from industrial to post-industrial production, *prognostika* reduced discourse on the future to a detailed economic forecasting of the "national economy" of state socialist Czechoslovakia. Both STR and *prognostika* challenged the reality of state socialism and used the activity of prognosis and prediction in order to show frictions in the communist system. In the 1960s, Richta criticized the post-Stalinist dictatorship for its inability to cope with the demands of social change caused by the interaction between labor, scientific knowledge and new technologies. He attempted to find possible paths for far-reaching reform based on the practical application of science combined with the participation of individuals in the decision-making process. Richta interconnected techno-optimism and a firm belief in the civilizational mission of science with serious concerns for the "human implications" of social transformations caused by post-industrial labor and the rise of "scientific civilization." In short, reform communist STR criticized the post-Stalinist regime in the name of a more scientific, more rational and more humanistic socialism. Richta's conscious ideological suppression of his own project after 1968 removed the humanistic and emancipatory appeals and transformed the STR into a developmental theory of technocratic governance and a legitimizing narrative for late socialist dictatorship.

Prognostika of the late 1980s offered a substantially different critique of the stagnating regime. Scholars criticized primarily the poor performance of the centrally planned economy. Regardless of conflictual concepts of economic reform, they were all convinced that the implementation of the market was crucial task of reformers. Their thinking about future was deep rooted in post-1968 forecasting expertise, first of all in its economism, which replaced more complex reform communist thinking about the future of socialism. Although *prognostika* had an indisputable critical potential and active nonconformist and critical-minded scholars, as well as pragmatic reformist technocrats present in its networks, the mental horizons of these actors remained fundamentally affected by economic determinism and geographical perspective limited by the borders of Czechoslovakia.[77] Reform concepts or heretical thoughts on the transformation of socialism into a different economic and political system were elaborated precisely in this specific intellectual framework. The rational and effective governance of the "national economy" was seen as the most important precondition for future social and political changes. This perspective, which put the economy above other social phenomena, paved the way for the neoliberal solution of the 1990s. In 1989, the future was reality and several *prognostika* experts became powerful members of the emerging post-socialist political elite. They had an unique opportunity to realize their concepts, which had been constructed in the laboratories of late socialist forecasting. Their policies were seriously affected by the significant intellectual limitations of their previous expertise. The over-simplification of social reality to calculative choices, and economism, resulting in narrow-minded fascination with economic performance, were political and cultural phenomena of the post-socialist era that had their origins also in the world of late socialist *prognostika*.

NOTES

1. The most accurate English translation of *prognostika* is "forecasting." In this chapter I also occasionally use the literal English translation *prognostics*. The terms *futurology* and *prognostika* are used in their concrete historical meanings, which is explained further in the text (most importantly in the case of the terminological shift in the early 1970s). The term *future studies* is used as a neutral label for future-oriented scholarship.
2. For an analysis of Czechoslovak *perestroika*, see Michal Pullmann, *Konec experimentu: přestavba a pád komunismu v Československu* (Prague: Scriptorium, 2011).
3. Miloš Zeman, "Prognostika a přestavba," *Technický magazín* 32, no. 8 (1989): 6–9. For Zeman's 1989 TV appearance, see " 'Jistý prognostik' si dovolil kritizovat komunistické hospodářství," *Czech TV* video, August 3, 2010, www.ceskatelevize.cz/ct24/exkluzivne-na-ct24/62717-jisty-prognostik-si-dovolil-kritizovat-komunisticke-hospodarstvi/.
4. Gordon L. Rocca, " 'A Second Party in Our Midst': The History of the Soviet Scientific Forecasting Association," *Social Studies of Science* 11 (1981): 199–247; Gil Eyal, *The Origins of Postcommunist Elites: From Prague Spring to the Breakup of Czechoslovakia* (Minneapolis: University of Minnesota Press, 2003).
5. This chapter discusses primarily central forecasting in the Czech part of Czechoslovakia and in Prague institutions of the CSAS. In the 1980s, a special forecasting team dealing with the prospects of the Slovak part of the federation was active in Slovakia. For the main outcome of this expertise, see Jozef Markuš, ed. *Slovensko na prelome tretieho tisícročia* (Bratislava: Veda, 1991).
6. Radovan Richta, ed., *Civilizace na rozcestí: společenské a lidské souvislosti vědeckotechnické revoluce* (Prague: Svoboda, 1966). For an English translation, see Radovan Richta, ed., *Civilization at the Crossroads: Social and Human Implications of the Scientific and Technological Revolution* (Prague: Svoboda, 1969).
7. The relationship between STR and the Prague Spring is elaborated in H. Gordon Skilling, *Czechoslovakia's Interrupted Revolution* (Princeton: Princeton University Press, 1976), 125–131. For the *Action Program*, see Robin Alison Remington, ed., *Winter in Prague: Documents on Czechoslovak Communism in Crisis* (Cambridge, MA: MIT Press, 1969), 88–137.
8. See Ota Šulc, *Futurologie: Přehl. literatury a informací* (Prague: St. knihovna ČSSR-Ústř. ekon. knihovna, 1968); Ota Šulc and Miloš Zeman, *Futurologický slovníček* (Prague: Státní knihovna ČSR—Ústřední ekonomická knihovna, 1969); Ota Šulc, *Problémy výzkumu budoucnosti: Názory na futurologii ve světě a u nás* (Prague: Institut poradenství při Čs. komitétu pro vědecké řízení,1970).
9. See the list of futurology projects in Šulc, *Futurologie: Přehl. literatury a informací*.
10. Rocca, " 'A Second Party in Our Midst,' " 205; Zeman, *Jak jsem se mýlil v politice* (Prague: Ottovo nakladatelství, 2005), 41. The Socialist Academy was an organization concerned with popular scientific education. The first CFS seminar took place in December 1968.
11. Zeman, *Jak jsem se mýlil v politice*, 22.
12. Miloš Zeman, "Futurologie a filosofie," *Filosofický časopis* 17 (1969): 531.
13. "Seminář českých futurologů," *Trend* 2, no. 2 (1970): 33. The "Seminar of Czech Futurologists," together with an "interdisciplinary team for futurology," was organized in February 1970.
14. The subtitle of the journal, which was published until 1989, was *Prognostické informace* (forecasting information). The concept of *Trend* was similar to the

German journal *Prognosen und Analysen: Über die Welt von Morgen*, which was published from 1968 in West Berlin by Zentrum Berlin für Zukunftsforschung.

15. Jan Adam, *Planning and Market in Soviet and East European Thought, 1960s–1992* (New York: St. Martin's Press, 1993); Martin Myant, *The Czechoslovak Economy, 1948–1988: The Battle For Economic Reform* (Cambridge, UK: Cambridge University Press, 1989).
16. Heinz-Dieter Haustein, *Wirtschaftsprognose: Grundlagen-Elemente-Modelle* (Berlin: Verl. Die Wirtschaft, 1969); Fritz Haberland and Heinz-Dieter Haustein *Die Prognostik als neues Element der Führungstätigkeit zur Meisterung der wissenschafltich-technischen Revolution* (Berlin: Dietz, 1968).
17. Jindřich Schermer, *Prognostika-věda-plánování* (Prague: Svoboda, 1968). Schermer was an assistant professor at the HSE in Prague. His research on forecasting was interrupted by his emigration to the United Kingdom after 1968.
18. Schermer, *Prognostika-věda-plánování*, 44.
19. *Futurologický sborník I* (Prague: Futurologická společnost, 1968).
20. Zeman, *Jak jsem se mýlil v politice*, 46.
21. The campaign against "bourgeois theories" was part of the general struggle against "revisionism" and "rightist opportunism." The canonic text of this campaign was Ladislav Hrzal and Jakub Netopilík, *Ideologický boj ve vývoji české filozofie* (Prague: Svoboda, 1975).
22. Ota Šulc, "Marxisticko-leninská prognostika v boji proti buržoazním koncepcím výzkumu budoucnosti," *Filosofický časopis* 19 (1971): 102–108. This text was written as a conference paper for the Third Czechoslovak-Soviet Symposium of Philosophy, which took place in October 1970.
23. See the following GDR works concerned with the critique of "bourgeois" future studies: Alfred Bönisch, *Futurologie: Eine kritische Analyse bürgerliche Zukunftsforschung* (Berlin: Akademie-Verlag, 1971); Jürgen Kuczynski, *Propheten der Wirtschaft: Studien zum Probleme der Wirtschaftsprognose im Kapitalismus* (Berlin: Dietz, 1970).
24. For example, see Jindřich Filipec, *Industriální společnost v sociologické diskusi* (Prague: Svoboda, 1967).
25. Even in 1988, a book was published concerned with the "critical analysis of non-Marxist future studies conceptions": Jana Duffková, *Zkoumání budoucnosti: kritická analýza nemarxistických koncepcí* (Prague: Svoboda, 1988). It was significant for the changed ideological context of the late 1980s that the book received a critical and overall negative review in the central forecasting journal. See Marian Timoracký, "Duffková, J.: Zkoumání budoucnosti," review of *Zkoumání budoucnosti: kritická analýza nemarxistických koncepcí*, by Jana Duffková, *Trend* 21 no. 5 (1989): 29–31. The author of the review wrote that "critical analysis" was, as usual, reduced to the construction of an "ideological bogeyman."
26. Archive of Masaryk Institute of Czech Academy of Sciences Prague (MUA), Collection Prezídium ČSAV, 25. 2. 1970, 64. zasedání, Zřízení Ústavu pro filozofii a sociologii ČSAV.
27. See for example MUA, Collection Základní dokumenty ČSAV, Ústav pro filosofii a sociologii ČSAV, Organizační řád ústavu 1970.
28. Ota Šulc, "Rozvoj výzkumu prognostiky v ČSSR," *Trend* 4, no. 6 (1972): 4–6.
29. Šulc, *Problémy výzkumu budoucnosti*, 24–25.
30. MUA, Collection Prezídium ČSAV, 29. 9. 1971, 21. zasedání, Účast ČSAV na vypracování prognóz.
31. Ibid. See also Ota Šulc, *Prognositka od A do Z* (Prague: SNTL, 1987), 174 and Josef Schorcht, "Založení systému prognostických prací v ČSSR," *Trend* no. 3 (1971): 5–8.
32. Šulc, *Prognostika od A do Z*, 174.

33. Schorcht, "Založení systému prognostických prací v ČSSR," 6–7.
34. MUA, Collection Prezídium ČSAV, 29. 9. 1971, 21. zasedání, Účast ČSAV na vypracování prognóz.
35. MUA, Collection Prezídium ČSAV, 2. 2. 1972, 27. zasedání, Informace o stavu prací na přípravě prognóz.
36. MUA, Collection Prezídium ČSAV, 27. 9. 1972, 37. zasedání, Informace o stavu prací na přípravě prognóz.
37. For the text of the report, see MUA, Collection Prezídium ČSAV, 27. 6. 1973, 50. zasedání, Materiál „Utváření socialistického způsobu života". The outcomes of the research were published as an internal publication of the Institute for Philosophy and Sociology. See *Vývoj způsobu života v souvislosti s perspektivním rozvojem společnosti. Studie sv. 8* (Prague: Ústav pro filosofii a sociologii ČSAV, 1973).
38. MUA, Collection Prezídium ČSAV, 16. 10. 1974, 69. zasedání, Projednávání materiálu „Socialistická koncepce formování potřeb pracujících, socialistický způsob života, vědeckotechnická revoluce a socialistický člověk."
39. MUA, Collection Prezídium ČSAV, 24. 5. 1974, 29. zasedání byra Prezídia, Zpráva o stavu prognostických prací v ČSAV.
40. For similar Soviet discussions about planning, see Pekka Sutela, *Economic Thought and Economic Reform in the Soviet Union* (Cambridge, UK: Cambridge University Press, 1991).
41. For example, in September 1972 the Polish Academy of Sciences (group "Poland 2000") organized a Summer School of Prognostics with eighty participants from Poland, the Soviet Union, GDR, Hungary, Romania, Czechoslovakia and Bulgaria. Among the participants was also the prominent Austrian future studies scholar Erich Jantsch. Czechoslovak scholars Ota Šulc and František Petrášek presented papers dealing with "prognostics of life-style" methodology, respectively with the "system approach to prognostics." See Ota Šulc, "Letní škola prognostických studií v Polsku," *Trend* 4, no. 6 (1972): 29. The conference papers were published in *Problemy prognozowania w krajach socjalistycznych* (Wrocław: Zakład Narodowy imienia Ossolińskich Wydawnictwo Polskiej Akademii Nauk, 1975).
42. In May 1973, Czechoslovakia hosted two international forecasting conferences: the CSAS Institute for Economics organized the Czechoslovak-Soviet symposium on methodological problems of socio-economic forecasting and the international colloquium on "social forecasting and planning" was organized by the CSAS Institute for Philosophy and Sociology. See Jiří Slíva, "Čs-sovětské symposium 'Metodologické otázky prognózování sociálně ekonomického rozvoje'," *Trend* 5, no. 4 (1973): 28–29 and Ota Šulc, "Mezinárodní kolokvium o sociálním předvídání a plánování," *Trend* 6, no. 1 (1974): 27–28.
43. Pavel Alexander, *K problematice prognostiky vědních objevů* (Prague, UVTEI, 1971); Miloslav Fejt, *Prognostika ve vojenství* (Prague: Horizont, 1971); Jiří Volf, *Aplikace prognostiky v řízení* (Prague: UVTEI, 1972; Milan Hromádka, *Kultura a prognostika: K základním otázkám koncepce prognostiky kultury* (Prague: Ústav pro výzkum kultury, 1974). For theory and methodology of forecasting, see Juraj Kolek and Jaroslav Vrchota, *Prognostické metódy a ich aplikácia* (Bratislava: Alfa, 1972); Ladislav říha, *Dlouhodobé prognózy a plány* (Prague: Práce, 1974); Alena Koutská, *Scénář jako prognostická technika* (Prague: UVTEI, 1973); Miloš Zeman, *Komplexní prognostické modelování* (Prague: Institut pro výchovu vedoucích pracovníků ministerstva průmyslu ČSR, 1975).
44. František Kutta, *Systém řízení sociálních procesů* (Prague: Svoboda, 1976); František Kutta and Miroslav Soukup, eds., *řízení v období vědeckotechnické revoluce: Principy socioekonomického řízení* (Prague: Svoboda, 1973);

František Kutta et al., *Teorie a praxe sociálního plánování a programování v ČSSR* (Prague: Svoboda, 1980).

45. Josef Goldmann, *Makroekonomická analýza a prognóza* (Prague: Academia, 1975). For a short biography of Josef Goldmann (1912–1984), see Hans-Jürgen Wagener, ed., *Economic Thought in Communist and Post-Communist Europe* (London and New York: Routledge, 1998), 365.
46. Goldmann, *Makroekonomická analýza a prognóza*, 8.
47. Valtr Komárek, *Aktuální otázky plánování a prognostiky* (Prague: Academia, 1976), 5; Valtr Komárek, *Hospodářské a vědeckotechnické prognózy* (Prague: SNTL, 1977).
48. MUA, Collection Prezídium ČSAV, 17. 12.1979, 20. zasedání, Informace o účasti ČSAV při rozvinutí prací na dlouhodobém výhledu ČSSR do roku 2000.
49. MUA, Collection Prezídium ČSAV, 11. 11. 1980, 29. zasedání, Prognóza vývoje socialistické společnosti, zejména socialistického způsobu života do roku 2000.
50. National Archives Prague (NA), Předsednictvo ÚV KSČ 1980–1989, svazek (sv.) P 80/83, bod (b.) 4a, 16. 9. 1983, Projednávání návrhu ústavního zákona o zřízení Státní komise pro vědeckotechnický a investiční rozvoj. Czech and Slovak "republic commissions" were also established. See NA, Předsednictvo ÚV KSČ 1980–1989, sv. P 88/83, b. 8, 10. 11. 1983, Návrh na složení Státní komise a republikových komisí pro vědeckotechnický a investiční rozvoj.
51. NA, Předsednictvo ÚV KSČ 1980–1989, sv., P 104/84), b. 13, 13. 4. 1984, Změny ve složení pléna Státní plánovací komise.
52. NA, Předsednictvo ÚV KSČ 1980–1989, sv. P 112/84, b. 1, 6. 7. 1984, Hlavní směry dalšího rozvíjení Souboru opatření ke zdokonalování soustavy plánovitého řízení národního hospodářství.
53. MUA, Collection Prezídium ČSAV, 47. zasedání, 4. 9. 1982, Informace o zabezpečení rozvoje prognostické činnosti v ČSAV.
54. Valtr Komárek and Jan Bauer, *Mé pády a vzestupy* (Prague: Nakladatelství dopravy a spojů, 1992), 34–41. For the establishing of the center, see MUA, Collection Prezídium ČSAV, 11. zasedání, 24. 1. 1984, Návrh na zřízení Kabinetu prognóz ČSAV.
55. MUA, Collection Prezídium ČSAV, 14. zasedání, 10. 4. 1984, Návrh projektu obsahového zaměření, organizačního a hmotného zabezpečení cyklických prací na souhrnné prognóze. Komárek was officially appointed as director in June 1985. See MUA, Collection Prezídium ČSAV, 26. zasedání, 11. 6. 1985, Jmenování ředitele Kabinetu prognóz.
56. MUA, Collection Prezídium ČSAV, 65. zasedání kolegia předsedy ČSAV, 16. 9. 1986, Návrh na přebudování Kabinetu prognóz ČSAV na Prognostický ústav ČSAV. For the organization of the Center and Institute, see MUA, Collection Sbírka základních dokumentů ČSAV, Prognostický ústav (Kabinet prognóz), box 179.
57. For annual reports of the Institute, see MUA, Collection Sbírka základních dokumentů ČSAV, Prognostický ústav (Kabinet prognóz), box 179.
58. Interview with Jan Mládek, May 10, 2013.
59. MUA, Collection Prezídium ČSAV, 14. zasedání, 10. 4. 1984, Návrh projektu obsahového zaměření, organizačního a hmotného zabezpečení cyklických prací na souhrnné prognóze. MUA, Collection Prezídium ČSAV, 67. zasedání kolegia předsedy ČSAV, 11. 11. 1986, Návrh opatření k zásadnímu zkvalitnění práce rady pro souhrnnou prognózu a jejích komisí. The head of the General Prognosis Board was Josef říman, who was the vice president and from 1986 the president of the CSAS.
60. MUA, Collection Prezídium ČSAV, 71. zasedání kolegia předsedy ČSAV, 10. 3. 1987, Zpráva o stavu prací na souhrnné prognóze vědeckotechnického, ekonomického a sociálního rozvoje ČSSR do roku 2010.

61. NA, Předsednictvo ÚV KSČ 1980–1989, sv. P 40/87, b. 3, 3. 7. 1987, Základní poznatky z dosavadních prací na souhrnné prognóze vědeckotechnického, hospodářského a sociálního rozvoje ČSSR do roku 2010 a doporučení pro konkretizaci strategie do roku 2000 ve druhém cyklu prací na dlouhodobém výhledu.
62. The Czech term for *perestroika* was *přestavba* (reconstruction, rebuilding). For *přestavba* discourse, see Pullmann, *Konec experimentu*. The emergence of the *přestavba* political program is described in Martin Štefek, "Komunistická strana Československa a proces přestavby v letech 1985–1989," in Jiří Kocian, Jaroslav Pažout, and Jakub Rákosník, eds., *Bolševismus, komunismus a radikální socialismus v Československu VII.*(Prague: Dokořán-ÚSD AV ČR, 2010), 92–138. For economic reform in the *perestroika* period, see Myant, *The Czechoslovak Economy*.
63. Valtr Komárek, ed., *Prognóza a program* (Prague: Academia, 1990).
64. Komárek and Bauer, *Mé pády a vzestupy*, 40–41. According to Komárek, parts of *General Prognosis* were read in the broadcasting of Radio Free Europe and Voice of America.
65. Interview by the author with Jan Mládek, May 10, 2013
66. Valtr Komárek, "Pojednání: Prognostická sebereflexe čs. společnosti," *Politická ekonomie* 37 (1989): 523–535. The English translation of this article was published in 1990; for quotations see Valtr Komárek, "Treatise: Prognostic Self-Analysis of Czechoslovak Society," *Eastern European Economics* 28 (1990): 11, 21, 27.
67. Tomáš Ježek and Otakar Turek, "Strukturální změny a hospodářský mechanismus," *Politická ekonomie* 37 (1989): 583, 584, 585.
68. Václav Klaus, "Imperativy dlouhodobé prognózy a dominantní charakteristiky současné ekonomiky," *Politická ekonomie* 37 (1989): 570–579. Cited according to the English translation published as Václav Klaus, "The Imperatives of Long-Term Prognosis and the Dominant Characteristics of the Present Economy," *Eastern European Economics* 28 (1990): 49.
69. Klaus, "The Imperatives of Long-Term Prognosis and the Dominant Characteristics of the Present Economy," 50.
70. Miroslav Flek, ed., *Metodologické problémy společenskovědního výzkumu budoucnosti* (Prague: Sportpropag, 1983).
71. Among the department's researchers were sociologists Jiří Kabele, Martin Potůček and Miloš Bárta. Kabele and Potůček gained influential positions in Czech sociology after 1989. Bárta was an active reform communist in 1968. External collaborators were, among others, the Slovak sociologist and leading future studies scholar Fedor Gál and Pavel Machonin, the most prominent reform communist sociologist and head of the most important reform communist sociological research team in the 1960s. See also Zeman, *Jak jsem se mýlil v politice*, 49–60.
72. Jiří Kabele, "Sportpropag—nepravděpodobné místo pro studium společnosti: Osobní pohled," *Sociální studia* 8 (2011): 17–35. Similar semi-official seminars concerned with the economy were organized in the Czechoslovak State Bank by Klaus. The group of Prague monetarists was described by Tomáš Ježek in his memoirs; see Petr Husák and Tomáš Ježek, *Budování kapitalismu v čechách: Rozhovor s Tomášem Ježkem* (Prague: Volvox Globator, 1997), 41–99.
73. Sociologist Josef Alan discussed the "myths of empirical sociology" and its retreat from social reality to abstract data. Klaus criticized contemporary official economics and called for research based on "economic subjects" (actors of exchange). Moreover, one chapter of the book was written by economist Karel Kouba, who was influential actor of the 1960s economic reform and was purged from academia after 1968. See Flek, *Metodologické problémy společenskovědního výzkumu budoucnosti*, 40, 60.

74. See Kabele, "Sportpropag," 22; Zeman, *Jak jsem se mýlil v politice*, 59–60.
75. Zeman, *Jak jsem se mýlil v politice*, 65.
76. For the conference papers, see *Aktuální problémy prognostiky: sborník statí* (Prague: Dům techniky ČSVTS, 1986); *Aktuální problémy prognostiky 88* (Prague: Dům techniky ČSVTS, 1988). The third conference took place in 1990, see *Aktuální problémy prognostiky 90* (Prague: Dům techniky ČSVTS, 1990).
77. Probably the most important exception from *prognostika* technocratic mainstream was the future-oriented research of Slovak non-conformist social scientists on the "problem-oriented participative forecasting," which was concerned with the interaction between society (public) and expertise. See Fedor Gál and Pavol Frič, "Problem-Oriented Participative Forecasting: Theory and Practice," *Futures* 19 (1987): 678–685.

ARCHIVES

Archive of Masaryk Institute of Czech Academy of Sciences Prague (MUA)
National Archives Prague (NA)

REFERENCES

Adam, Jan. *Planning and Market in Soviet and East European Thought, 1960s–1992*. New York: St. Martin's Press, 1993.

Aktuální problémy prognostiky 88. Prague: Dům techniky ČSVTS, 1988.

Aktuální problémy prognostiky 90. Prague: Dům techniky ČSVTS, 1990.

Aktuální problémy prognostiky: sborník statí. Prague: Dům techniky ČSVTS, 1986.

Alexander, Pavel. *K problematice prognostiky vědních objevů*. Prague: UVTEI, 1971.

Bönisch, Alfred. *Futurologie: Eine kritische Analyse bürgerliche Zukunftsforschung*. Berlin: Akademie-Verlag, 1971.

Duffková, Jana. *Zkoumání budoucnosti: kritická analýza nemarxistických koncepcí*. Prague: Svoboda, 1988.

Eyal, Gil. *The Origins of Postcommunist Elites: From Prague Spring to the Breakup of Czechoslovakia*. Minneapolis: University of Minnesota Press, 2003.

Fejt, Miloslav. *Prognostika ve vojenství*. Prague: Horizont, 1971.

Filipec, Jindřich. *Industriální společnost v sociologické diskusi*. Prague: Svoboda, 1967.

Flek, Miroslav, ed. *Metodologické problémy společenskovědního výzkumu budoucnosti*. Prague: Sportpropag, 1983.

Futurologický sborník I. Prague: Futurologická společnost, 1968.

Gál, Fedor and Pavol Frič. "Problem-Oriented Participative Forecasting: Theory and Practice." *Futures* 19 (1987): 678–685.

Goldmann, Josef. *Makroekonomická analýza a prognóza*. Prague: Academia, 1975.

Haberland, Fritz and Heinz-Dieter Haustein. *Die Prognostik als neues Element der Führungstätigkeit zur Meisterung der wissenschafltich-technischen Revolution*. Berlin: Dietz, 1968.

Haustein, Heinz-Dieter. *Wirtschaftsprognose: Grundlagen-Elemente-Modelle*. Berlin: Verl. Die Wirtschaft, 1969.

Hromádka, Milan. *Kultura a prognostika: K základním otázkám koncepce prognostiky kultury*. Prague: Ústav pro výzkum kultury, 1974.

Hrzal, Ladislav and Jakub Netopilík. *Ideologický boj ve vývoji české filozofie*. Prague: Svoboda, 1975.

Husák, Petr and Tomáš Ježek. *Budování kapitalismu v čechách: Rozhovor s Tomášem Ježkem*. Prague: Volvox Globator, 1997.

Ježek, Tomáš and Otakar Turek. "Strukturální změny a hospodářský mechanismus." *Politická ekonomie* 37 (1989): 580–591.

"'Jistý prognostik' si dovolil kritizovat komunistické hospodářství." *Czech TV* video, August 3, 2010. www.ceskatelevize.cz/ct24/exkluzivne-na-ct24/62717-jisty-prognostik-si-dovolil-kritizovat-komunisticke-hospodarstvi/.

Kabele, Jiří. "Sportpropag—nepravděpodobné místo pro studium společnosti: Osobní pohled." *Sociální studia* 8 (2011): 17–35.

Klaus, Václav. "The Imperatives of Long-Term Prognosis and the Dominant Characteristics of the Present Economy." *Eastern European Economics* 28 (1990): 39–52.

Klaus, Václav. "Imperativy dlouhodobé prognózy a dominantní charakteristiky současné ekonomiky." *Politická ekonomie* 37 (1989): 570–579.

Kolek, Juraj and Jaroslav Vrchota. *Prognostické metódy a ich aplikácia*. Bratislava: Alfa, 1972.

Komárek, Valtr. *Aktuální otázky plánování a prognostiky*. Prague: Academia, 1976.

Komárek, Valtr. *Hospodářské a vědeckotechnické prognózy*. Prague: SNTL, 1977.

Komárek, Valtr. "Pojednání: Prognostická sebereflexe čs. společnosti." *Politická ekonomie* 37 (1989): 523–535.

Komárek, Valtr. "Treatise: Prognostic Self-Analysis of Czechoslovak Society." *Eastern European Economics* 28 (1990): 7–23.

Komárek, Valtr, ed. *Prognóza a program*. Prague: Academia, 1990.

Komárek, Valtr and Jan Bauer. *Mé pády a vzestupy*. Prague: Nakladatelství dopravy a spojů, 1992.

Koutská, Alena. *Scénář jako prognostická technika*. Prague: UVTEI, 1973.

Kuczynski, Jürgen. *Propheten der Wirtschaft: Studien zum Probleme der Wirtschaftsprognose im Kapitalismus*. Berlin: Dietz, 1970.

Kutta, František. *Systém řízení sociálních procesů*. Prague: Svoboda, 1976.

Kutta, František and Miroslav Soukup, eds. *řízení v období vědeckotechnické revoluce: Principy socioekonomického řízení*. Prague: Svoboda, 1973.

Kutta, František, ed. *Teorie a praxe sociálního plánování a programování v ČSSR*. Prague: Svoboda, 1980.

Markuš, Jozef, ed. *Slovensko na prelome tretieho tisícročia*. Bratislava: Veda, 1991.

Myant, Martin. *The Czechoslovak Economy, 1948–1988: The Battle For Economic Reform*. Cambridge, UK: Cambridge University Press, 1989.

Problemy prognozowania w krajach socjalistycznych. Wrocław: Zakład Narodowy imienia Ossolińskich Wydawnictwo Polskiej Akademii Nauk, 1975.

Pullmann, Michal. *Konec experimentu: přestavba a pád komunismu v Československu*. Prague: Scriptorium, 2011.

Remington, Robin Alison, ed. *Winter in Prague: Documents on Czechoslovak Communism in Crisis*. Cambridge, MA: MIT Press, 1969.

Richta, Radovan, ed. *Civilizace na rozcestí: společenské a lidské souvislosti vědeckotechnické revoluce*. Prague: Svoboda, 1966.

Richta, Radovan, ed. *Civilization at the Crossroads: Social and Human Implications of the Scientific and Technological Revolution*. Prague: Svoboda, 1969.

Rocca, Gordon L. "'A Second Party in Our Midst': The History of the Soviet Scientific Forecasting Association." *Social Studies of Science* 11 (1981): 199–247.

říha, Ladislav. *Dlouhodobé prognózy a plány*. Prague: Práce, 1974.

Schermer, Jindřich. *Prognostika-věda-plánování*. Prague: Svoboda, 1968.

Schorcht, Josef. "Založení systému prognostických prací v ČSSR." *Trend* 3 (1971): 5–8.

"Seminář českých futurologů." *Trend* 2 (1970): 33.

Skilling, H. Gordon. *Czechoslovakia's Interrupted Revolution*. Princeton: Princeton University Press, 1976.

Slíva, Jiří. "čs-sovětské symposium 'Metodologické otázky prognózování sociálně ekonomického rozvoje.'" *Trend* 5 (1973): 28–29.
Sutela, Pekka. *Economic Thought and Economic Reform in the Soviet Union*. Cambridge, UK: Cambridge University Press, 1991.
Štefek, Martin. "Komunistická strana Československa a proces přestavby v letech 1985–1989." In *Bolševismus, komunismus a radikální socialismus v Československu VII.*, edited by Jiří Kocian, Jaroslav Pažout, and Jakub Rákosník, 92–138. Prague: Dokořán-ÚSD AV ČR, 2010.
Šulc, Ota. *Futurologie: Přehl. literatury a informací*. Prague: St. knihovna ČSSR-Ústř. ekon. knihovna, 1968.
Šulc, Ota. "Letní škola prognostických studií v Polsku." *Trend* 4 (1972): 29.
Šulc, Ota. "Marxisticko-leninská prognostika v boji proti buržoazním koncepcím výzkumu budoucnosti." *Filosofický časopis* 19 (1971): 102–108.
Šulc, Ota. "Mezinárodní kolokvium o sociálním předvídání a plánování." *Trend* 6 (1974): 27–28.
Šulc, Ota. *Problémy výzkumu budoucnosti: Názory na futurologii ve světě a u nás*. Prague: Institut poradenství při Čs. komitétu pro vědecké řízení,1970.
Šulc, Ota. *Prognositka od A do Z*. Prague: SNTL, 1987.
Šulc, Ota. "Rozvoj výzkumu prognostiky v ČSSR." *Trend* 4 (1972): 4–6.
Šulc, Ota, and Miloš Zeman. *Futurologický slovníček*. Prague: Státní knihovna ČSR—Ústřední ekonomická knihovna, 1969.
Timoracký, Marian. "Duffková, J.: Zkoumání budoucnosti." Review of *Zkoumání budoucnosti: kritická analýza nemarxistických koncepcí*, by Jana Duffková. *Trend* 21 (1989): 29–31.
Vývoj způsobu života v souvislosti s perspektivním rozvojem společnosti. Studie sv. 8. Prague: Ústav pro filosofii a sociologii ČSAV, 1973.
Volf, Jiří. *Aplikace prognostiky v řízení*. Prague: UVTEI, 1972.
Wagener, Hans-Jürgen, ed. *Economic Thought in Communist and Post-Communist Europe*. London and New York: Routledge, 1998.
Zeman, Miloš. "Futurologie a filosofie." *Filosofický Časopis* 17 (1969): 522–532.
Zeman, Miloš. *Komplexní prognostické modelování*. Prague: Institut pro výchovu vedoucích pracovníků ministerstva průmyslu ČSR, 1975.
Zeman, Miloš. "Prognostika a přestavba." *Technický magazín* 32 (1989): 6–9.

7 Official and Unofficial Futures of the Communism System

Romanian Futures Studies between Control and Dissidence

Ana-Maria Cătănuş

This chapter focuses on the analysis of official and unofficial futures research in Communist Romania during the 1970s and 1980s. It begins with an analysis of the historical and political context in which futurology emerged and developed in Romania, and it emphasizes the importance of the relative liberalization of regime control in the early years of Nicolae Ceauşescu's regime. The beginning of the Ceauşescu years saw not only forms of internal liberalization but also attempts of opening up Romanian society to the Western world. The second part of the chapter deals with official future research in Romania. The field was created under direct supervision of the Party and it benefited from the support of institutions controlled by Ceauşescu. Official future, or *prospective* research, as it was also called, focused mainly on the setting up of methodologies for the elaboration of economic and social development forecasts. In 1972, Ceauşescu welcomed the third International Conference on Futures Research to Bucharest. However, this finest hour of a relatively liberal climate for futures research would not last. As Ceauşescu's rule developed into a dictatorship, he also identified the future as a mere instrument to justify the policies of his regime. This put Romanian futurists in a difficult position. The third part of the chapter focuses on the personal story of Mihai Botez, one of the leading Romanian futurologists, who at the end of the 1970s crossed over the border of official forecasting into the production of unofficial futures of the communist system. He would use these unofficial futures as the basis of his dissidence, a conduct that he would rather describe as a "scientific and moral commitment."[1]

FROM LIBERALIZATION TO REPRESSION: THE POLITICAL REGIME OF NICOLAE CEAUŞESCU

The mid 1960s brought a critical change at the highest level of political communist leadership in Romania. In 1965, Gheorghe Gheorghiu-Dej, the first secretary of the Romanian Workers' Party (RWP) died suddenly. The appointed successor was Nicolae Ceauşescu. A son of peasants, Ceauşescu had studied only four years in school. At the age of eleven he left his native village, Scorniceşti, to come to Bucharest, where he became the apprentice of a shoemaker. In the interwar period, Ceauşescu had been close to the illegal communist movement. Consequently he was arrested several times and imprisoned. This provided him the opportunity to meet future communist leaders, including Gheorghiu-Dej.[2] His career in the Romanian Communist Party took off in 1944, when Romania became an ally of the Soviet Union. In the years to come, Ceauşescu steadily climbed in the nomenklatura. He was a born politician and a true communist. In 1955, he became a full member of the Politburo.[3] Ten years later, he became the head of the Romanian Communist Party (RCP), a position he kept to his death, in 1989.[4]

Contemporary historiography has concentrated mainly on the second part of Ceauşescu's rule in the 1980's. This has contributed to a dominant picture of Ceauşescu as the hardened dictator that he would indeed become, and of his regime as a form of neo-Stalinism. But Nicolae Ceauşescu's regime began and developed in his first years of power under very different auspices. Ceauşescu chose not only to continue Gheorghiu-Dej's "enlightened" policies from the first part of the 1960s, in which Soviet cultural influence was kept at bay and national traditions emphasized, while Romania also sought relations with the West. The liberalization reforms implemented by Ceauşescu in his first years led to an improvement of living conditions in Romania, and a relaxation in its relationship to the West. The period from 1965 to 1971 was one of diminishing repression. People were no longer arrested in the streets, prosecuted, and sent to prison without trial, as did happen in the 1950s. Instead, the regime appealed more and more to "prophylactic measures." That meant that harsh repression was replaced by increased surveillance by the secret police, or Securitate, through informants, control of correspondence, records of phone conversations, and listening devices at work or at home.[5] Meanwhile there was increased freedom of expression in art and literature, within ideological boundaries. The opening up toward the West enabled new forms of circulation: foreign translations and foreign press arrived in Romania and Romanian artists and scientists could travel more freely.[6] Romanian historiography refers to this period as the "*liberalizare*" (liberalization), and it brought the regime wide popular support. The highpoint of Ceauşescu's popularity came during the 1968 Czechoslovakian crisis, when the Romanian regime refused to join the Warsaw Treaty Organization troops marching to crush the Prague Spring. Ceauşescu's refusal helped position him as a national leader, and he won the sincere backing of his people—for the first and last time. *National-communism* was a political construct linking nationalism and communism, and it resulted in increased legitimacy for the regime both within the country and abroad. Many intellectuals, some of them former political prisoners under Gheorghe Gheorghiu-Dej' s regime, joined the RCP. That was the case of writers Alexandru Ivasiuc and Paul Goma. As young students, they were sentenced to prison in the 1950s for manifesting solidarity with the 1956 Hungarian revolution. After being released in the

1960s, they were allowed to resume their studies and continue in professional life.[7] As he distanced Romanian foreign policy from the Soviet Union, Ceauşescu improved Romanian relations with the West,[8] and gained political recognition and even economic and financial support.

Ceauşescu was nevertheless a firm adherent to the postulates of Marxism-Leninism. According to doctrine, there could only be one future, in the shape of communist society with an intermediary stage—socialist society—as predicted. But realizing this future demanded political action, so Ceauşescu' s regime embarked on a series of political reforms aimed at the development of the Romanian economy. At the Ninth Congress of the RCP in 1965, industrialization was set as the central objective of economic policy, with an emphasis on metallurgy and chemical industries. Subsequently, it was emphasized that science and technology would constitute the foundation of industrial development.[9] In accordance with this direction, Ceauşescu stated: "Science is the basis for the building of socialism. It has to make an active contribution in raising our society to even higher levels. It is necessary to harmoniously combine general fundamental research with applied research related to the practical concrete needs of building socialism."[10] The message was that the Party was fully committed to sustain the development of all scientific domains and in particular the technical sciences. The direct lead came from Ceauşescu himself, who personally met with scientists and academicians presenting them his plans and assuring them of all his support.[11]

On the institutional level, science and research were to be coordinated by a central organism—the National Council of Scientific Research that was to function under the authority of the Council of Ministers.[12] The creation of the Council was announced at the Ninth RCP congress, and judging from its objectives, the subordination of science to the political goals of the regime was clear:

> [The National Council of Scientific Research] will have as its task to ensure the fulfillment of Party and government policy regarding the development of scientific research in SRR [Socialist Republic of Romania], in strong relation to the concrete needs of our economy and culture; [another task was the] organization of the active participation of scientists and researchers to the solving of important problems raised by the construction of socialism.[13]

Starting with the five year plan for 1966–1970, the program of scientific research elaborated by the National Council of Scientific Research was to constitute a part of each five year plan.[14] This centralization of scientific life was in accordance with the desired objectives set by the communist leadership, namely, to trigger a process of industrialization of the Romanian economy. But that was not enough. Ceauşescu's interest in the development of Romanian science and research was part of a plan to ensure social progress and reduce Romanian dependence on other countries.[15] In other words, he embedded the idea of a scientific and technological revolution in his project of national communism.

Although ideologically motivated, regime interest in science and research actually proved beneficial to the development of these sectors. The number of researchers sent abroad for training and research increased. Also, there was more space for debate and innovation, especially in technical sciences. The regime facilitated access to foreign scientific literature, including journals. Moreover, the regime became receptive to, and encouraged, the development of new areas such as management or forecasting that it understood as tools for rationalizing the Romanian economy. Consequently, the regime concentrated its efforts on the active creation of these fields through the education of Romanian management theorists and forecasters and the development of methods in these sectors. The Romanian government signed bilateral cultural and scientific protocols with Western countries such as Belgium, Holland, and the Federal Republic of Germany. The last one, signed in January 1968, gives an idea of the areas of interest for the Romanian leadership: the mathematical sciences, mechanics, atomic and nuclear physics, and, importantly, the methods and techniques of short-, medium-, and long-term forecasting.[16]

During liberalization, Romania also increased its contacts with international organizations such as the United Nations. One result of these contacts was the creation in Romania of a training center in management techniques. This was the *Centrul de Perfecţionare a Cadrelor de Conducere din Întreprinderi* (the Center for Management Education), better known under the acronym CEPECA; it was set up with financial and technical support from the UN and the International Labor Bureau in Geneva. In the negotiations that led up to the creation of the CEPECA in 1967, the mathematician Mircea Maliţa, at that moment deputy of the Foreign Ministry, had an important role. Trained in philosophy and mathematics, Maliţa held high ranking scientific and diplomatic positions during the communist regime.[17] A member of the RCP since 1946, Maliţa was seen as “a rare flower in the garden of the nomenklatura.”[18] The CEPECA was built as a cutting-edge institution. It had modern training facilities, including an imported IBM computer, international experts, and it offered studies abroad for Romanian university staff.[19]

The CEPECA project was a success. Classes were taught on the basis of a syllabus drafted by UN experts, which included modern methods of management and marketing. CEPECA was also central for the creation in Romania of the new professions of systems analyst or programmer. CEPECA established relations with over 70 university institutions and training centers in the world, including the London Business School, Pennsylvania University—Marketing Faculty, the Ford Foundation, and the Young Men’s Christian Association (YMCA).[20] As successful as it was, however, CEPECA tested the limits of the regime’s acceptance of intellectual exchange with the West. In 1971, CEPECA collaboration's with the UN was terminated and the institution was absorbed by the Ştefan Gheorghiu Academy. The Academy was a Party school, an ideological training institution for leading cadres of the Party. This change signified the new authoritarian orientation of Ceauşescu's regime. In July 1971, upon returning from a tour of the communist states in Asia, Ceauşescu announced what became known as the July Theses. The July Theses was a set of propositions intended to improve the Marxist/Leninist education of the people.[21] It marked the return in force of communist ideology.

The essence of Ceauşescu's miniature cultural revolution, as it has been called in the Romanian historiography, was the strengthening of the role of the Party in Romanian society and the subordination of all domains of social activity under Marxist-Leninist ideology. This applied especially to the fields of culture and education. As Ceauşescu firmly stated: "the leading role of the Party applies in education, literature, art, music or science, and the RCP has the right to accept only what serves the objectives of socialism."[22] As harsh as this ideological turn was at the level of political discourse, it is worth mentioning that changes happened gradually. In the first years of the 1970s, ideological tightening affected primarily the humanities and culture. Liberalization had also introduced tensions in Romanian society. A new strata of technical intellectuals had emerged and been encouraged by the regime. These professionals now had experience from the Western world. Some of them accepted the official vision of the socialist future and continued working with the regime. However, among these intellectuals were also those that developed and publicly affirmed a different vision on the future Romanian society. In the 1970s and 1980s, these alternative visons would collide with the official future.

OFFICIAL FUTURES OF COMMUNISM

The Communist Party understood the future as already set in the past. According to Marxism-Leninism, the final goal of a stage-driven historical process was accomplishing communism, embodied in the idea of the classless society.[23] Whereas this objective was derived from doctrine, the implementation of the new order relied on the communist leadership and was to be carried out through planning. In the ideological literature, planning was described as "the main method of unitary organizing and conducting by the socialist state of the entire economic and social activity using a unique national plan, with the purpose of building socialism and communism."[24] This "national unique" plan[25] was drafted by the government in strict accordance with the Party program, aiming at the realization of the main directions settled for the period by the Party Congresses. Based on synthesis studies and forecasts, the national plan encompassed all economic and social areas. Initially the plan was drafted for one year, in 1949 and 1950.

That changed in 1951 when the *cincinal* (a five-year plan) was instituted.[26] The institution in charge of drafting the plans was *Comitetul de Stat al Planificării* (the State Planning Committee), similar to the Soviet's Gosplan. Besides the five-year plan, prospective plans were made for a longer period of fifteen to twenty-five years. After approval by the communist parliament, *Marea Adunare Naţională* (the Great National Assembly), the plan became mandatory.[27]

Affirming that "the socialist society needs to organize its entire life on the basis of the newest scientific results,"[28] Ceauşescu manifested a growing personal interest toward scientific areas that could help speed up the process of changing Romania from a developing to a developed nation. A series of conferences on the topic of the scientific and technological revolution were organized at the *Academia de Ştiinţe Sociale şi Politice* (the Academy of Social and Political Sciences), an institution created by the Party in 1970 to control the research activities in social and political sciences. Here, scholars from different areas discussed how science could be applied to the Romanian economy and society. Economic and social forecasts occupied a particular space, as did demography, international relations, and the relationship between science and production. These works were published in several volumes at the *Editura Politică* (the Political Publishing House), in a special collection named *Idei Contemporane* (Contemporary Ideas). During the 1970s and 1980s, *Contemporary Ideas* was the place where all the important literature related to scientific and technological revolution, and, importantly, futurology, was published. *Contemporary Ideas* published texts both written by Romanian authors and translations. The Romanian public could thus read, in translation, futurist literature consisting of Alvin Toffler's *Future Shock*, Radovan Richta's *Civilisation at the Crossroads*, or John Naisbit's *Megatrends*, and get in touch with the future related theories of Edward Cornish, Herman Kahn, or Willis Harman.[29] That these Western futurists, as well as the critical work of the Czech Richta group, was translated, demonstrates Ceauşescu's interest in the new methods of anticipating and exploring the future. The development of future studies in Romania began at the end of the 1960s under the direct benediction of Ceauşescu. Maliţa was central intellectual in the shaping of this field in Romania.

Maliţa's futurological adventure began with a series of articles published in the mid-1960s in the newspaper *Contemporanul* (Contemporary). In 1969, they were gathered in a book named *Cronica anului 2000* (Chronicle of the Year 2000) that proved to be a success. A series of essays on future, Maliţa's book imagined the changes society will suffer within a generation. The *Chronicle* displayed a wide range of futuristic views about life on Earth—housing, communication, transportation, education, diplomacy and international cooperation.[30] There were recurring themes in his paper such as the computer—"the new instrument of assisting the human thought," mathematics, which "will become the new universal language", progress, and education.[31]

Maliţa' s next step was the setting up of a multidisciplinary team to study problems related to the future. But that was not an easy task. Research into the future had to be performed against the backdrop of awareness that in socialism, there was only one future, and bearing in mind that even officially sanctioned explorations into the future could be interpreted as subversive and find themselves in the unfortunate space of the unofficial. Maliţa's solution to this problem was to link his proposal for futures research to the problem of planning. As Maliţa retrospectively confessed, "I thought: we had five years plans, didn't we? So that was already an act on the future."[32] Maliţa's efforts were rewarded when, at the beginning of the 1970s, the *Laboratorul de Studii Prospective* (Laboratory of Prospective Research) at the University of Bucharest was born. The prospective research laboratory was a place where researchers from different fields, including mathematics, sociology, and economics, could gather. Mihai Botez, a mathematician, the youngest ever to have a PhD in mathematics, was chosen by Maliţa to direct research.[33] Maliţa had worked with Botez in CEPECA, where Botez was in charge of teaching the course on mathematic applications to the economy.

The purposes of the laboratory for prospective research were summarized in an article written by Maliţa in the review *Lupta de clasă* (the Class Struggle). The aim of the laboratory was to develop methods for the new field of "prospective research," a term coined by Maliţa, inspired by the French *prospectiviste* philosopher Gaston Berger. It was also to organize seminaries, publish a bulletin, and establish international contacts.[34] Besides these concrete tasks, the beginning of the article allows a glimpse into the ideological theory that stood behind all Romanian futurological studies. The first and most important assumption was that futurology was not at all a new field. Its roots were to be found in Marx's thinking, Marx being the first "to undertake, a century ago, integral prospective research, verified by facts, and as a result of these researches he foresaw a future model of society, that became socialist society."[35] According to Marxist researchers, unlike the capitalist societies, where "futurology was an odd graft on a philosophical corpus unaccustomed to rigorous planning and the rational organization of the future," socialist societies could offer futurology a more fertile ground, based on the scientific theory of Marxism.[36] That was to show the righteousness of socialist researches on the future; unlike the Western futurology, which was seen not only as new but ideologically impregnated, and so unacceptable.

But assuming ideological primacy in terms of future-oriented researches was only one element in the theoretical construction guiding futurology in socialist countries. Another came from regime interpretations of a much larger debate in futures research, gaming, and modeling, around the prediction of one systemic future, or the possibility of several plural outcomes. Marxist ideologists tried to overcome what they called a Western ideological preconception about Marxist theory as projecting a one-dimensional future, or better said, the socialist future. The matter was tackled in several works by philosopher Pavel Apostol, who in the 1970s had an ascending career in official futurology. A member of the RCP since 1945, Apostol experienced his ups and downs. In 1952, he was accused of espionage and ideological mistakes in his activity. Although the first accusation was dismissed, Apostol was excluded from the RCP, and was confined in a working colony for two years without being subjected to a trial. He was released in 1954 and was rehabilitated in 1968 under Ceauşescu's rule.[37] As Pavel Apostol explained in his works:

> The expression "socialist (or communist) future" does not mean the creation of a standard future society but that of a set of future societies (social systems) having a common structure. Diversity in reality is not cancelled out by structural oneness. The diversification of the roads to socialism and communism does not only concern the possibility of different national roads, but also the multiple possibilities of realizing them within the framework or each national community or state.[38]

Moreover, in search of common grounds with Western futurology, communist ideologists spoke about the plural, probabilistic vision on the future in Marxist theory, and emphasized its similarities with the French liberal philosopher Bertrand de Jouvenel's theory on possible futures, the *futuribles*.[39] This short digression into Marxist future theory gives us an idea of the theoretical framework in which future related research took place in Romania. In short, it had to be performed on the theoretical grounds of Marxism-Leninism, in the concrete conditions of the socialist society (meaning in officially sanctioned institutions), and for the benefit of socialist society.[40] In spite of this ideological straitjacket, the first years of the 1970s brought a real development of the field of future research in Romania. Research was directed in two ways: a first strand was oriented toward theoretical and methodological research, while a second strand aimed at producing forecasts for Romania's economic and social development for the year 1990—and in some areas for the year 2000. In 1972, the laboratory produced the first manual of forecasting. The book was the result of the efforts of a research team coordinated by Mihai Botez, who by that time had already published several works on economic and social forecasting, including *Introducere în studiul prospectivei* (Introduction to the study of prospective), in 1971.

The successes of Romanian futurology in the 1970s drew on the many contacts that were established with Western institutions and researchers. We need to take into account that this happened with the approval of the regime, and more specifically the approval of Nicolae Ceauşescu himself, who not only allowed but encouraged such contacts. Not only Romanian researchers could travel abroad, but foreign scientists came to Romania, giving lectures or working on common projects.[41] As a result, in September 1972 the Third International Conference of Futures Research took place in Bucharest

Figure 7.1 The Third International Conference of Futures Research, Bucharest, Romania; September 1972. Courtesy of the National History Museum of Romania.

(see Figure 7.1). The conference was a success as it brought together important researchers from both ideological camps in an attempt to overcome differences and to work together for the *common future of human beings*, as the title of the conference suggested.[42] Among them were Bertrand de Jouvenel, Johan Galtung, Ossip K. Fleichteim, Igor Bestuzhev-Lada, and Romanians Mircea Malit¸a, Mihai Botez, and Pavel Apostol. Besides disseminating scientific knowledge, the conference proposed the setting up of an international center for methodological prospective studies. The proposition was made by Mihai Botez, with the idea to create a framework for multidisciplinary future research, drawing on multinational teams working together for the development of future methodologies.[43] *Centrul International de Metodologie a Studiilor asupra Viitorului şi Dezvoltării* (International Center of Methodology for Futures and Development Studies) became a reality two years later, in

March 1974, when its creation was approved by the General Assembly of World Future Studies Federation (WFSF) in Paris. The Center functioned as an autonomous organization within the University of Bucharest, and it was led by Botez. The setting up of the Center was approved by Nicolae Ceauşescu, to whom Johan Galtung, president of the WFSF, addressed a letter on May 10, 1974. In this letter, Galtung expressed his vision of the Center as a place to coordinate future studies, publish monographs, and hold seminars.[44] The two discussed the matter personally on May 25, 1974, when Ceauşescu received Galtung in the presence of Maliţa.[45] The Center functioned for only three years; in 1977, it was combined with the Department of Systems Study within the University of Bucharest and Botez lost his position. This measure was part of a wider institutional reorganization as a result of the ideological tightening after the end of liberalization.

Building up methodologies of future research and taking part in an international future field were activities that benefited from the regime's approval. However, what interested the Romanian communist leadership even more were the concrete activities of forecasting. Forecasting was understood as a tool that could accelerate the realization of communist society. In an article concerning forecast and the theoretical basis of future study, the author, Sergiu Tămaş, a prolific writer of books and articles on the communist future, synthesizes the role of forecast thusly: one of the features of the activity of studying the future within the socialist conditions is the total commitment to the finding of new resources of progress for the realization in a shorter historic term of the communist society. Tămaş also speaks about "mastering the social time by the working people as a way of deliberated approaching the desired future."[46]

Several institutions were set up by the Party for the elaboration of forecasts regarding the economic and social development of Romania. The first was *Comisia Centrală de Prognoză* (the Party and State Central Commission for the Elaboration of Long-Term Forecasts), created in July 1971. Within it, work was divided on field commissions (for different branches of the industry, raw materials, agriculture etc.) and problem commissions (demography, living standard, education, science, health). There were three types of forecasts: long-range (fifteen to thirty-five years), medium-range (five to ten years) and short-range (one to two years or less) forecasts.[47] The Central Commission was headed by Ceauşescu.[48] A year later, the commission was integrated into a new organism, *Consiliul Suprem al Dezvoltării Economice şi Sociale* (the Supreme Council for Economic and Social Development of Romania), which was a unique Party and state organization in charge of the current development of the country as well with the elaboration of forecasts.[49] This was also headed by Ceauşescu.

The gathering under the same roof of all activities regarding the future development of Romania was motivated ideologically by the need to concentrate planning resources toward the general goal of communism. Moreover, the Council, counting 250–260 members representing the Party, the state, scientists, artists, and workers, so the entire society, was to act like an economic parliament where decisions were made in deliberation. In ideological terms, that was proof of the continuous development of "socialist democracy." But as Ceauşescu centralized power in his own hands, these institutions were emptied of content.

The centralization of decision-making in Romania was intended to match a new goal set by the party leader, the construction of the *multilaterally developed socialist society*. This new phase was theorized by Ceauşescu as a superior stage of socialism, yet intermediary to communism. The new short-term destination of communist society was announced at the Tenth RCP Congress in 1969. *Programul Partidului Comunist Român de făurire a societăţii socialiste multilateral dezvoltate şi înaintare a României spre comunism* (the RCP program for building the multilaterally developed socialist society and for advancement toward communism) became public in 1975; it was defined as the "fundamental ideological, theoretical and political chart of the Party, establishing the general strategy and tactical orientations of the Party and state internal and foreign policy for twenty to twenty-five years and the fundamental directions of Romanian society development towards the socialism and communism."[50] The directions set for development were based on long-term forecasts. The program also reaffirmed the emphasis on industrial planning and announced the intention to intensify the activity of prospective forecasts in this area. This profound turn toward industrialization explains the reaction of the regime toward the first report of the Club of Rome, *Limits to Growth*, in 1972. *Limits to Growth* provoked serious reactions in the socialist countries. Most of the variables discussed in *Limits*, population growth, for instance, were the subject of planning in Romania, hence of nonnegotiable political decisions. At the end of the 1960s, the regime began controlling population growth and banned abortions. The program of industrialization relied heavily on the exploitation of natural resources. As a result *Limits to Growth* was dismissed by the regime. "Western European technocrats have failed and are projecting their failure on the whole world."[51] As in the West (see Vieille Blanchard in this volume) ideological critique was entangled with discussion of the methodology of the World model used in *Limits*.[52]

The Romanian Communist regime had a strong vision about the future and about how this future should be built. There was no room for flexibility. However, among Romanian futurolgists there were few that dared oppose this vision.

UNOFFICIAL FUTURES OF COMMUNISM: THE DISSIDENCE OF MIHAI BOTEZ

Futurology was a field in which international contacts were encouraged by the regime. Consequently, Romania sent representatives into professional futurist organizations, such as the World Futures Studies Federation. Scholars from Romania also contributed to the drafting of reports of the Club of Rome, following the debate on *Limits*; Maliţa's 1979 report, co-authored with James W. Botkin and Mahdi Elmandjra, titled "No Limits to Learning: Bridging the Human Gap," was an example. However, as with every other area of social life in communist society, futurology was subjected to Party control, and the development of futurology was also highly influenced by ideology. The interest the Party showed in the field of futures research created the preconditions for the formation of a new category of professionals who trained in an international milieu. Futurology developed mainly via modeling, and many of the researchers were mathematicians. Others were economists, sociologists, or philosophers. For them, the opening to the West manifested during the period of liberalization meant the chance to travel abroad, and to produce and exchange ideas in a free world. Some of these researchers opted for a normal professional life (meaning the permission to publish and to travel). In return, they did not challenge the regime. There were others, however, who by challenging the regime jeopardized their careers, sometimes even having their lives threatened. This was the case of Botez, the futurologist who became a dissident.

Botez's cold, rational, and unemotional critique of the communist system developed in a series of studies about communism in general and the Romanian regime in particular. Unlike other dissidents who rejected the Stalinism of Ceauşescu, Botez's struggle was not against Ceauşescu personally. His opinion was, rather, that the communist system needed to change, and that in order to change it, it needed to be analyzed and understood.

Botez was a highly interdisciplinary mathematician, a "butterfly researcher" who saw futurology as a way of making connections between mathematics and social science.[53] He was fundamentally interested in the study of probability. Botez graduated in 1963 from the Faculty of Mathematics at the University of Bucharest and three years later defended a PhD thesis on "*Probability Theory, Mathematical Statistics and Operational Research*."[54] The PhD was the beginning of an exceptional career. During the 1960s, Botez worked successively at the Institute of Mathematics of the Romanian Academy, the Center for Mathematical Statistics, and became the director of the Department for Operational Research and Mathematics in Management and Data Processing at CEPECA. In the beginning of the 1970s, Botez took charge of the Laboratory of Prospective Research, and eventually became the director of the International Center of Methodology for Future and Development Studies.[55] Botez's shift to the study of prospective happened at the end of the 1960s. He traveled abroad, for specialized trainings in Federal Republic of Germany, Denmark, Poland, Greece, and the US. Also, he went for study tours in Belgium, France, and Austria, focusing mainly on *operation research* and *decision theory*, the interest which brought Botez closer to the new field of the systems dynamics. In his travels, Botez met Jay W. Forrester, C. West Churchmann, and Erich Jantsch.[56]

In 1976, after meeting Martin Diamond in Bucharest, Botez was able to leave Romania for a fellowship in the Woodrow Wilson International Center (WWIC). He was the first Romanian fellow of WWIC, and at the end of his study tour he concluded a paper exploring the general conditions in future science from a sociological perspective.[57] Botez contribution in Romanian official futurology ranged from methodology handbooks, to papers detailing the future of economy, technology, tourism, and literature. By participating in international conferences and acting as member of the editorial board of international reviews, such as *Technological Forecasting and Social Change*, he became internationally known in the field of futures research. He was also a member of the Scientific Board of the World Future Studies Federation and of the Center for Logic and Comparative Sciences, Bologna, Italy.[58] This international prestige protected him, for a while, as it pushed the regime to exercise constraint in punishing a dissident backed by Western scientific groups.

In the light of this promising career, Botez made a surprising choice at the end of the 1970s as he embarked on dissidence. Publicly challenging the regime meant suffering professional marginalization, social isolation, and constant harassment from the secret police, sometimes outright death threats. Importantly, Botez assumed the same principles of dissent as his more famous colleagues in Eastern Europe and the Soviet Union, namely, *legality*—acting within the limits of the communist laws, and acting with *transparency* and the *refusal of clandestinity*, assuming his stands publicly.[59]

What makes someone become a dissident? In the context of the communist regimes of Eastern Europe, usually the answer lies in personal backgrounds, parents or relatives who had been members of "bourgeois" political parties who became victims of repression. Thus, the traumatic past acts as a trigger for future dissidence. In other cases, it is possible to identify a critical point, a fracture, in the professional or personal evolution of a dissident. But this is not the case of Mihai Botez. Botez came from an educated family. His father was a violinist who had studied in France, his mother was a teacher. Botez parents lived in the "old world," but their children grew up in the new one. Memories of the former regime were present, even if spoken about cautiously, and bourgeois habits were cultivated (Mihai took piano classes) but Botez parents did not raise their children in hostility toward communism. As Botez confessed in an article, in 1988, after he had left Romania:

> For me, Communism was not an option, and I never felt obliged to justify it. Rather, it was the unchallengeable political system I was born into. Therefore, my problem is not to cry, to claim, or to blame, but to understand this new world and do my best to improve it. As a futurist, I see the present Communist world not only as the result of an evolutionary process, but also—and may be more fundamentally—as the starting point for the future.[60]

It was his training as a mathematician and a futurist that made Botez aware of the internal flaws of the system. Confronted by democracy in the West, he began seeing social dialogue as a critical element in establishing notions of what was good or bad for society as a whole. Through his probabilistic studies of the future of the communist system, Botez arrived at the conclusion that the road chosen by the political leadership was wrong and would lead to crisis. From the beginning of the 1970s, he started to raise questions about Party economic policy. As he would later remember: "Every year I began my classes at Ştefan Gheorghiu Academy with a question: we are now building industry, but what are we going to do with it later? It was a way to justify the need for correct forecasts."[61] Decisions were based on the leadership directives, so forecasts were ignored unless they forecasted what the Party had decreed. No answer from the Party came, not even to the dozens of reports officially sent in which Botez warned in futurological terms about a future crisis in energy, raw materials, and the environment.[62]

Botez was also disappointed by the end of liberalization. At the beginning of his regime, Ceauşescu appeared to be shifting from hard notions of ideology to a notion of professional efficiency. He announced that past understandings of the "healthy social origin" of a person would no longer be allowed to determine this person's ability to participate in the building of socialist society."[63] This signaled to non-Party intellectuals that they could join the ranks of the RCP and be allowed to pursue professional careers without political intervention. In this context, Botez joined the RCP in 1970. He was mistaken to do so and he admitted it in 1984 when he officially left the party. With the end of liberalization, complete loyalty to the Party was reinstated as the basis for societal position. As Ceauşescu centralized control over Romanian institutions in his own hands, the forecasting institutes would also be directed by him personally. This suffocated intellectual exchange in the field. As Botez concluded in a study written in the 1980s, the crisis in Romanian society was also caused by "the poor intellectual quality of the leading class."[64]

Mihai Botez's dissidence lasted from 1979 to 1987, when he left Romania. It can be divided into two major parts. The first consisted in taking a public stand on errors in planning and Party management and their subsequent consequences. In an effort to create the foundations for a future civil society, Botez conceptualized and tried to disseminate the theory of "solitary dissidence." Solitary dissidence meant individual dissidence, outside of any organization, and it encouraged individuals to step out into the public sphere and speak their mind regarding problems of the Romanian society. The second part of his dissidence saw him draw up studies regarding the future of the communist system and the Romanian regime. These studies were elaborated employing the methods of futures research.

Botez's dissidence began in 1979 with an open letter addressed to the Party, in which he drew attention to the negative developments in Romania. He noticed not only the growing lack of food and the rise in foreign public debt, but also the fact that "the strategy of Romania's planned development focused on industries totally depending on imports of raw materials and on foreign markets at the expense of agriculture led to different results than the expected ones." Industrialization was leading to the depopulation of entire villages and "the weakening of peasantry [. . .] without forging an authentic working class."[65] Finally, Botez's letter brought into focus the precarious situation in education, culture, and science, as well as the growing number of people wishing to leave the country.[66]

The 1979 letter was followed by dozens of memorandums regarding the economic and social situation of Romania. They were sent to Romanian institutions, but also to the West, and became known to the Romanian public especially through the broadcastings of Radio Free Europe. There was a recurrent topic in Botez's memorandums. It was the crisis in energy generated by the huge investments in industrial branches, such as metallurgy and the petrochemical industry. Growth in these sectors put a hamper on the economy as a whole because of their energy demand and their dependence on imported raw materials. These industries were not profitable. Overdevelopment of industry had disruptive social effects in terms of urbanization. Imports were paid for by exporting goods from other economic branches than industry. This situation created shortages among the population in food, fuel, and public services.[67] Botez kept challenging the Party:

> The general strategy of development in Romania is wrong and is having devastating implications for Romania's future. This situation is generated by absolutely unrealistic economic forecasts and incorrect estimations of the country's resources and needs. The development plans are wrong and fulfilling them can only worsen the economic crisis. What I consider most dangerous is that the government is ignoring these facts and we continue to live in an atmosphere of unreality.[68]

What was to be done in order to correct things? In Botez's thinking, the rulers had to accept public dialogue, responsibility for mismanagement, and start working with specialists in order to find solutions.

That was a naive request, to say the least. The growing personality cult of Ceauşescu did not leave room for dialogue. Questioning economic policy meant nothing less than questioning the Party's vision of the future. The Party could not accept this. Multilaterally developed socialism was to be accomplished regardless of social costs. Botez's acts of dissidence did not generate reactions among his fellow intellectuals; however, they did raise the interest of the institutions of repression. He was constantly harassed and interrogated by the Securitate. Professionally, he was downgraded and could no longer travel abroad for scientific purposes. The Party offered him the possibility of exile, which he refused. In 1987 he was sent on internal exile to a little town called Tulcea. The reason was to cut him off from his contacts with foreign journalists and to have him under permanent surveillance.

The public acts of Botez were however just one side of his dissidence. Behind these stood his futurological studies of the structure and mechanisms of the communist system. Botez approached the inner flaws of communist system in several papers. Some of these were presented publicly, drafted, usually, in a highly codified, scientific language of mathematics, probabilistic, and systems analysis. Others, speaking more straightforwardly, were subversive and could only be addressed to a quasi-clandestine public. An example of the former was a paper on modeling in international affairs called "Cooperative Management of Force-Induced International Situations: An Exercise in Formal Modeling." It was published in the review *Policy Science* in 1977.[69] Botez worked on this paper with Sergiu Celac, one of his closest friends. The article was divided in two parts, one of more narrative language, while the latter consisted of range of mathematical formulas that expressed the critical theses that could not be put in words. The article developed the following thesis: the more participants there are in decision making, the bigger the probability of reaching a rational decision. The reverse was also demonstrated: the fewer participants there were in decision making, the smaller the probability of arriving at a rational conclusion. In the case of only one decision maker, the models predicted a perfectly irrational outcome, such as dictatorship. Sergiu Celac later confessed to writing the first section, while Botez was behind the mathematic demonstration of what could not be said in words.[70] The article, even if codified in scientific language, was a clear hint to Western readers of the Romanian regime's turn to dictatorship. Amazingly, Botez and Celac managed to publish the article abroad, by first presenting a very short version to the Romanian Academy conference.

An example of Botez's more openly critical writings was an extensive study on the desirable society. *Undesirable versus Desirable Societies* was the result of a research project, carried out in collaboration with the WFSF and the United Nations University on the modeling of desirable societies.[71] In its preparation, Botez worked with Mariana Celac, his onetime companion and collaborator, as members of a team from the Bucharest GPID (Goals, Processs and Indicators of Development), headed by Solomon Marcus. The report made too direct references to the Romanian leadership, and could not be presented as the result of a project developed within a Romanian institution. In consequence, it was assumed by the two researchers personally. In order to define what was a desirable society, the authors embarked on a demonstration on what was undesirable and of how undesirable social settings could be avoided. They applied a terminology from GPID that differentiated between the desired society, referred to as *beta type society*, and current society, called *alpha type society*. Beta was an abstract construction, described as horizontally integrated and participatory whereas Alpha was "the real societal forms currently in existence," described as large, vertically integrated, and growth orientated.[72] After this demonstration, the authors embarked on a critical analysis of how to realize desirable societies, and discussed individual versus group images of future societies and how they were distributed on the level of class, nation, and world. Concerning the communist image of the future, the authors stated:

> It is often considered that beta society can be reached—may be during our lifetime. For one third of the people of this planet, communist society—theoretically, a beta type one—is no longer a utopian abstract construct but a goal to be attained: the journey into the future is already in motion on a one-way road. For those who are engaged in this "battle" it is evident that their children will share their ideal and will proceed toward the goal projected by their parents (or even grandparents) animated with the same enthusiasm. Apparently, in this long perspective the problems of efforts (or costs) to be paid by the community lose their meaning [. . .]. [73]

The study also emphasized the cult of personality in national-communism, as a "natural outcome of the structure of power, and not an accident."[74] The authors had to codify the conclusions of their research. Undesirable developments would be those that led to "more alpha," as in the case of dictatorship is. A desirable society would be that with "less alpha," that meant increased social dialogue and the possibility for the citizens to make their own choice for the future.[75]

Botez's unofficially conducted studies concerned the structure and mechanism of the communist system. This work was part of his efforts of understanding and explaining how the communist world functioned and how it might evolve. Botez used a special denomination for the countries under communism. That was the "Second World" and it consisted of Eastern Europe countries, the Soviet Union, Yugoslavia, but also China, Vietnam, Cambodia, Cuba, Ethiopia, Mongolia, Afghanistan, Angola, Yemen, Laos, North Korea, and Guinea-Bissau.[76]

The first of his papers regarding the functioning of the communist world was entitled Lumea a doua: Introducere în comunismologia structurală (*The Second World: An Introduction to Structural Communismology)*. Botez tried to establish new methodologies based on mathematics that could model the evolution of the communist system as it was transplanted onto societies with different national traditions and in different stages of development. "Communist societies are new social structures with particular mechanisms of decision and self-regulation, with new definitions of ideological, political and social cost and benefits, elements that are generating new schemes of political thinking, new rules of conduct, also introducing a new ethic and a new logic."[77] By contradicting Marxist Leninist postulates, Botez stated that communist societies were not the result of a historically law-bound process, but rather, the result of a conscious implementation of large social projects that led to brutal interruptions in the history of a nation.[78]

In his *Second World* study, Botez constructed an abstract/canonical model of communism. Consequently he analyzed how this model functioned in reality at the level of the societies it was implemented. The structural features of the abstract/canonical model of communism were the following: Marxist/Marxist-Leninist/modern variants of scientific socialism official ideology; state-controlled economies; political power held by the communist party and the leader; and from the social point of view, societies that express their egalitarian calling and in which critical intellectuals are almost absent.[79] Subsequently, Botez argued that the implantation of this model on societies with different stages of economic, political, social, and cultural development generated a series of perturbations. The resulted societies, although similar at the level of structures, registered a wide range of evolutions "within the limits of communist society species."[80]

In this context, Mihai Botez discussed *national-communism* as an ideological construct resulting from the alliance between nationalism and communism. In Botez's opinion, "nationalism became a major source of stability for the communist order" and it brought benefits to the communist leadership both within and abroad.[81]

At the end of his analysis, Botez drafted two scenarios regarding the development of the real communism: a canonical and an uncanonical one. The canonical scenario would result into maintaining the existing communist society.[82] Even if chances of turning into reality were seen as quite small, Botez developed also a noncanonical scenario of real communism. The starting point of this scenario would be a popular rebellion, leading to social chaos and permanent instability.[83] Botez conclusion was that "the second world was a stable one and its stability resulted from its ability to absorb all perturbation and to conserve for the long term its canonical structure."[84] The communist system could not change.

The *Second World* referred to the communist system in general. Botez developed another study to the Ceauşescu regime, *Românii despre ei înşişi. O cercetare de comunismologie prospectivă* (Romanians on Themselves: A Research of Prospective Communismology). He prepared this with the aim to present it to an international conference in Paris in 1986 on the topic "Romania after Ceauşescu."[85] Botez was not allowed to leave the country for the conference. But he conducted a comprehensive study of the images of Romanian society held by different social groups in Romanian society. He wanted to find out what people thought about economic, social, and cultural life, and their opinion on Ceauşescu's growing personality cult. He interviewed professors, diplomats, party activists, waiters, and even cab drivers.[86] This was tremendously risky research, both for Botez and for his interview subjects. In order to protect his research, Botez imagined a complex system of codifying information. He used stars, crosses, and points to mark the people he had talked to and their answers.[87] The result was grim, showing social fatigue, intellectual apathy, and a state of fatalistic resignation in the Romanian population. Botez concluded that Romanian society would accept government policy, regardless of its "long-term catastrophic implications."[88] The most likely of the scenarios that he evaluated, making use of prospective methods, was an up to ten-year period of status quo. Disturbances might appear, Botez predicted, as a result of a series of systemic crises: a crisis in leadership and of the Party apparatus, a crisis of the Romanian economy, and a crisis of the young population. As a result of these crises, variations where possible, including a Romanian alignment to Gorbachev's *Perestroika*, but Botez saw as most likely that no significant change would occur before the year 2000.[89]

After he left Romania for the US in 1987, Botez continued his researches on the communist societies. He devoted special attention to the relationship between communist regimes and intellectuals. His ideas were developed in conference papers and articles and benefited from wide dissemination.[90] Botez identifies four types of social conduct: *positive support, positive acceptance, negative acceptance, total and active rejection.*[91] *Positive support* came with high moral cost. It ensured professional and social success and access to the nomenklatura. It also allowed for the creation of "islands of competence" within the communist system.[92] *Positive acceptance* referred to those who accepted the existing situation and showed willingness to take part in gradual improvement. In return, they were allowed to have a normal professional life, including going abroad for professional purposes.[93] *Negative acceptance* meant denial of social implication. The individual was concerned with surviving and due to his lack of implication he was marginalized. However, he could still have a normal professional life within the country.[94] In contrast, *total and active rejection* of the regime led to professional and social isolation of the individual, internal exile and finally external exile.[95] In analyzing the relationship between the political regime and the intellectuals, Botez put a special emphasis on the small social segment of critical intellectuals[96] seen as having an important role in developing autonomous conducts and as "carriers of a negative regulatory feedback in society."[97] Critical intellectuals could not exert influence over the policies of the communist power. However, they could monitor the abuses of the power and take public stands against it. That proved that even in the hardest conditions social reflexes of a normal society were still alive.

CONCLUDING REMARKS

Botez, like the majority of prominent Western analysts, did not foresee the disintegration of communist rule in Eastern Europe, which occurred only a few years after his departure from Romania. Ceauşescu's Romania is a perfect example of how communist regimes held the future captive. Trying to predetermine the future of Romanian society, Ceauşescu invented a new stage of development toward communism, the multilaterally developed socialist society. Instead of bringing the anticipated prosperity and equality for the Romanian people, efforts to realize this vision only brought equality in poverty and hard life.

Mihai Botez was a rare example of a public dissident in Romanian life. Like Winston Smith, the hero of Orwell's *1984*, Botez could not agree that "two plus two equals five."[98] Similar to others Romanian scholars, Botez recognized the gap between what the regime presented as reality and the actual state of things. However, unlike the majority of his peers he did not keep his views to himself. Nevertheless, Botez did not share the tragic fate of Winston Smith as he did not allow himself to be broken by the communist regime. Although he had a promising career ahead, Botez embarked on dissidence in the 1980s. It was a scientific rooted critique. Botez not only criticized the regime, but moreover he studied it and tried to anticipate its futures. According to Mihai Botez, a fair future could have been built by those leaders who would accept the errors and strive to correct the path leading to the future. This implied seeing things "as they were" or, metaphorically, accepting the reality that two plus two *always* equals four. In other words Botez's critique did not aim to discard the role of ideology, but rather to enhance the role and authority of autonomous science in the construction of a communist future.

ACKNOWLEDGMENT

The author would like to thank Mariana Celac, Jenny Andersson and Eglė Rindzevičiūtė for reading this chapter and making useful comments.

NOTES

1. Interview with Mariana Celac, December 14, 2013.
2. Cristina Diac, "Nicolae Ceauşescu (1918–1989)," *Arhivele Totalitarismului* 1–2 (2008): 236–237.
3. Ibid., 238.
4. Ceauşescu also assumed the position of head of the state in 1967, when he was elected president of the State Council. In 1974, when the institution of presidency was created, Ceauşescu became the president of the Socialist Republic of Romania.
5. Ana-Maria Cătănuş, *Vocaţia libertăţii. Forme de disidenţă în România anilor 1970–1980* (Bucureşti: Institutul Naţional pentru Studiul Totalitarismului, 2014), 65.
6. Anneli Ute Gabanyi, *Literatura şi politica în România după 1945* (Bucureşti: Editura Fundaţiei Culturale Române, 2001), 150, 156.
7. Ana-Maria Cătănuş, *Vocaţia libertăţii*, 75.
8. In 1967, Willy Brandt visited Bucharest. He was soon followed by General de Gaulle, in May 1968, and the American president, Richard Nixon, in 1969.
9. See *Congresul al IX-lea al Partidului Comunist Român, 19–24 iulie 1965* (Bucureşti: Editura Politică, 1965).
10. Arhivele Naţionale ale României, Fond CC al PCR., Cancelarie, 63/1965, 4. Hereafter A.N.I.C.-C.C. P.C.R.-Cancelarie.

11. Ibid., 83/1966, 2–30.
12. In 1971, the institution changed its name into the National Council of Science and Technology and in the 1980s it was coordinated by Elena Ceauşescu. Ana-Maria Cătănuş, "Consiliul Naţional pentru Ştiinţă şi Tehnologie," in Dan Cătănuş, coord., *România 1945–1989. Enciclopedia regimului comunist. Instituţii de partid, de stat, obşteşti şi cooperatiste* (Bucureşti: Institutul Naţional pentru Studiul Totalitarismului, 2012), 188–192.
13. Nicolae Ceauşescu, *România pe drumul desăvârşirii construcţiei socialiste. Rapoarte, cuvântări, articole, iulie 1965–septembrie 1966* (Bucureşti: Editura Politică, 1968), 231–233. Passage translated into English by the author.
14. Ibid.
15. *Congresul al IX-lea al Partidului Comunist Român*, 60.
16. Romania was the first communist state to resume diplomatic relations with Federal Republic of Germany, in February 1967. A.N.I.C.-C.C. P.C.R.-Cancelarie, 109/1968, 2.
17. During 1962–1970, Maliţa was deputy minister on the problems related to United Nations and other international organizations. Further, during 1970–1972 he occupied the seat of Minister of Education. Maliţa worked in the proximity of Ceauşescu, as minister at the State Council, beginning with 1974, and as Ceauşescu's councilor from 1979–1981. In the 1980s, he returned to diplomatic activity, serving among other positions as the Romanian ambassador to the US (1982–1985). Florica Dobre, coord., *Membrii C.C. al P.C.R., 1945–1989. Dicţionar* (Bucureşti: Editura Enciclopedică, 2004), 371.
18. Solomon Marcus, Intervenţie la "Colocviul dedicat lui Mihai Botez" (Târgovişte, 1997). Manuscript. Private Archive of Mihai Botez.
19. Remember CEPECA—Centrul de Perfecţionare a Cadrelor de Conducere din Întreprinderi. Accessed September 21, 2014. www.marketwatch.ro/articol/4130/Remember_CEPECA_-_CEntrul__de_PErfectionare_a_Cadrelor_de_Conducere_din_Intreprinderi/pagina/1.
20. Ibid.
21. Nicolae Ceauşescu, *România pe drumul construirii societăţii socialiste multilateral dezvoltate. Rapoarte, cuvântări, articole, mai 1971–februarie 1972* (Bucureşti : Editura Politică, 1972), 185–195.
22. Ibid., 224.
23. *Statutul Partidului Comunist Român* (Editura Partidului Comunist Român, 1945), 6.
24. *Dicţionar de economie politică* (Bucureşti: Editura Politică, 1974), 554.
25. The official complete name was "the unique national plan for economic and social development of Romania."
26. Brânduşa Costache, "Comitetul de Stat al Planificării," in Dan Cătănuş, coord., *România 1945–1989. Enciclopedia regimului comunist. Instituţii de partid, de stat, obşteşti şi cooperatiste*, 108–114.
27. Manea Mănescu, "Planul naţional unic de dezvoltare economico-socială a României," *Era socialistă* 14 (1973): 9.
28. Academia de Ştiinţe Sociale şi Politice a Republicii Socialiste România, *Revoluţia ştiinţifică-tehnică şi progresul. Lucrările sesiunii ştiinţifice din 5 aprilie 1974* (Bucureşti: Editura Politică, 1974), 7.
29. See Ioniţă Olteanu, ed., *Dialoguri despre viitor cu Edward Cornish, Herman Kahn, Willis Harman, Alvin Toffler* (Bucureşti: Editura Politică, 1982).
30. See Mircea Maliţa, *Cronica anului 2000* (Bucureşti: Editura Politică, 1969).
31. Ibid.
32. Interview with Mircea Maliţa, December 20, 2013.
33. Ibid.

34. Mircea Maliţa, "Linii directoare ale cercetărilor prospective," *Lupta de clasă* (August 1970): 20.
35. Mihai Botez, coord., *Cursul de prognoză* (elaborat de membrii Laboratorului de cercetări prospective al Universităţii din Bucureşti) (Bucureşti : Centrul de Informare şi Documentare în Ştiinţele Sociale şi Politice, 1972), 7.
36. Mircea Maliţa, "Linii directoare ale cercetărilor prospective," 13.
37. Arhiva Consiliului Naţional pentru Studierea Arhivelor Securităţii, Fond Documentar, dos. 19, vol. 5, 1–40. Hereafter ACNSAS.
38. Pavel Apostol, *Calitatea vieţii şi explorarea viitorului* (Bucureşti: Editura Politică, 1975), 12, 98–99. See also Pavel Apostol, *Omul anului 2000* (Iaşi: Junimea, 1972); Pavel Apostol, "Marxism and the Structure of Future," *Futures* (September 1972): 201–210; Pavel Apostol, *Viitorul* (Bucureşti: Editura Ştiinţifică şi Enciclopedică, 1977).
39. Pavel Apostol, *Calitatea vieţii şi explorarea viitorului*, 98–99. (See Andersson's chapter in this volume.)
40. Mihai Botez, *Introducere în prospectivă* (Bucureşti: Centrul de Informare şi Documentare în Ştiinţele Sociale şi Politice, 1971), 51.
41. In 1972, Johan Galtung gave lectures to the University of Bucharest. In May 1976, Alvin Toffler visited Romania.
42. Mihai Botez, and Mircea Ioanid, coord, Viitorul comun al oamenilor. Comunicări prezentate la cea de a III-a Conferinţă mondială de cercetare a viitorului, Bucureşti, septembrie 1972 (Bucureşti: Editura Politică, 1976).
43. Ibid., 58.
44. A.N.I.C.—C.C. P.C.R.-Relaţii Externe, 86/1974, 6–7.
45. Ibid, 1–8.
46. Sergiu Tămaş,"Prognoza şi bazele teoretice ale cercetării viitorului," *Era socialistă* 4 (1973): 26–29. See also Sergiu Tămaş, *Cercetarea viitorului* (Bucureşti: Editura Politică, 1976).
47. *Dicţionar de conducere şi organizare* (Bucureşti: Editura Politică, 1985), 642. See also A.N.I.C.-C.C. P.C.R, Secţia Propagandă şi Agitaţie, 36/1973, 1–30.
48. A.N.I.C.-C.C. P.C.R.-Cancelarie, 80/1971, 2–13.
49. A.N.I.C.-C.C. P.C.R.-Cancelarie, 140/1972.
50. *Programul Partidului Comunist Român de făurire a societăţii socialiste multilateral dezvoltate şi înaintare a României spre comunism* (Bucureşti: Editura Politică, 1975).
51. A. Pora, "Clubul de la Roma se înşală," *Era socialistă* 2 (1973): 55.
52. For debates on the *Limits* in Romania, see Mihail Florescu, "Limitele unei analize a creşterii," *Era socialistă* I, 3 (1973): 44–47; Mihail Florescu, "Limitele unei analize a creşterii," *Era socialistă* II, 6 (1973): 47–50; Victor Calcan, "Anti-Limitele creşterii," *Era socialistă* I, 8 (1974): 39–43; Victor Calcan, "Anti-Limitele creşterii," *Era socialistă* II, 9 (1974): 38–41; Virgil Ioanid, "Un punct de vedere cu privire la limitele creşterii," *Era socialistă* 17 (1974): 44–48.
53. Solomon Marcus, Intervenţie la "Colocviul dedicat lui Mihai Botez," 1997; Solomon Marcus, "30 de ani cu Mihai Botez," *România literară* 15 (2000): www.romlit.ro/30_de_ani_cu_mihai_botez. Accessed September 21, 2014
54. HU OSA Records of Radio Free Europe/Radio Liberty Research Institute, 300–60–3-box 6. Hereafter HU OSA.
55. Ibid.
56. Interview with Mariana Celac, December 14, 2013. An architect and a dissident herself, Celac was also interested in futurology and co-authored several papers with Botez, such as *Undesirable versus Desirable Societies* (Japan: United Nations University, 1983). Accessed September 21, 2014 www.transcend.org/galtung/papers/HSDR-GPID66.PDF.

57. Ibid.
58. HU OSA 300–60–3-box 6.
59. Ana-Maria Cătănuş, *Vocaţia libertăţii*, 25–30.
60. "Romania: Stalinism in One Country. Interview with Mihai Botez," *Uncaptive Minds* (June–July–August 1988): 21.
61. Mihai Botez, *Scrisori către Vlad Georgescu* (Bucureşti: Editura Fundaţiei Culturale Române, 2003), 85.
62. Ibid.
63. Nicolae Ceauşescu, *România pe drumul desăvârşirii construcţiei socialiste. Rapoarte, cuvântări, articole, ianuarie 1968–martie 1969* (Bucureşti : Editura Politică, 1969), 111.
64. Mihai Botez, *Românii despre ei înşişi. O cercetare de comunismologie prospectivă* (Bucureşti : Editura Litera, 1992), 65.
65. Mihai Botez, *Scrisori către Vlad Georgescu*, 13.
66. Ibid.
67. HU OSA-300–5–190-box 1.
68. Ibid.
69. Mihai Botez and Sergiu Celac, "Cooperative Management of Force-Induced International Situations: An Exercise in Formal Modeling," *Policy Science* 8 (1977): 455–468.
70. Interview with Sergiu Celac, April 10, 2012. Sergiu Celac worked on the Ministry of Foreign Affairs. In the 1970s, he had been Ceauşescu's both English and Russian interpreter. At the end of the 1970s, he was removed from this position and sent to work at a publishing house. Celac believes this article was one of the reasons he and Botez became undesirables. Under the influence of Mihai Botez, he became interested in future studies and became a researcher on the field. Currently he is member of the Romanian Association for the Club of Rome.
71. Mihai Botez and Mariana Celac, *Undesirable versus Desirable Societies*.
72. Ibid., 4.
73. Ibid, 6.
74. Ibid., 10.
75. Ibid., 58–59.
76. Mihai Botez, *Lumea a doua. Introducere în comunismologia structurală* (Bucureşti : Editura Du Style, 1997), 84.
77. Ibid., 26.
78. Ibid., 28.
79. Ibid., 82.
80. Ibid., 89.
81. Ibid., 127.
82. Ibid., 165–168.
83. Ibid., 169–172.
84. Ibid., 173.
85. Mihai Botez, *Românii despre ei înşişi. O cercetare de comunismologie prospectivă* (Bucureşti: Editura Litera, 1992), 7.
86. Ibid., 12.
87. Ibid., 12–13.
88. Ibid., 63.
89. Ibid., 74.
90. Mihai Botez, "East-European Intellectuals and the National-Communist State: A View from Bucharest," *PRAXIS International* 3 (1988): 350–359; Mihai Botez, "Intellectuals and the Communist State: A View from Bucharest," in Vladimir Tismăneanu and Judith Shapiro, eds., *Debates on the Future of Communism*, (Basingstoke: Macmillan, 1991), 170–181.

91. Mihai Botez, *Intelectualii din Europa de Est (Intelectualii est-europeni şi statul naţional comunist)—un punct de vedere românesc* (Bucureşti: Editura Fundaţiei Culturale Române,1993), 54.
92. Ibid., 56.
93. Ibid., 58.
94. Ibid., 59.
95. Ibid., 54.
96. Critical intellectuals were those persons whose opinions were blocked by the regime in their way to the public. If they expressed these opinions in unofficial press from within the country (*samizdat*) or abroad, they could be subjected to repression by the communist regime. Ibid., 82.
97. Ibid., 86.
98. George Orwell, O mie nouă sute optzeci şi patru (Bucureşti: Univers, 1991), 221.

ARCHIVES

Arhiva Consiliului Naţional pentru Studierea Arhivelor Securităţii. File D 19/5. www.cnsas.ro/ancheta_1968_dosare.html.

Arhivele Naţionale ale României. Fond CC al PCR, Cancelarie, files 63/1965; 68/1965; 83/1966; 109/1968; 80/1971; 140/1972.

Arhivele Naţionale ale României. Fond CC al PCR, Secţia Propagandă şi Agitaţie, file 36/1973.

Arhivele Naţionale ale României. Fond CC al PCR, Sectia Relaţii Externe, file 86/1974.

HU OSA Records of Radio Free Europe/Radio Liberty Research Institute. Fonds 300, Subfonds 60, box 6; subfonds 5, 190, box 1.

Private Archive of Mihai Botez.

REFERENCES

Academia de Ştiinţe Sociale şi Politice a Republicii Socialiste România, *Revoluţia ştiinţifică-tehnică şi progresul. Lucrările sesiunii ştiinţifice din 5 aprilie 1974*. Bucureşti: Editura Politică, 1974.

Apostol, Pavel. *Calitatea vieţii şi explorarea viitorului*. Bucureşti: Editura Politică, 1975.

Apostol, Pavel. "Marxism and the Structure of Future." *Futures* (September 1972): 201–210.

Apostol, Pavel. *Omul anului 2000*. Iaşi: Junimea, 1972.

Apostol, Pavel. *Viitorul*. Bucureşti: Editura ştiinţifică şi Enciclopedică, 1977.

Botez, Mihai, coord. *Cursul de prognoză* (elaborat de membrii Laboratorului de cercetări prospective al Universităţii din Bucureşti). Bucureşti: Centrul de Informare şi Documentare în Ştiinţele Sociale şi Politice, [1972].

Botez, Mihai. "East-European Intellectuals and the National-Communist State: A View from Bucharest." *PRAXIS International* 3 (1988): 350–359.

Botez, Mihai. *Intelectualii din Europa de Est (Intelectualii est-europeni şi statul naţional comunist)—un punct de vedere românesc*. Bucureşti: Editura Fundaţiei Culturale Române,1993.

Botez, Mihai. "Intellectuals and the Communist State: A View from Bucharest." In *Debates on the Future of Communism*, edited by Vladimir Tismăneanu and Judith Shapiro, 170–181. Basingstoke: Macmillan, 1991.

Botez, Mihai. *Introducere în prospectivă.* Bucureşti: Centrul de Informare şi Documentare în Ştiinţele Sociale şi Politice, 1971.

Botez, Mihai. *Lumea a doua. Introducere în comunismologia structurală.* Bucureşti: Editura Du Style, 1997.

Botez, Mihai. *Românii despre ei înşişi. O cercetare de comunismologie prospectivă.* Bucureşti: Editura Litera, 1992.

Botez, Mihai. *Scrisori către Vlad Georgescu.* Bucureşti: Editura Fundaţiei Culturale Române, 2003.

Botez, Mihai and Mariana Celac. *Undesirable versus Desirable Societies.* Japan: United Nations University, 1983. Accessed September 21, 2014. www.transcend.org/galtung/papers/HSDR-GPID66.PDF.

Botez, Mihai and Mircea Ioanid, coord. *Viitorul comun al oamenilor. Comunicări prezentate la cea de a III-a Conferinţă mondială de cercetare a viitorului. Bucureşti, septembrie 1972.* Bucureşti: Editura Politică, 1976.

Botez, Mihai and Sergiu Celac. "Cooperative Management of Force-Induced International Situations: An Exercise in Formal Modeling." *Policy Science* 8 (1977): 455–468.

Calcan, Victor. "Anti-Limitele creşterii." *Era socialistă* I, 8 (1974): 39–43.

Calcan, Victor. "Anti-Limitele creşterii." *Era socialistă* II, 9 (1974): 38–41.

Ceauşescu, Nicolae. *România pe drumul desăvârşirii construcţiei socialiste. Rapoarte, cuvântări, articole, iulie 1965–septembrie 1966.* Bucureşti : Editura Politică, 1968.

Ceauşescu, Nicolae. *România pe drumul desăvârşirii construcţiei socialiste. Rapoarte, cuvântări, articole, ianuarie 1968–martie 1969.* Bucureşti : Editura Politică, 1969.

Ceauşescu, Nicolae. *România pe drumul construirii societăţii socialiste multilateral dezvoltate. Rapoarte, cuvântări, articole, Mai 1971–Februarie 1972.* Bucureşti : Editura Politică, 1972.

Cătănuş, Ana-Maria. "Consiliul Naţional pentru Ştiinţă şi Tehnologie." In *România 1945–1989. Enciclopedia regimului comunist. Instituţii de partid, de stat, obşteşti şi cooperatiste*, coordinated by Dan Cătănuş, 188–192. Bucureşti: Institutul Naţional pentru Studiul Totalitarismului, 2012.

Cătănuş, Ana-Maria. *Vocaţia libertăţii. Forme de disidenţă în România anilor 1970–1980.* Bucureşti : Institutul Naţional pentru Studiul Totalitarismului, 2014.

Congresul al IX-lea al Partidului Comunist Român, 19–24 iulie 1965. Bucureşti: Editura Politică, 1965.

Costache, Brânduşa. „Comitetul de Stat al Planificării." In *România 1945–1989. Enciclopedia regimului comunist. Instituţii de partid, de stat, obşteşti şi cooperatiste*, coordinated by Dan Cătănuş, 108–114. Bucureşti : Institutul Naţional pentru Studiul Totalitarismului, 2012.

Diac, Cristina. "Nicolae Ceauşescu (1918–1989)." *Arhivele Totalitarismului* 1–2 (2008): 234–239.

Dicţionar de economie politică. Bucureşti: Editura Politică, 1974.

Dicţionar de conducere şi organizare. Bucureşti: Editura Politică, 1985.

Dobre, Florica, coord. *Membrii C.C. al P.C.R., 1945–1989. Dicţionar.* Bucureşti: Editura Enciclopedică, 2004.

Florescu, Mihail. "Limitele unei analize a creşterii." *Era socialistă* I, 3 (1973): 44–47.

Florescu, Mihail. "Limitele unei analize a creşterii." *Era socialistă* II, 6 (1973): 47–50.

Gabanyi, Anneli Ute. *Literatura şi politica în România după 1945.* Bucureşti: Editura Fundaţiei Culturale Române, 2001.

Ioanid, Virgil. "Un punct de vedere cu privire la limitele creşterii." *Era socialistă* 17 (1974): 44–48.

Maliţa, Mircea. *Cronica anului 2000.* Bucureşti: Editura Politică,1969.

Maliţa, Mircea. "Linii directoare ale cercetărilor prospective." *Lupta de clasă* (August 1970): 13–20.

Mănescu, Manea. "Planul naţional unic de dezvoltare economico-socială a României." *Era socialistă* 14 (1973): 6–11.

Marcus, Solomon. "30 de ani cu Mihai Botez." *România literară* 15 (April 19–25, 2000). Accessed September 21, 2014. www.romlit.ro/30_de_ani_cu_mihai_botez

Olteanu, Ioniţă, ed. *Dialoguri despre viitor cu Edward Cornish, Herman Kahn, Willis Harman, Alvin Toffler*. Bucureşti: Editura Politică, 1982.

Orwell, George. *O mie nouă sute optzeci şi patru*. Bucureşti: Univers, 1991.

Pora, A. "Clubul de la Roma se înşală." *Era socialistă* 2 (1973): 55.

Programul Partidului Comunist Român de făurire a societăţii socialiste multilateral dezvoltate şi înaintare a României spre comunism. Bucureşti: Editura Politică, 1975.

Remember CEPECA—Centrul de Perfecţionare a Cadrelor de Conducere din Întreprinderi. Accessed September 21, 2014. www.marketwatch.ro/articol/4130/Remember_CEPECA_-_CEntrul__de_PErfectionare_a_Cadrelor_de_Conducere_din_Intreprinderi/pagina/1.

"Romania: Stalinism in One Country. Interview with Mihai Botez." *Uncaptive Minds* (June–July–August 1988): 20–25.

Statutul Partidului Comunist Român. Editura Partidului Comunist Român, 1945.

Tămaş, Sergiu. *Cercetarea viitorului*. Bucureşti: Editura Politică, 1976.

Tămaş, Sergiu. "Prognoza şi bazele teoretice ale cercetării viitorului." *Era socialistă* 4 (1973): 26–29.

8 Virtually Nigeria

USAID, Simulated Futures, and the Politics of Postcolonial Expertise, 1964–1980

Kevin T. Baker

In early 1967, Nigeria was on the brink of civil war. The previous year had seen two military coups, widespread intercommunal violence, and a major crisis of confidence in Nigeria's federal system. With secessionist sentiments escalating in the Eastern Region, Glenn L. Johnson was undergoing his own crisis. He was the director of Michigan State University's Consortium for the Study of Nigerian Rural Development (CSNRD), a USAID-funded group of social scientists and other experts working on problems of rural development in Africa's largest country. This project, despite propitious beginnings, was on the verge of collapse. Already marred by a series of scandals in faraway Vietnam that damaged the group's credibility, the unrest in Nigeria made further data collection nearly impossible. With the future of the Consortium—and, indeed, the future of Nigeria—in doubt, Johnson began searching for ways to salvage the group's research project. Familiar with a farm management computer game being developed at Michigan State called Simfarm, he began exploring the possibility of creating a simulation model of the Nigerian economy from CSNRD data. As Nigeria collapsed into civil war, Johnson and a team of programmers, agricultural economists, and engineers endeavored to create a new version of that country, a Virtual Nigeria.

This chapter explores how Glenn Johnson and his colleagues at Michigan State University (MSU) sought not only to predict, but also shape Nigeria's future. In a time before oil revenues came to dominate economic life in the country, Johnson and his team were dedicated to modernizing Nigeria's dominant industry: agriculture. When conventional methods failed, they designed an interactive computer simulation called the Nigerian Agricultural Sector Simulation (NASS), which would enable its users to generate and select from competing agricultural policy futures for the country. Because much of the Nigerian documentation from this period is still closed to researchers, this chapter focuses on the story of the modelers, exploring how they sought to create credible knowledge about the future in a climate of postcolonial suspicion. Johnson and his colleagues worked during a period of transition, one in which American development experts no longer seemed credible and their visions of the future, embodied most famously in the “modernization theory” of W. W. Rostow, no longer seemed appropriate or desirable to many leaders in the developing world. Consequently, MSU’s simulation aimed not only at creating knowledge about Nigeria’s potential futures, but it was consciously committed to producing credibility and trust in those futures.

Theodore Porter’s *Trust in Numbers* has shown that the quantification of bureaucratic decision-making in the nineteenth and twentieth centuries was not primarily due to a desire to produce scientifically rigorous data.[1] Instead, Porter argues, quantification emerged because trust in elites had eroded in the face of a growing, democratizing society. “Trust in numbers” and notions of objectivity came to stand in for trust in people. This chapter asks what happens when numbers are no longer enough to produce trust. Instead of relying on objective numbers to produce trust, Johnson and his team brought subjectivity—the subjective experience of decision makers—to bear. With its interactive, responsive design, the Nigerian Agricultural Sector Simulation (NASS) involved the user directly in the research process, allowing a policy-maker to create multiple, experimental futures. By allowing a decision-maker to experiment and to bring their own experience to bear, Johnson and his team hoped their simulation could produce Nigerian futures that were credible and could be translated into actual policy. An exploration of the design of the NASS simulation suite offers important insights into the politics of trust in the postcolonial period. Similarly, this chapter argues that computer simulations like the NASS encouraged policy planners to imagine the future as a potential site of political intervention, choice, and even control.

After the decline of modernization theory’s credibility after the Vietnam War, simulation approaches like Michigan State’s NASS seemed for a time to be a useful way to repackage Washington’s international development agenda. In a climate in which leaders in developing countries were beginning to view Western-oriented theories of modernization with great suspicion, NASS’s designers incorporated a number of experiential and affective features into the simulation environment. These were intended to enhance the model’s credibility among Nigerian officials. By allowing decision-makers to manipulate policy variables at will, the interactive nature of NASS gave officials the impression of total control, while the underlying assumptions of the economic model guided their actions and shaped their political intuition.

Modernization theory was a political vision that purported to provide “pre-modern” nations a path to Western-style, capitalist modernity. Unlike other, earlier theories of economic development, modernization theory denied the relevance of racial or geographic barriers to capitalist development. In the work of Rostow and his colleagues, some countries may be “ahead” of others in terms of economic “progress,” but there were no intractable barriers that would prevent a nation like Nigeria from becoming an urbanized, industrial, and commercial society like the US.

Although the scholars at MIT's Center for International Studies were undoubtedly important architects of development policy in the Kennedy and Johnson administrations, a single-minded focus on modernization theorists in the historiography of international development homogenizes what was, in fact, an ideologically plural community of development workers. Daniel Immerwahr has argued that much of the US's international development work stressed not the large scale, urban, and industrial vision of Rostow and MIT's Center for International Studies, but instead devoted effort to small scale, rural, and community development.[2] Further, Helen Tilley's work on colonial development in Africa has shown that, when exposed to the realities of field work, development experts often complicated or subverted metropolitan ideology.[3]

In many ways, an examination of the work of Glenn Johnson and the CSNRD provides a useful corrective to a development historiography that has been overly focused on high theoretical perspectives emanating from Cambridge and Washington. First, Johnson and the Consortium (like many of their peers in USAID) were not primarily concerned with urban or industrial problems; they were animated by smaller-scale issues of rural life. Johnson and his team of scholars were concerned with increasing agricultural productivity and improving the quality of life in rural Nigeria. Second, whereas many of the scholars associated with modernization theory in the US had backgrounds in macroeconomics, Glenn Johnson was trained as an agricultural economist. Marion Fourcade has characterized the macroeconomics profession in the US as highly mathematized and abstract, and insulated from both other disciplines and state control.[4] The picture of agricultural economists is much more complicated. Agricultural economists like Glenn Johnson often began their careers in the employ of the federal government, were interdisciplinary in outlook, and were sensitive to the embeddedness of economic life. The interdisciplinary orientation of agricultural economists like Johnson led them to search for methodological alternatives when conventional approaches failed. Attentive to the complex social nature of economic life, Glenn Johnson and his colleagues in the Consortium saw in the systems sciences a way to unify their observations of rural Nigerian life into one coherent vision. For these scholars, the systems sciences provided an important coordinating function, ordering and consolidating knowledge garnered from a variety of disciplinary backgrounds.

Johnson's team was also attracted to the possibility of designing an interactive simulation of the Nigerian economy, which they hoped would offer Nigerian policy makers and USAID officials a way to experiment with policy alternatives and gain experience in running a national economy. Paul Edwards, in his article "The World in the Machine," has argued that systems models, like the famous World 3 model of *The Limits to Growth* publication of 1972 or the climatic General Circulation Models (GCMs) of the 1960s and 1970s, not only helped to establish computer simulation's reputation as a technique of policy analysis; they also shaped the "background assumptions for a large subset of the world policy community."[5] By bringing the world into a unified cybernetic control scheme, World 3 and GCMs presented the globe as a unitary system and suggested possible interventions into long-term international processes. Similarly, simulation models like NASS allowed planners to manipulate policy variables interactively and at will; this gave them a sense of having a level of control over political and economic circumstances not available in real life.

By assimilating Nigeria into a unified cybernetic control model, the simulation models of development experts opened up the possibility of intervention into long-term economic and social processes. Moreover, these development simulations were not just important for what they explained about a specific development context, but for what sort of aspirations they *evoked* in decision makers. Writing about a very different kind of simulation—disaster preparedness drills—the anthropologist Andrew Lakoff has identified the production of an "affect of urgency" as an important function of disaster simulations.[6] Producing this "affect of urgency" with scenario drills gives participants a *feeling* of how an emergency might unfold in the absence of an actual emergency event.[7] Building on the work of Lakoff, this chapter shows how a group of designers attempted to operationalize a user's simulation experience to build trust in the NASS simulation software and the Nigerian futures it produced.

This chapter begins with a brief look at the state of economic planning in Nigeria at the time of the country's independence, paying close attention to Wolfgang Stolper's account of his time at the head of Nigeria's Economic Planning Unit. It continues with a discussion of the field work of the CSNRD in the context of widespread disorder and civil war in Nigeria. The next two sections explore the CSNRD's research program's curious afterlife as a source of development data for Nigeria, and adds a (brief) look at how the NASS's codebase was also used to produce simulations for the South Korean and Brazilian economies.

PLANNING WITHOUT FACTS

American involvement in the Nigerian national economy began soon after Nigeria gained its independence from the United Kingdom. From an early stage, American officials and development experts saw Nigeria as providing the setting for a "significant historical demonstration" of the quality of American technical assistance.[8] Wolfgang Stolper, a Harvard PhD and former student of Joseph Schumpeter, played an important role in defining the economic policy of the newly independent state. With Ford Foundation backing, the economist was dispatched to Nigeria to head its Economic Planning Unit.

Arriving in Lagos in 1961, Stolper was tasked with leading a team of Nigerian economists in crafting the National Development Plan for 1962–1968. Initially, he expressed a great deal of enthusiasm for the project, exclaiming "I have the chance to weld this territory [Nigeria] into a nation," but he soon soured on the prospects for the national development plan and for Nigeria more generally. Stolper (who once described himself "the best economist in West Africa") remained confident in the value of his personal expertise, but doubted the ability of Nigerian elites, who he saw as corrupt and overly entangled in patronage systems, to carry out the plan's recommendations. Furthermore, he lamented the "lack of facts" available from the colonial period while drafting the national development plan, arguing that the lack of statistical and social scientific knowledge about the country made planning exceedingly difficult.[9]

Out of this frustration came *Planning without Facts*, a book Stolper hoped would provide guidance to future planners and theorists working on problems of economic development in Nigeria and in other national contexts. The book was published in 1966, but circulated in manuscript form for several years before that. It argued that in the absence of statistical and social scientific "facts" about development, planners should decentralize decision-making, make use of indirect market-based incentives, and not interfere with price signaling mechanisms.[10] In short, Stolper argued that in development situations where the planner has low levels of information about a specific national context, authorities must relinquish central controls and keep long-term plans flexible and pragmatic.

The Consortium for Nigerian Rural Development (CSNRD) emerged, in large part, to redress the paucity of statistical and scientific information that Wolfgang Stolper had identified. The group was comprised of teams from four universities—Michigan State University (MSU), Colorado State, Kansas State, and the University of Wisconsin—plus three institutional partners, the Research Triangle Institute, the US Department of Agriculture, and the Department of the Interior. It was commissioned to appraise existing Nigerian and USAID programs for agricultural development, analyze "development potentials," and make recommendations for improving agricultural programs. In order to accomplish these goals, USAID tasked the CSNRD with studying the Nigerian agricultural sector "in its broadest context."[11] Glenn L. Johnson, an economist experienced with Nigerian agricultural development, was made the project's director.

Trained as an agricultural economist at the University of Chicago, Johnson conducted his earliest research at Kentucky State University's Agricultural Experiment Station. There, he created an econometric model of the US's burley tobacco industry.[12] In 1955, after receiving an appointment for a professorship in MSU's agricultural economics department, Johnson became convinced that the field of economics was too insular and that future research needed to account for institutional, technological, and social change. Recalling these years, Johnson lamented that "though we did not realize it then, we were trying to compress a relatively productive multidisciplinary, problem-solving field of work into production economics . . . at the expense of destroying its technological, institutional, and humanistic breadth."[13]

In 1961, the US State Department sent Johnson to Thailand as part of an economic and military advisory team, his first experience with international development work. Even though the majority of Thailand's economic activity was agrarian, Johnson was the only agriculturalist on the mission. By the end of the project, he was very dissatisfied with the group's final product—a national econometric model—which, owing to the overemphasis of the "disciplinary" concerns of economists, he would later describe as "oversimplified, too macro, and naïve."[14] This first experience of development planning increased his wariness of narrowly disciplinary approaches that reduced the complexity of social systems to simple time-series data and maximization models. At the same time, he became convinced of the superiority of informal, ad-hoc projections that borrowed insights from decision-makers themselves and from "technical, institutional, and humanistic disciplines."[15] This interest in disciplinarily eclectic, pragmatic projection methodologies eventually endeared him to the interdisciplinary features of the systems simulation approach.

Johnson took this perspective with him to his first USAID contract in Nigeria, where he was named the first director of the Economic Development Institute at the University of Nigeria. During his two-year term at the helm of the Institute, Johnson maintained a methodologically eclectic approach and kept the group deliberately "unspecialized on any source or type of information or any computational technique."[16] By Johnson's own account, the Institute's research program gained the attention and respect of the country's policy officials, who drew on the group's publications when drafting Nigeria's national development plan. It was on the basis of this reputation that Glenn Johnson was named the head of the CSNRD.

MAKING DEVELOPMENT FACTS

Glenn L. Johnson left for Nigeria on March 23, 1965 to prepare for the arrival of the rest of the CSNRD team. He was met there by Vernon Johnson and Francis Jones, both from the USAID Mission in Lagos, who briefed Johnson on the agricultural programs of the mission. The following week, the three men toured the East and Northern Regions of Nigeria, making contact with Nigerian and USAID officials in those regions. According to a report of the forward planning team of the Consortium, both the meetings in Lagos and the consultations in the regions revealed that the team was aware of serious misgivings by Nigerian government officials about the activities of the CSNRD. Although the report noted that the fear of "external, overseas planning of Nigerian affairs" tended to disappear when the project and its mandate were explained to skeptical officials, these anxieties continued to dog the Consortium throughout its period of fieldwork in Nigeria, and the Consortium's lack of credibility with policy-makers eventually jeopardized the continuation of the project.[17]

Although the Consortium's report from their tour of the country's regions makes no note of it, Nigeria had been undergoing a period of increasing sectional and interethnic tensions.[18] At the time of the advance team's arrival, Nigeria had just come through an unsettled, hotly contested election in December 1964. The political campaign had been marked by arbitrary arrests, intimidation, and ethnic violence. In the Northern Region, home to the ruling party's power base, opposition candidates had been arrested, denied entry to rallies, and imprisoned. In the face of widespread intimidation, one of the opposition parties had called for a boycott of the December elections. The boycott was only partially successful and the ruling party coalition, the Nigerian National Alliance, won an overwhelming victory. Following the certification of election results, President Azikiwe, a founder of the opposition National Council of Nigerian Citizens, refused to invite the formation of a National Alliance government by the reelected Prime Minister Abubakar Tafawa Balewa. Out of the stalemate, the two men reached a compromise that would require the formation of a unity government, and allow seats that had been boycotted in the December election to be recontested the next year.

An election for the Western Regional assembly was scheduled for October. This election essentially followed the pattern established during the December general election, with widespread charges of violence and voter suppression. On election day, October 11, 1965, regional Premier Samuel Akintola decreed that the election results were to be announced only at the central headquarters in Ibadan, instead of the normal practice of declaring vote tallies at local polling places. When the results were announced two days later, the official tallies put the ruling NNDP ahead, but both sides claimed victory. In response to widespread accusations of fraud, protests and riots took place across Nigeria's Western Region throughout November and December. Nigeria's first military coup began on the morning of January 15, 1966. Led by a group of officers that would later be called the "Five Majors," the plotters arrested all of Nigeria's regional premiers and executed several federal officials. In the chaos, John Aguiyi-Ironsi, the General Officer Commanding of the Nigerian Army, rose to be the Head of State of the Nigerian military government. After securing power, the new military government quickly outlawed political parties, deposed regional governors and replaced them with military governors.

About this time on a tape-recorded message to East Lansing, Glenn Johnson issued a report on the new political landscape in Nigeria and the implications of the change in leadership for the ongoing CSNRD. He noted that, in spite of the coup, things were generally quiet in Lagos. Despite administrative reshuffling in the Nigerian government and general political uncertainty, Johnson remained optimistic about the CSNRD's progress, noting that the team members were "not letting minor difficulties serve as an excuse for not getting things done."[19] Around that time, Johnson was made aware of the military government's establishment of a National Advisory Committee for Agriculture earlier that year. Johnson suspected, judging from the committee's organizational structure, that its working groups were acquainted with and attentive to the CSNRD's research agenda. He accorded the highest priority to gaining the trust of this group of advisors:

> I found myself saying today that [the] CSNRD may have a real opportunity to relate itself to the National Advisory Committee, but that if it is unable to do so, that it probably should be abandoned. With this National Advisory Committee established, it seems to me that the CSNRD should not continue if its work is not to be used by this committee.[20]

Despite these high stakes, Johnson remained optimistic about the Consortium's ability to influence the commission.

His optimism, however, was tempered by embarrassing news from the US. The New Left magazine *Ramparts* had published reports in its April 1966 issue concerning Michigan State University's far-reaching involvement in covert CIA operations in Vietnam. The article, which launched a firestorm of criticism of US activities in Southeast Asia among members of the nascent anti-war movement, alleged that academics at MSU, under contract to the International Cooperation Administration (USAID's predecessor agency), had been training South Vietnamese police forces, purchasing firearms, and helping to write a new South Vietnamese constitution.[21] Johnson's recent optimism was "offset," as he put it, "by the most untimely publicity being given to the involvement of MSU with [the] CIA. I fully expect that sometime during this trip I will be charged with fronting for the CIA in an attempt to infiltrate the Nigerian planning organization, false as an accusation as that would be."[22] The contents of the article quickly became known among Nigerian policy officials, who did indeed become suspicious of USAID and the Consortium's activities in particular. Although he was not directly connected to the CIA's covert actions, Johnson later claimed that the "*Ramparts* article continue[s] to cause me considerable embarrassment" and hoped "that it [would] not lead to serious problems for the CSNRD."[23]

Later that summer, on July 29, 1966, Nigeria would see its second military coup of the year. A group of Northern NCOs and officers, dissatisfied with what they saw as an "Igbo conspiracy," captured and killed Head of State Johnson Aguiyi-Ironsi in Idaban. For three days, Nigeria had no head of government until a group of elite Northern officers selected thirty-one-year-old Yaubu "Jack" Gowan. The renewed political upheaval in the country sparked massive population movements and widespread political violence in the end of 1966 and into early 1967.

On December 15, 1966, in the Consortium's first meeting following the countercoup, the group began to discuss ways of putting the pieces of their research proposal together into a final set of recommendations to guide USAID and Nigerian agricultural planning. Noting the presence of what they called a "synthesis gap," the group struggled to show how the research group's activities fit into the Nigerian rural economy.[24] Several months earlier, Johnson had requested that the group's data be made available to researchers at MSU for use in the construction of computer simulation models. Drawing on the work of the simulation team and the consortium more broadly, Johnson announced that he was developing a flow chart that would show how USAID and Consortium activities fit into the Nigerian national economy.[25]

In March, the military governor of the Eastern Region of Nigeria, C. Odumegwu Ojukwu, declared that the regional government would take over all federal functions in the region, effectively declaring the autonomy of the Eastern Region. Upon returning to the US in early May, Johnson reported to the Consortium that "[p]rogress in general was very favorable and that work underway in Nigeria was not being impeded by internal conditions" but that staff recruiting had become difficult due to the political circumstances.[26] Furthermore, he noted that it was becoming apparent that the project's researchers needed to recognize regional differences within Nigeria, both because the recent upheavals had made ignoring regional concerns impossible, but also because Johnson thought that greater attention to these matters would result in more robust policy recommendations.[27]

The new Republic of Biafra, formerly the Eastern Region of Nigeria, was formally declared on May 30, 1967. This event marked a serious change in the tone in Consortium meetings. Johnson had returned to the US from Nigeria two days earlier to report that the political situation in Nigeria was "grim."[28] In order to respond to the changing political situation, the group developed a number of contingency plans for consortium activity in case of further deterioration. Meanwhile, Robert Brown at the USAID office in Washington reported that the USAID's policy for Nigeria was to "continue to the best of [our] ability," while acknowledging that as the "situation" there worsened, the Agency's range of action would become more restricted.[29] Johnson and Brown agreed that all future study activities would need to be "regionalized" as much as possible to benefit USAID missions in West Africa and elsewhere, a goal that would eventually come to fruition in MSU's simulation efforts.[30] The political realities continued to worsen until July 6, 1967, when civil war within Nigeria finally broke out.

Later, at the December 13, 1967 meeting of the Consortium in Chicago, it was noted with relief that internal disorder in Nigeria seemed to be "quieting down."[31] Furthermore, transit restrictions in the Western Region had been ended and consortium researchers had returned to work there. Researchers were also working again in the Northern Region, but public officials there were preoccupied with immediate "government problems" and were therefore not focused on longer-range development issues.[32] As a result of the problems associated with doing research in a war zone—plus the inability to get data from the Republic of Biafra and the unwillingness of Northern administrators to cooperate with researchers—the Consortium now focused its efforts on the Western state, although discussion lingered on ways to gather information from the Northern, Eastern, and Midwestern Regions. It was further decided that several of the subprojects should be merged due to "Nigerian contingencies." These included USAID restrictions on movement, and the need to "round out" works in progress.[33]

In the Consortium's final meeting on September 6, 1968, Jim Emerson of USAID in Washington reported that the agency was pleased with the Consortium's activities and was incorporating CSNRD reports into program materials for the Lagos mission. For reasons beyond the Consortium's control, however, political circumstances in Nigeria made implementation of the group's recommendations impossible. In addition to the fallout from the civil war, there had been a number of problems with the CSNRD's research program in Nigeria. The group had been inattentive to the politics of sectionalism in the country and this had compromised their ability to interact with regional decision-makers and administrators. This problem had been compounded by the negligible participation of Nigerians in an authorial or consultative capacity, an omission which had further undermined the project's credibility in Nigeria.

On a more methodological level, the "synthesis gap" that the group had identified had never been completely resolved, and consequently the CSNRD's final report had no real unifying theoretical or disciplinary perspective. Finally, internal strife in the country had exposed the very idea of "Nigeria" as an artificial, colonial-era political creation. As the CSNRD wound down its operations and Nigeria dissolved into a bloody civil war, researchers at MSU set about programming a software suite that would allow Nigerian and USAID officials to experiment with policy proposals that would unfold within a simulated, virtual Nigeria. This simulation was constructed around Consortium data and offered essentially the same policy recommendations as the CSNRD. However, this American-crafted model would gain enough credibility with the Nigerian military government for them to use the software to draft classified economic development planning documents in the early 1970s.

PROGRAMMING A VIRTUAL NIGERIA

Glenn L. Johnson, even before the conclusion of the CSNRD's research activities, had expressed serious doubts about the Consortium's ability to meaningfully synthesize its work using conventional methods. Early in his tenure as CSNRD director, he began to look for ways to systematize and present the group's data and recommendations. Shortly after the start of the Consortium's field work, Johnson attended a conference where Albert

Halter, an economist based at Oregon State, gave a paper presenting his work on watershed management and development using a systems simulation method. Johnson was immediately taken with the flexibility and cost effectiveness of the approach. Even more, he was impressed with the way computer simulation conveyed complex policy information to public officials, noting with excitement "[t]he print-outs look more like the menus one inspects when he goes out to a restaurant than the solution to a linear programming problem."[34] The value of this approach, aside from its low cost and high speed, was that it allowed Oregon officials to imaginatively enact and experiment with different policy alternatives and to see their administrative domain as an apolitical, unified, and closed system—all with the ease of ordering an entree from a restaurant menu. Simulations also enabled these officials to think of the future in plural terms. In a sense, the Oregon watershed simulation allowed officials to generate a slate of alternative future scenarios and identify and optimize attractive ones. By bringing present and future into one cybernetic control scheme, computer simulations aided policy planners in imagining the future as a political site of intervention and even control.

After hearing Albert Halter's conference presentation, Johnson began to investigate the possibility of using computerized systems simulation methods in the work of the Consortium. With a grant from MSU, he organized a conference to explore the feasibility of employing this methodology in the Nigerian context. Although the conference was well attended, with Wolfgang Stolper and several systems experts from the space industry in the audience, it was concluded that the simulation approach was not sufficiently developed for the Consortium to make use of it.[35] Consequently, the CSNRD continued using conventional pencil and paper methods of sector analysis.

Although the conference concluded that software suites and modeling methodologies were not ready for generalized, national-level simulations, Johnson was granted a separate USAID contract in 1967 to develop this approach. Because the Research Advisory Committee of USAID was skeptical about computerized economic analysis, it was decided that the simulation team would spend a year constructing a subsector model and that subsequent funding would be awarded only if the approach was deemed successful. The team chose to create a model of the northern Nigerian beef industry, because, as Johnson noted, "simulation requires that something from the real world be modeled . . . [and] we knew a great deal about Nigeria as a result of the CSNRD study."[36] In other words, they had the "development facts" to work with. However, there was another reason for this decision as well: the team chose to model the northern beef industry specifically because the earlier beef subproject of the CSNRD study had been canceled due to the "secessionist difficulty" in Biafra and because they wanted to make some use of data that had already been collected.[37]

In spelling out the significance of their simulation approach for USAID and Nigerian policy officials, Johnson and his simulation team pointed to a farm management simulation called "Simfarm,"[38] which had been developed by MSU's Warren Vincent to teach farm management as a dynamic, interactive game.[39] This early form of "gamification" enabled students to experiment and gain experience with management strategies without the normal computational burdens of pencil-and-paper projections. In this game, the student made management decisions, which were inputted into the computer, after which Simfarm immediately presented the students with the consequences of the user's decisions. Johnson and his coauthors saw their beef subsector model as a macro-level version of this simulation, one that would allow USAID and Nigerian officials to experiment with different policy proposals, giving them hard-to-come-by experience with policy interventions in a complex economic system.[40] Indeed, at the time of the construction of the model the Nigerian Civil War was still raging; thus the "virtual Nigeria" of the modelers enabled users to experience a unified Nigeria that did not exist in the real world.

The simulation team succeeded in convincing USAID of the value of the approach and in 1968, the agency extended the duration of the contract and tasked the group with creating a sector-wide simulation of the Nigerian agricultural economy. Thomas Manetsch, a professor of electrical engineering and systems science at MSU was an important figure in the adaptation of the Consortium's research with the systems perspective. Manetsch had written his dissertation on a simulation model of the US plywood industry, based on Jay Wright Forrester's industrial dynamics approach to systems modeling. Manetsch's model simulated the activities of several thousand firms involved in the production, retail, and wholesale sectors of forestry, incorporating a simulated price mechanism and a system-wide distributed delay process to model industrial production and information-response lags.[41] Manetsch, along with several MSU graduate students working in systems science, brought their considerable modeling expertise to the production of the Nigerian Agricultural System Simulation (NASS). Gloria Page—who was a programming instructor at MSU and, incidentally, the mother of Google cofounder Larry Page—was one of the NASS's lead programmers.

The completed national model, initially hosted on a CDC 6500 computer on MSU's campus, was capable of analyzing and investigating the consequences of various agricultural policy options and was able to take account of the implications of non-agricultural economic developments as well as interactions with the international economy. In total, the model was composed of some 3,000 equations. A forty-two-year simulation run required about a minute to execute.[42] To ensure that it would be adaptable for a variety of scenarios, the model allowed users to evaluate policy alternatives by altering an array of performance variables, including nutritional levels, total value added, export amounts, and GDP. This model for Nigeria was composed of two regional submodels, a northern one and a southern one (albeit with regions based on ecological, rather than political boundaries) and a noneconomic sector submodel. In this way, the NASS was able to sidestep the thorny problems of regional and ethnic strife by dividing Nigeria into ecological regions.[43] This ecological regionalization—while still dividing the country—did so in an "apolitical" way. These were not intended to be administrative units, rather they were imagined as reflections of different patterns of agriculture in different ecological zones.

The team that created the NASS had two sets of goals in mind. First, the team wanted to create a computer simulation that would serve as a training and experimental environment for economic development officials. In essence, they wanted the NASS to work something like a disaster drill or a war game where participants were called upon to imaginatively enact the assumptions of a military strategy or a disaster plan. The NASS was intended to work in a similar way, allowing new Nigerian and international development officials to gain experience in managing a national economy and to internalize the economic logic of the underlying model. Second, the MSU simulation team intended to establish a body of source code that could be used in other national contexts. In this way, the group hoped that their simulated Nigeria could, with slight tweaks to the model, stand in for any "Least Developed Country."

By presenting decision makers with an easily controllable interface with simplified summaries of simulation data, the MSU simulation team expected the model would gain credibility in the Nigerian policy community. To deal with this, the manufacture of credibility was built into the model construction procedure itself, with the simulation team creating an algorithmic meta-model for the design process. The designers of the NASS model imagined the construction of a system simulation as an iterative process; in their view, the model was never considered "finished," as policy-maker inputs, and new, more complete real-world data could be used to fine-tune or radically adjust the model at any point. As Figure 8.1 shows, the NASS simulation designers viewed the construction of the model as a phase in a larger decision-making process. This flowchart shows the MSU team's imagination of ideal simulation model production, with feedback from advanced stages of the process informing every earlier stage of the process in model adjustment. By integrating decision makers into the model construction process, and the simulation model into a larger policy choice process, the model builders hoped to make decision-makers invested in future model construction, evaluation, and experimentation. In the words of Glenn Johnson and George Rossmiller, by ensuring their patrons' participation throughout the design phase, "formal models can become institutionalized directly into the decision structure as part of the investigative capacity. Hence, the credibility gap often observed among decision-makers, professional analysts, and modelers is greatly diminished."[44]

In a comparable manner, the MSU simulation team also sought to appeal to decision-makers through thoughtful interface design. The group emphasized the necessity of clear variable input prompts and easily readable simulation outputs.[45] To this end, the team developed a front-end simulation

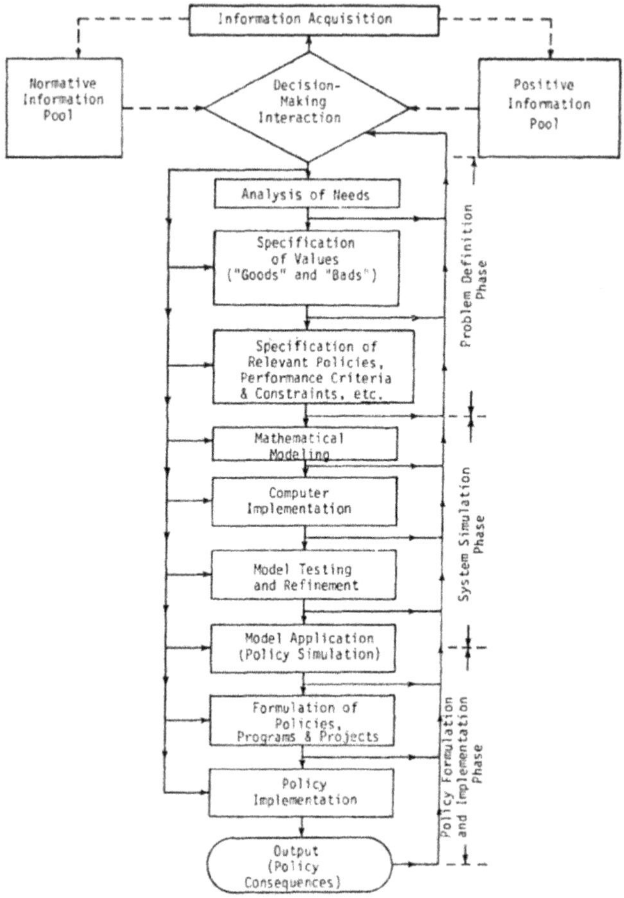

Figure 8.1 System simulation and the decision-making process, from Glenn Leroy Johnson and George E, Rossmiller, "Improving Agricultural Decision Making: A Conceptual Framework," 48.

program called the Policy Analysis Language. Designed to make computer simulation models more easily accessible to novice users, Policy Analysis Language offered a "conversational" terminal interface that would communicate with the model's FORTRAN back-end. The interface offered policy-makers two separate tracks of command entry. For the nontechnical user, Policy Analysis Language presented a series of natural-language questions, each with instructions for entry. At any point, the end-user could enter an "E" for a more detailed explanation about a specific question. Conversely, the conversational interface could be overridden and commands could be used to execute model functions if the user was familiar with the intricacies of the program's back-end.[46] On the output end, the team stressed the importance of clear and easily understandable policy data, asserting that "considerable effort often must go into the design of special tables and graphs that will readily communicate with decision-makers and evaluators."[47] By presenting decision-makers with an easily controllable interface and simple outputs, the MSU simulation team expected that the simulation would gain credibility in the Nigerian policy community.[48]

With the credibility of the modeling approach defended to the team's satisfaction, they could move on to the more phenomenological aspects of software design. Arguably, the most important affective response simulation modeling provided planning officials was the experience of experimentation. Michael Abkin, a MSU graduate student who wrote the southern submodel of the NASS for his doctoral dissertation, framed the problem facing development officials this way: "Policy making is a process immersed in uncertainty because it concerns the future. Development policy making is submerged to uncertainty's darkest depths due to the immensely complex (and thus still imperfectly understood) process of economic development."[49] By giving policy planners this experience of experimentation, regardless of the accuracy of the model, some of this perception of uncertainty is removed. The MSU simulation team justified the high cost of developing simulation models in these terms. As Kwon-Yaun Chong, one of the NASS's designers, noted, these costs were "quite small compared to the price which society will pay for mistaken policies and programs in designing the development strategy."[50] In the face of potentially disastrous social costs, this experimental enactment of policy alternatives served an important defensive function.

In order for the NASS to be successful as an experimental platform, the model builders first had to re-imagine the Nigerian agricultural economy as a unified cybernetic control system. As mentioned above, the design of the NASS incorporated two regional submodels that corresponded to ecological divisions rather than to Nigeria's ever-changing political or ethnic boundaries. This allowed policy-makers to imagine a national economy in which conflicts over federal and regional administration of agricultural policy and sectional strife could be ignored. In essence, the NASS—with its presentation of Nigeria as a diverse-but-unified country that operated as a closed, rational system—allowed policy makers to experience an idealized version of their country.

PROGRAMMING PLACELESSNESS

The systems sciences, with their focus on structural and functional isomorphies across systems, served as a perfect shuttle for transporting CSNRD-collected knowledge from locale-specific, discipline-oriented information to abstract, universal forms. The work of the sociologist of knowledge Thomas F. Gieryn speaks to the strategies that social scientists like the consortium members have used to manage issues of site specificity and generality. He distinguishes between the "field site" and the laboratory as two contrasting but interrelated frameworks by which social scientists understood their object of study as a "truth spot," Gieryn's term for a "delimited geographical location that lends credibility to claims." By "field site," Gieryn means a specific location that can be treated as possessing the qualities of a "natural object." Viewed through the frame of a field site, the location being studied becomes a natural thing. In an important sense, social scientists see field sites as not constructed, but found and observed in their "original and unsullied state." By contrast, Gieryn describes a laboratory as an artificial and controlled environment that allows for easy generalizations outside of the fixed "placeness" of a unit of analysis. He does not see these two varieties of "truth spots" as mutually exclusive—rather he argues that social scientists shifted between them depending on the sort of claim being made:

> On some occasions, the city assumes the qualities of a lab: a restricting and controlling environment, whose placelessness enables generalizations to 'anywhere', and which demands from analysts an unfeeling detachment. On other occasions, the same city becomes a field-site, and assumes different qualities: a pre-existing reality discovered by intrepid ethnographers who develop keen personal sensitivities to the uniquely revealing features of this particular place. As Chicago-the-city is textually shuttled back and forth between laboratory and field-site, the claims about metropolitan life by Chicago School authors take on credibility by being situated in the complementary legitimating languages of both truth-spots—lab and field.[51]

In a similar fashion, experts affiliated with the CSNRD strategically alternated between treating Nigeria as a field site and a laboratory. The early work of the consortium was aimed at studying and transforming the agricultural conditions of Nigeria itself. But the volatility of Nigerian national politics during the civil war made it unlikely that their recommendations would be implemented any time soon. Faced with uncertain research and professional prospects, these development experts attempted to take the lessons they learned from the Nigerian study and apply them to problems of rural development in other national contexts.

To make such claims plausible, however, they were obliged to create a new theoretical framework whereby "development facts" could be wrenched from their local, field-site specific context into a general, synthetic framework that could be applied—with some calibration—to any arbitrarily chosen development environment. The systems perspective, which stressed the importance of identifying homologies across different types of systems, was a useful way to assimilate site specific knowledge into a more universal framework. In essence, NASS transformed Nigeria from a "field site" into a development "laboratory." After designing the Nigerian simulation model, the MSU team designed two more systems simulations based closely on its experience: one for the South Korean agricultural sector, and one for that of Brazil. In designing these new simulation models, the MSU team drew upon the work begun in the Nigerian simulation, incorporating their knowledge of the Nigerian agricultural economy and their impressive software library.

From the very beginning of the NASS project—and the collapse of their data-gathering efforts in Nigeria—the MSU simulation team imagined their project as the first step to the construction of a more generic model that could be modified to suit various local conditions. Kwon-Yuan Chong noted after the completion of the NASS national model that the team's initial concern was "to develop system simulation computing features that are applicable to the analysis of the planning problems of other economies." Given MSU's considerable expertise in Nigerian agricultural economics, the team decided to use that country as a case study to test and validate the functionality of the general systems approach.[52]

Alongside the construction of South Korean and Brazilian models, USAID commissioned the systems simulation team to develop a generic software library, CLASS, or the Computer Library for Agricultural Systems Simulation. The library, which was constructed from "genericized" components from the Nigerian simulation, contained model subroutines that, with some modification, could be usable in any national context. Among the first subroutines developed for the library were a basic demographic model, a system for national agricultural accounting, and an input-output macroeconomic model, which linked the agricultural and non-agricultural sectors.[53]

The use of Nigeria as a "development laboratory" to create South Korean, Brazilian, and generic development simulation environments is just one example of experts attempting to make development knowledge portable in the face of a political crisis. The political upheavals associated with the Cold War and decolonization destabilized the "professional geographies" of many development experts. As Donna C. Mehos and Suzanne M. Moon have noted, technical experts were often forced to apply highly localized knowledge amassed after decades of research to radically different occupational environments. Consequently, many Cold War technical experts developed methods for translating location specific, "place-based" knowledge, to more "portable," generic forms of knowledge.[54] In a similar way, James Ferguson has shown how constantly rotating cadres of international development experts have applied fairly uniform policies across widely divergent development contexts, regardless of the social, political, or economic realities on the ground. Never in one place for too long, these experts lacked concrete knowledge of any specific place, but they did develop a generic, placeless kind of development expertise. As Ferguson puts it, "Tanzania may be very different from Lesotho on the ground, but, from the point of view of a 'development' agency's head office, both may be simply 'the Africa desk.' "[55] Thus, the Cold War and decolonization produced the placeless development expert, and the placeless expert produced the generic "Least Developed Country" as an object of knowledge and intervention. The generic CLASS library, which could be adapted to stand in for any developing country, promised to make the simulation knowledge developed for Nigeria placeless and portable.

CONCLUSION

As this chapter has explored, Glenn Johnson and his colleagues, anxious about the status of their professional authority, "delegated" the problem of credibility generation to an interactive computer simulation. Although the use of a computer program to build trust with an end-user might have been a new strategy, the strategy of replacing expert authority with that of the authority of a methodology was not. Theodore Porter has written extensively about the gradual displacement of expert authority and discretion by numerical measures of value. Over the course of the nineteenth and twentieth centuries, trust in an expert's knowledge and skill has gradually been replaced by trust in quantitative, rule-governed methodologies. This shift had less to do with a story of scientific or methodological "progress," and more to do with a gradual democratization of expert professions, especially in the civil service. These newer bureaucratic actors lacked the traditional, "gentlemanly" trappings of authority, and thus were more vulnerable to the scrutiny of outsiders. "The appeal of numbers is especially compelling," Porter argues, to officials "who lack the mandate of a popular election, or divine right."[56] For these experts, a decision based on explicit, "objective" measures can insulate these actors from claims of bias or partiality. "Scientific objectivity," Porter claims, "provides an answer to a moral demand for impartiality and fairness. Quantification is a way of making decisions without seeming to decide. Objectivity lends authority to officials who have very little of their own."[57]

Glenn Johnson and his colleagues faced a similar challenge. Like Porter's experts, they could not rely on the authority of their professional expertise to justify their claims. But neither could they rely on "objective" quantitative measures to augment their limited credibility. The designers of the NASS simulation recognized this and were aware of the fact that they were faced with the unenviable task of making policy recommendations in an atmosphere of mistrust with an unproven methodology. With this in mind, they sought to build the production of credibility explicitly into their simulation methodology. Thus, in many ways, the most important product of the Nigerian simulation might have been user "trust," not the quantitative data it generated.

Part of what makes interactive policy simulations so compelling to their users is that they provide the experience of choice and control in a way that deciding among traditional policy proposals does not. Users are themselves enrolled in the data production and policy choice process. They are empowered to adjust inputs and alter the settings of a simulation, in other words they are enabled to produce and experiment with possible futures. But not all futures in policy simulations are created equal. By altering the context of a decision-making process, computer simulations can lead users to arrive at different decisions without necessarily changing their underlying preferences. Simulations like the NASS circumscribe the spectrum of possible decisions and tend to privilege the "default" settings chosen by the developers.

Unlike the experts Porter describes, however, Glenn Johnson and his colleagues did not attempt to appeal to decision-makers through recourse to impersonal, "objective" numbers, at least not primarily. Instead, they drew on the subjective experience of the simulation's users to add credibility to the model. Although the research and recommendations of the CSNRD were considered by most observers (including the project's funders at USAID) to be of very high quality, the project failed to fulfill its patrons' hopes. The NASS model, conversely, enjoyed a great deal of success in Nigerian policy circles. Indeed, a long-term agricultural planning document produced by the Nigerian military government in 1974 used the NASS model to understand the prospects of the national economy.[58] As I have suggested, the two projects were based on virtually identical data gathered by the same personnel, who came to the same policy recommendations. The key to the NASS project's success in persuading Nigerian policy-makers lay more in the experiential qualities of simulation than in its empirical superiority. These affective qualities of the simulation, its designers hoped, would allow Nigerian policy makers to experiment with policy alternatives, imagine themselves in possession of an unrealistic level of political control, and think of Nigeria as a harmonious, rationally-organized, closed system. In combination, the designers of the NASS hoped that these features would make its users feel that they were, so to speak, "in the driver's seat," so that they could imagine the future as an object of experimentation, intervention, and, ultimately, control.

ACKNOWLEDGMENTS

The author would like to thank Adam Plaiss, Alexandra Lindgren-Gibson, Blake Smith, Christina C. Chiknas, Daniel Immerwahr, Deborah Cohen, Helen Tilley, Jared Poley, Joe Perry, Judith Baker, Ken Alder, Lori Baker, Lukas Rieppel, Margarita Rayzberg, Sally Olson, Steve Epstein, and Thomas Baker, as well as the members of the Science Studies Doctoral Colloquium, and the staff of the Michigan State University Archives for their advice, assistance, and support throughout this project.

NOTES

1. Theodore Porter, *Trust in Numbers: The Pursuit of Objectivity in Science and Public Life* (Princeton, NJ: Princeton University Press, 1996).
2. Daniel Immerwahr, *Thinking Small: The United States and the Lure of Community Development* (Cambridge, MA: Harvard University Press, 2014)
3. Helen Tilley, *Africa as a Living Laboratory: Empire, Development, and the Problem of Scientific Knowledge, 1870–1950* (Chicago: University Of Chicago Press, 2011)
4. Marion Fourcade, *Economists and Societies: Discipline and Profession in the United States, Britain, and France, 1890s to 1990s* (Princeton, NJ: Princeton University Press, 2009).

5. Paul N. Edwards, "The World in a Machine," in Agatha C. Hughes and Thomas P. Hughes, eds., *Systems, Experts, and Computers: The Systems Approach in Management and Engineering, World War II and After*, reprint ed. (Cambridge, MA: MIT Press, 2011), 245.; Donella H. Meadows and Club of Rome, *The Limits to Growth: A Report for the Club of Rome's Project on the Predicament of Mankind* (New York: Universe Books, 1972).
6. Andrew Lakoff, "From Population to Vital System: National Security and the Changing Object of Public Health," in Andrew Lakoff and Stephen J. Collier, eds., *Biosecurity Interventions: Global Health & Security in Question* (New York: Columbia University Press, 2008), 35.
7. For more on this from a performance studies perspective, see Tracy C. Davis, *Stages of Emergency: Cold War Nuclear Civil Defense* (Durham, NC: Duke University Press Books, 2007).
8. Larry Grubbs, *Secular Missionaries: Americans and African Development in the 1960s* (Amherst: University of Massachusetts Press, 2009), 110.
9. Wolfgang Frederick Stolper, *Planning without Facts: Lessons in Resource Allocation from Nigeria's Development: With an Input-Output Analysis of the Nigerian Economy, 1959–60* (Cambridge, MA: Harvard University Press, 1966), 6; See also James Ferguson, *Anti-Politics Machine: Development, Depoliticization, and Bureaucratic Power in Lesotho* (Minneapolis: University of Minnesota Press, 1994), 25–75.
10. Stolper, *Planning without Facts*, 6.
11. Michigan State University. Consortium for the Study of Nigerian Rural Development Records, 1964–1971. 22.9.5.15-259-16, 1st Meeting, December 18–19, 1964. Hereafter "MSU."
12. Glenn Leroy Johnson, *Burley Tobacco Control Programs: Their Over-All Effect on Production and Prices, 1933–50* (Lexington, KY: Kentucky Agricultural Experiment Station, University of Kentucky, 1952).
13. Glenn Leroy Johnson, "General, Systems, Simulation Models for Sector Analysis" (paper presented at the Seminar on the Evaluation of the Korean Agricultural Sector Simulation Model, Arlie House, Arlie, VA, 1974), 14.
14. Ibid., 16.
15. Ibid., 10.
16. Ibid., 17.
17. MSU-259-43—Report of Phase I CSNRD Planning Team, 1965.
18. For the role official figures played in this conflict, see Morten Jerven, *Poor Numbers: How We Are Misled by African Development Statistics and What to Do About It* (Ithaca, NY: Cornell University Press, 2013), 57–75.
19. MSU-263-14—CSNRD Personnel—Johnson, Glenn L—Correspondence—March–April 1966.
20. Ibid.
21. Stanley K. Sheinbaum, "The University on the Make," *Ramparts*, April 1966.
22. MSU-263-14—CSNRD Personnel—Johnson, Glenn L—Correspondence—March-April 1966.
23. Ibid.
24. MSU-259-25-10th Meeting, December 14, 1966.
25. MSU-259-24-9th Meeting, June 30, 1966.
26. MSU-259-26-11th Meeting, May 3, 1967.
27. Ibid.
28. MSU-259-27-12th Meeting, June 29, 1967.
29. Ibid.
30. Ibid.
31. MSU-259-28-13th Meeting, December 11, 1967.
32. Ibid.
33. Ibid.

34. Johnson, "General, Systems, Simulation Models for Sector Analysis," 13.
35. For more on the influence of defense and space industry intellectuals in Cold War policy making, see Jennifer S. Light, *From Warfare to Welfare: Defense Intellectuals and Urban Problems in Cold War America* (Baltimore, MD: Johns Hopkins University Press, 2005); David R. Jardini, "Out of the Blue Yonder: The Transfer of Systems Thinking From the Pentagon to the Great Society, 1961–1965," in Agatha C. Hughes and Thomas P. Hughes, eds., *Systems, Experts, and Computers: The Systems Approach in Management and Engineering, World War II and After*, reprint ed. (Cambridge, MA: MIT Press, 2001), 311–357.
36. Glenn Leroy Johnson, *A Simulation Model of the Nigerian Agricultural Economy: A Progress Report to the Agency for International Development* (East Lansing, MI: Michigan State University, 1968), 13.
37. Johnson, "General, Systems, Simulation Models for Sector Analysis," 13.
38. This simulation has no connection to the 1993 Maxis game.
39. Vincent later published his results in Warren H. Vincent, *Simfarm 1: A Farm Business Simulator and Farm Management Game* (East Lansing: Michigan State University, 1970).
40. Johnson, *A Simulation Model of the Nigerian Agricultural Economy*.
41. This was later published in article form as T. J. Manetsch, "The United States Plywood Industry—Systems Study," *IEEE Transactions on Systems Science and Cybernetics* 3, no. 2 (November 1967): 92–101, doi:10.1109/TSSC.1967.300088.
42. Michael H. Abkin, *Policy Making for Economic Development: A System Simulation Model of the Agricultural Economy of Southern Nigeria* (PhD diss., Michigan State University, 1972), 20.
43. Michigan State University Agricultural Sector Simulation Team, "System Simulation of Agricultural Development: Some Nigerian Policy Comparisons," *American Journal of Agricultural Economics* 55, no. 3 (1973): 404–405, doi:10.2307/1239121.
44. George E. Rossmiller, ed., *A Systems Approach to Agricultural Sector Development Decision-Making: Building and Institutionalizing an Investigative Capacity* (East Lansing, MI: Agricultural Sector Analysis and Simulation Projects, Michigan State University, 1980), 49–50.
45. Ibid., 119–122.
46. Claudia S. Winer and Chris Wolf, *Policy Analysis Language Version 2.3: Guide to Using a PAL Programs* (East Lansing, MI: Michigan State University, 1976), 1.
47. Rossmiller, ed., *A Systems Approach to Agricultural Sector Development Decision-Making*, 120.
48. Ibid., 443.
49. Abkin, "Policy Making for Economic Development," 221.
50. Kwon-Yuan Chong and the Agricultural Simulation Team, Michigan State University, "Some Illustrative Policy Runs Using the Nigerian Agricultural System Simulation Model for Development Planning" (paper presented at the Sixth Hawaii International Conference on Systems Science, Honolulu, HI, 1973), 5.
51. Thomas F. Gieryn, "City as Truth-Spot Laboratories and Field-Sites in Urban Studies," *Social Studies of Science* 36, no. 1 (February 1, 2006): 7.
52. Chong and the Agricultural Simulation Team, Michigan State University, "Some Illustrative Policy Runs Using the Nigerian Agricultural System Simulation Model for Development Planning," 1.
53. Agricultural Sector Analysis and Simulation Projects, Department of Agricultural Economics, and Center for International Studies, *Annual Report for Period, July 1, 1973–June 30, 1974 of the Project on Adapting and Testing of*

Agricultural Simulation Models to Sector Analysis (East Lansing, MI: Michigan State University, 1974), 67–70.

54. Donna C. Mehos and Suzanne M. Moon, "The Uses of Portability: Circulating Experts in the Technopolitics of Cold War and Decolonization," in Gabrielle Hecht, ed., *Entangled Geographies: Empire and Technopolitics in the Global Cold War* (Cambridge, MA: MIT Press, 2011), 43–44.
55. Ferguson, *Anti-Politics Machine*, 258.
56. Porter, *Trust in Numbers*, 8.
57. Ibid.
58. Rossmiller, *A Systems Approach to Agricultural Sector Development Decision-Making*, 592.

REFERENCES

Archival Sources

Michigan State University. Consortium for the Study of Nigerian Rural Development Records, 1964–1971.

Printed Sources

Abkin, Michael H. *Policy Making for Economic Development: A System Simulation Model of the Agricultural Economy of Southern Nigeria*. PhD diss., Michigan State University, 1972.

Agricultural Sector Analysis and Simulation Projects, Department of Agricultural Economics,, and Center for International Studies. *Annual Report for Period, July 1, 1973–June 30, 1974 of the Project on Adapting and Testing of Agricultural Simulation Models to Sector Analysis*. East Lansing, MI: Michigan State University, 1974.

Chong, Kwon-Yuan, and the Agricultural Simulation Team, Michigan State University. "Some Illustrative Policy Runs Using the Nigerian Agricultural System Simulation Model for Development Planning." Paper presented at the Sixth Hawaii International Conference on Systems Science, Honolulu, HI, 1973.

Davis, Tracy C. *Stages of Emergency: Cold War Nuclear Civil Defense*. Durham, NC: Duke University Press, 2007.

Edwards, Paul N. "The World in a Machine." In *Systems, Experts, and Computers: The Systems Approach in Management and Engineering, World War II and After*, edited by Agatha C. Hughes and Thomas P. Hughes, 221–254. Cambridge, MA: MIT Press, 2011.

Ferguson, James. *Anti-Politics Machine: Development, Depoliticization, and Bureaucratic Power in Lesotho*. Minneapolis: University of Minnesota Press, 1994.

Fourcade, Marion. *Economists and Societies: Discipline and Profession in the United States, Britain, and France, 1890s to 1990s*. Princeton, NJ: Princeton University Press, 2009.

Gieryn, Thomas F. "City as Truth-Spot Laboratories and Field-Sites in Urban Studies." *Social Studies of Science* 36, no. 1 (2006): 5–38, doi:10.1177/0306312705054526.

Grubbs, Larry. *Secular Missionaries: Americans and African Development in the 1960s*. Amherst: University of Massachusetts Press, 2009.

Immerwahr, Daniel. *Thinking Small: The United States and the Lure of Community Development*. Cambridge, MA: Harvard University Press, 2014

Jardini, David R. "Out of the Blue Yonder: The Transfer of Systems Thinking From the Pentagon to the Great Society, 1961–1965." In *Systems, Experts, and*

Computers: The Systems Approach in Management and Engineering, World War II and After, edited by Agatha C. Hughes and Thomas P. Hughes, 311–357. Cambridge, MA: MIT Press, 2001.

Jerven, Morten. *Poor Numbers: How We Are Misled by African Development Statistics and What to Do About It.* Ithaca, NY: Cornell University Press, 2013. Johnson, Glenn Leroy. *Burley Tobacco Control Programs: Their Over-All Effect on Production and Prices, 1933–50.* Lexington, KY: Kentucky Agricultural Experiment Station, University of Kentucky, 1952.

Johnson, Glenn Leroy. "General, Systems, Simulation Models for Sector Analysis." Paper presented at the Seminar on the Evaluation of the Korean Agricultural Sector Simulation Model, Arlie House, Arlie, VA, 1974.

Johnson, Glenn Leroy. *A Simulation Model of the Nigerian Agricultural Economy: A Progress Report to the Agency for International Development.* East Lansing: Michigan State University, 1968.

Johnson, Glenn Leroy and George E. Rossmiller. "Improving Agricultural Decision Making: A Conceptual Framework." In *A Systems Approach to Agricultural Sector Development Decision-Making: Building and Institutionalizing an Investigative Capacity*, edited by George E. Rossmiller. East Lansing, MI: Agricultural Sector Analysis and Simulation Projects, Michigan State University, 1980.

Lakoff, Andrew. "From Population to Vital System: National Security and the Changing Object of Public Health." In *Biosecurity Interventions: Global Health & Security in Question*, edited by Andrew Lakoff and Stephen J. Collier, 33–60. New York: Columbia University Press, 2008.

Light, Jennifer S. *From Warfare to Welfare: Defense Intellectuals and Urban Problems in Cold War America.* Baltimore: Johns Hopkins University Press, 2005.

Manetsch, T. J. "The United States Plywood Industry—Systems Study." *IEEE Transactions on Systems Science and Cybernetics* 3, no. 2 (November 1967): 92–101, doi:10.1109/TSSC.1967.300088.

Meadows, Donella H., Dennis L. Meadows, Jørgen Randers, and William L. Behrens. *The Limits to Growth: A Report for the Club of Rome Project on the Predicament of Mankind.* New York: Universe Books, 1972.

Mehos, Donna C. and Suzanne M. Moon. "The Uses of Portability: Circulating Experts in the Technopolitics of Cold War and Decolonization." In *Entangled Geographies: Empire and Technopolitics in the Global Cold War*, edited by Gabrielle Hecht, 43–74. Cambridge, MA: MIT Press, 2011.

Michigan State University Agricultural Sector Simulation Team. "System Simulation of Agricultural Development: Some Nigerian Policy Comparisons." *American Journal of Agricultural Economics* 55, no. 3 (1973): 404–419, doi:10.2307/1239121.

Porter, Theodore. *Trust in Numbers: The Pursuit of Objectivity in Science and Public Life.* Princeton, NJ: Princeton University Press, 1996.

Rossmiller, George E., ed. *A Systems Approach to Agricultural Sector Development Decision-Making: Building and Institutionalizing an Investigative Capacity.* East Lansing, MI: Agricultural Sector Analysis and Simulation Projects, Michigan State University, 1980.

Sheinbaum, Stanley K. "The University on the Make." *Ramparts*, April 1966.

Stolper, Wolfgang Frederick. *Planning without Facts: Lessons in Resource Allocation from Nigeria's Development: With an Input-Output Analysis of the Nigerian Economy, 1959–60.* Cambridge, MA,: Harvard University Press, 1966.

Tilley, Helen. *Africa as a Living Laboratory: Empire, Development, and the Problem of Scientific Knowledge, 1870–1950.* Chicago: University Of Chicago Press, 2011.

Vincent, Warren H. *Simfarm 1: A Farm Business Simulator and Farm Management Game*, East Lansing: Michigan State University, 1970.

Winer, Claudia S., and Chris Wolf. *Policy Analysis Language Version 2.3: Guide to Using a PAL Programs.* East Lansing, MI: Michigan State University, 1976.

9 Pan-Africanism, Socialism and the Future

Development Planning in Ghana, 1951–1966

Jeff Grischow and Holger Weiss

In 1964, the Government of Ghana published the Seven-Year Development Plan designed to propel the nation into economic self-sufficiency. Ghana's Plan unfolded around the same time as the Nigerian experience with Wolfgang Stolper's ideas about development planning during the 1960s, presented by Baker in this volume, but the process and outcomes of planning in the two countries differed significantly. Stolper was the first in a series of American-trained development experts recruited by the Nigerian Government during the 1960s. In the absence of reliable statistics and facts on the ground, Stolper advocated a process of decentralized and flexible planning, an approach that he formalized in his 1966 book entitled *Planning Without Facts*.[1] In Nigeria this process would be directed primarily at rural development and the capitalist modernization of small-scale agriculture. In contrast, the Ghanaian government turned to socialist development economists including the Hungarian Marxist Joszef Bognar. The Ghanaian chair of the National Planning Commission, J.H. Mensah, was an avowed Marxist, as was most of the inner circle of the governing Convention People's Party, including President Kwame Nkrumah. Under these planners, Ghana's 1964 Plan laid out a centralized, socialist blueprint for rapid industrialization, although Nkrumah also negotiated with Western governments to develop a hydroelectric dam on the Volta River as a foundation for Ghanaian industry.

The Ghanaian Seven Year Plan was anchored in Nkrumah's vision of the future of Africa as a united federation of independent (and preferably socialist) republics. Behind this vision lay a radical form of the ideology of Pan-Africanism, which Nkrumah developed during the 1940s and 1950s as protection against an alternative future of neocolonial domination by the US and/or the Soviet Union. Nkrumah's Pan-Africanism emerged out of his links to a trans-Atlantic network of Pan-Africanists. Among this group, Nkrumah was influenced particularly by the Jamaican Marcus Garvey and Trinidadians C.L.R. James and George Padmore. Nkrumah never met Garvey, who radicalized Pan-Africanism in the early twentieth century through the idea of "Africa for the Africans," but he encountered Garveryism during the 1920s as a student and teacher in the Gold Coast (colonial Ghana).[2] The other two Nkrumah knew personally as a result of his sojourns to the US and England during the 1930s and 1940s. Both James and Padmore were more radical than Garvey, having come to socialism in the wake of the Russian Revolution. Garvey, James and Padmore provided an important intellectual foundation for Nkrumah's "future vision" of a united Africa that underpinned Ghana's Seven-Year Development Plan, and we focus, in this chapter, on the way that this ideological vision of the future shaped the Plan.

PAN-AFRICANISM AND ITS VISIONS OF THE FUTURE

Nkrumah considered national economic development and industrialization as a necessary stepping stone to continental independence and unity. In this manner the exercise of planning exerted a performative function in achieving Nkrumah's quest to "forge the future" of a Pan-Africanist continent free of neocolonial interference by the superpowers. Because the ideology of Pan-Africanism lay behind Nkrumah's project, we will begin with an analysis of the "future visions" of Pan-Africanist doctrines before moving on to Ghana's experience with development planning.

Pan-Africanism calls for the unification of all Africans into a single African state to which those in the African diaspora can return.[3] It promotes transnational solidarity, with a desire to use political, cultural, economic and religious means to achieve the goal of unifying all Africans.[4] Although the doctrine can be traced back to antislavery activism in the 1700s, its founding as a coherent ideology occurred between the creation of the African Association in London, England, in 1897 and the first Pan-African Congress in London in 1900. At this time, the most prominent Pan-Africanists (including Henry Sylvester Williams[5] and W.E.B. DuBois[6]) espoused a moderate, liberal vision of racial equality, an end to the colour bar, and economic opportunities for all people of African descent.[7] On the more radical side, some early Pan-Africanists such as Marcus Garvey (1887–1940) added a strong dose of anticolonialism to the mix. Born in Jamaica, Garvey began to speak out against colonialism while living in England in 1912–1913, lobbying for independent African states controlled by leaders of African descent. After returning to Jamaica, Garvey founded the United Negro Improvement Association in 1914 to promote his vision of Africa for the Africans.[8] During the 1920s and 1930s, other Pan-Africanists radicalized Garveyism as they adopted communist and socialist doctrines after the emergence of the Soviet Union.[9] In fact, most Africans who were converted to Pan-Africanism during the 1920s and 1930s also embraced socialism. In Britain, the Labour Party and the Fabian Society lent their support to Pan-Africanism, whereas many Pan-Africanists were exposed to the writings of socialist scholars such as Harold Laski. In the US and the European capitals, African students expanded the frontiers of Pan-Africanism with ideas and mobilization. Many of them rejected the colonial doctrine of assimilation and advocated radical political movements and independence for all African countries.[10] Three of the most important figures from this movement, for the purposes of our chapter, were George Padmore, C.L.R. James and I.T.A. Wallace-Johnson.

The Trinidadian-born Padmore (1902–1959) had moved to the US in 1924 and joined the American Communist Party in 1927. Padmore subsequently relocated to Moscow and, before a rift with the Comintern in 1933/1934, served as the main architect of a radical African Atlantic. As the secretary of the International Trade Union Committee of Negro Workers from 1931 to 1933, Padmore is remembered as the most influential Black Communist during this period. Known for being an ardent critic of both African American and African intellectuals, Padmore's first major book-pamphlet, *The Life and Struggles of the Negro Toiler* (1931), projected a communist-led, class-conscious revolutionary movement for the independence of the African colonies. At this time, he criticized reformist Garveyism as the most reactionary expression in African Atlantic bourgeois nationalism.[11] By 1933, though, Padmore had become disillusioned with the Communists and after his break with the Comintern started to seek collaboration with W.E.B. DuBois (1868–1963), the then leader of the Pan-African movement in the US. Padmore's outlook on the colonial problem also changed and he rejected his earlier negative stance toward the African American and African intellectuals as "bourgeois-reformist" nationalists.[12] Although he denounced a communist solution for the African future, Padmore remained a convinced socialist.[13] The change in Padmore's aspirations is best viewed through the international congresses he had planned in the 1930s and 1940s. While the First World Conference of Negro Workers, held in Hamburg in 1930, articulated a class-conscious approach and directed itself only to the workers in the African Atlantic, his (aborted) joint call with Garan Kouyaté's for a "World Negro Conference" in late 1933 was already echoing a Pan-African vision including a common platform for all anticolonial agitators, be they radical or bourgeois.[14]

When Padmore moved to London in 1934, he connected with another radical Pan-Africanist, C.L.R. James, whom he had known in Trinidad. The Trotskyist James had become involved in radical Pan-Africanism after moving to Britain in 1932. In 1935, he became chair of the International Friends of Abyssinia, established to support Ethiopian independence after the Italian invasion of Ethiopia. In 1937, Padmore and James helped to found the International African Service Bureau, an organization dedicated to promoting self-government for the African colonies.[15] The Bureau's founders also included I.T.A. Wallace-Johnston, a Sierra Leonian Marxist who had been a close associate to George Padmore and even studied for a short period in 1932/1933 in Moscow. Wallace-Johnson had moved to the Gold Coast in late 1933 and established the West African Youth League, a radical political platform, in 1934.[16] By the 1940s the IASB attracted a cluster of radical Pan-Africanists whose vision of the future focused on anticolonialism and socialism for the independent African states.

While Garvey, James and Wallace-Johnson developed their ideologies, the younger Kwame Nkrumah was introduced to Pan-Africanism between 1926 and 1935 while a student and then a teacher in the Gold Coast. During these years, Nkrumah personally knew several prominent nationalist/Pan-Africanist activists, including James Aggrey (his teacher at the Achimota Teaching College), the Nigerian nationalist Nnamdi Azikiwe, and Wallace-Johnston, who at the time lived in the Gold Coast. Although Garvey's United Negro Improvement Association did not have a branch in the Gold Coast, it did operate in the US, and Nkrumah would have been familiar with Garvey's ideas through people such as Aggrey and Azikiwe who had studied at Lincoln University in Pennsylvania.[17] Nkrumah followed in their footsteps, moving to the US in 1935 to study at Lincoln and then enrolling in graduate studies at the University of Pennsylvania. While at Pennsylvania, Nkrumah joined Garvey's Negro Improvement Association in New York (albeit after Garvey had been deported from the US), and he developed a close relationship with C.L.R. James between 1943 and 1945.[18] During this period, Nkrumah developed his political vision, focused on creating a federated union of independent West African states "where Africans can rule and govern themselves without outside interference."[19]

By late 1943, Nkrumah also had absorbed some very radical anticolonial, Marxist and socialist ideas. In an article published in November 1943, for instance, he looked toward a world free of imperialism, referring to "the masses of African peoples . . . rising in consciousness and solidarity" to secure a future with "a more balanced development of Africa's economy on the basis of the political, economic, social and educational needs of the African population." Nkrumah dreamed of a new Africa born out of a dialectical process of development in which opposite cultures collided and produced "a new thing—a new culture, a new education, or a new national life."[20] Radical anticolonial Marxism also permeated Nkrumah's dissertation proposal at Pennsylvania, which was rejected by the committee members as "too pro-communist" for the University. It is possible, perhaps even likely, that Nkrumah's radical turn grew out of his association with James and the American communist Claudia Jones.[21] In any case, by the time Nkrumah left the US for England in 1945, he was firmly committed to a vision of Africa's future in the form of a united, independent federation of African socialist states.[22]

Nkrumah arrived in England in May 1945, originally to study law, but quickly became caught up in the Pan-African circles surrounding George Padmore in London. Nkrumah met Padmore through James, who gave him a letter of invitation before Nkrumah left the US.[23] Nkrumah and Padmore became very close, and initially Padmore seemed to have a moderating influence on Nkrumah's more radical beliefs. By 1945, both men appeared to regard Pan-Africanism as an ideology in its own right, neither communist nor capitalist, but rather, a political force for uniting Africa against colonialism and imperialism.[24] In other words, anticolonialism emerged as the dominant theme. This idea influenced the resolutions of the Fifth Pan-African Congress in Manchester in October 1945, which Padmore asked Nkrumah to organize. The resolutions passed by the Congress reflected Padmore's "middle path" ideology. The main resolutions called for immediate decolonization for some colonies and increased political rights, the abolition of forced labour, and the accumulation of wealth from Africa into African hands.[25] Although Nkrumah played a prominent role in the Congress, he came away criticizing the resolutions as too moderate. In response, after the Congress Nkrumah established the West African National Secretariat, which included radical socialists and communists like Wallace-Johnson and Kweku Bankole Awooner-Renner, an outspoken communist from the Gold Coast. At this time, Nkrumah's socialism was more radically Marxist-Leninist than Padmore's ideology, which was more Trotskyist-socialist. This upset Padmore, who believed that that the Secretariat was linked closely with Moscow and the radical Marxist and communist left in Britain.[26]

In fact, it is likely that at this time the communist Awoonor-Renner influenced Nkrumah more strongly than Padmore, as shown, for instance, in Awooner-Renner's pamphlet *West African Soviet Union* published by the West African National Secretariat. The pamphlet advocated complete and absolute independence and a Federated People's State of West Africa, which in Awooner-Renner's mind comprised all regions from the Atlantic and the Congo Basin to the Nilotic Sudan and Kenya (including former British, French, Spanish and Portuguese colonies). This state was to be modelled upon the Soviet example: "All means of production would be in the hands of the state. And all forms of industries would be encouraged and nationalized, except small enterprises."[27] For his part, Nkrumah recruited supporters from the West African National Secretariat into another organization, the Circle, a secret society dedicated to achieving an independent West African federation of socialist states. Nkrumah envisioned the Circle as a kind of vanguard party along Leninist lines, and he required its members to swear an oath accepting him as the leader.[28] In the unpublished Circle manifesto, Nkrumah's went one step further than Awoonor-Renner in his vision of Africa's future. For Nkrumah, the Federated State of West Africa was only the first step in the eventual creation of a Union of African Socialist Republics.[29]

Nkrumah's time in London ended in November 1947, when he returned to the Gold Coast to take up a post as the general secretary of the United Gold Coast Convention, a new, moderate nationalist organization.[30] Awoonor-Renner followed in 1948 and threw his support behind Nkrumah's activities, culminating in a break with the United Gold Coast Convention, the founding of the quasi-socialist Convention People's Party and a campaign of "positive action" against colonial rule in late 1949 and early 1950. In 1951, however, Awoonor-Renner broke with Nkrumah and the Convention People's Party to found the Moslem Association Party and for six years contested elections, ostensibly seeking to unseat the Convention People's

Party and bring more effective representation for Ghana's Muslims. However, according to Sean Hanretta, Renner's split with Nkrumah may have reflected deeper tensions. Renner had spent the period from 1948 to 1950 lobbying Nkrumah to take a more orthodox communist position instead of the populism toward which he was drifting.[31] But that populism allowed Nkrumah to win a string of national elections. In 1951, he became the leader of government business under an arrangement of semi-independence. Further elections in 1954 and 1956 paved the way for Nkrumah to become prime minister of independent Ghana in 1957.[32]

After the break with Renner, George Padmore became Nkrumah's main political advisor and source of inspiration. In December 1957, Nkrumah nominated George Padmore as his special advisor on African affairs. His role was to carry through Nkrumah's vision of Pan-Africanism. Padmore was first and foremost to carry through Nkrumah's policy for the emancipation of those parts of Africa still under foreign rule and to work with nationalist movements and parties which would be inappropriate for the Ghanaian Foreign Ministry officially to engage in. In addition, Padmore guided the National Association of Socialist Students Organizations, created as a socialist association for Gold Coast students in London.[33] Padmore wrote several articles for Nkrumah's newspaper, the Accra Evening News, and analysed the road to independence in Ghana in his book *The Gold Coast Revolution* (1953). Three years later, Padmore published his blueprint for African independence, *Pan-Africanism or Communism?* (1956), where he presented his vision for a non-aligned, unified and socialist union of independent states in Africa—an idea that would become Nkrumah's raison d'etat. African socialism was thereafter to emerge as the label of a political philosophy propagated by several African statesmen, including Leopold Senghor of Senegal, Sekou Touré of Guinea, Julius Nyerere of Tanzania, Modibo Keita of Mali and Hastings Banda of Malawi. Being proposed as an alternative to both Western capitalism and European and Chinese socialism, the term was to encompass a range of different views and idea to economic and political development but all based on the idea that African socialism derived from ancient, uniquely African traditions. The vision was to modernize traditional collectivism and to apply it to contemporary society.

Invited by Nkrumah, Padmore moved to Ghana in 1957 and acted as Nkrumah's special advisor on African affairs until his death in 1959. However, Padmore's nomination would lead to contestation, and this contestation is indicative of the power struggle over the visions of Pan-African socialism in the Ghanaian context. Padmore's nomination was objected to especially by two groups. Some top civil servants in the Foreign Ministry resented the nomination as they—rightly—felt that his new office was out of the control of the Foreign Ministry, whereas some members of the Convention People's Party criticized his role in the National Association of Socialist Students Organizations as he mocked some Convention People's Party members over their phony socialism. However, such criticism had little impact on Nkrumah, and Padmore became an important voice in Ghana.[34] In April 1958, Nkrumah—with Padmore's help—organized the Conference of Independent African States in the Ghanaian capital, Accra. With Padmore's guidance, the aim of the conference was to promote the unity of the independent African nations, including co-ordinated financial policies and united stances on global issues. The delegates came from the handful of African states that were independent as of 1958: Morocco, Tunisia, Libya, Sudan, Ethiopia and the United Arab Republic (Egypt and Syria).

Under the guiding hand of George Padmore, the resolutions of the April 1958 Conference echoed the Manchester Pan-African Congress of 1945, including decolonization and continental co-operation in development.[35] A few months later, Nkrumah and Padmore organized a broader All-African People's Conference as a follow-up to the Conference of Independent African States. The All-African People's Conference included representatives from countries still under colonial rule to visit Accra to discuss strategies for decolonization and African unity.[36] If the 1945 Manchester Conference endorsed the dogma of Pan-African socialism,[37] the 1950s also witnessed a rehabilitation of Garveyism. Whereas Padmore during his communist period rejected Garveyism, Nkrumah claimed in his autobiography that Garvey's writings influenced Padmore more than anything else during his stay in the US, and the 1958 All-African People's Conference coupled him as an outstanding contributor to the spread of the idea of Pan-Africanism. Nkrumah's rehabilitation of Garvey marked a shift of the Pan-African movement from the US to Africa: as long as it remained a largely African American dogma for combatting the colour line, Garvey's 1920 plea "Africa for the Africans" was rejected as a petty-bourgeois concept. Nevertheless, the idea of Africa for Africans was found useful in the anticolonial visionary rhetoric of the African nationalists that were to spearhead the Pan-African movement after 1945.[38] Padmore's short sojourn marked, in Nkrumah's words, "the move of Pan-Africanism to Africa."[39] Indeed, Nkrumah publicized his Pan-Africanist views on African liberation in his autobiography, strongly inspired by Padmore,[40] and he ended the book with the declaration that Ghana had a "duty as the vanguard force" to act for the liberation of the entire African continent.[41] In Nkrumah's mind, this liberation project required national economic development in Ghana in order to lay the foundation for a strong federation of African states.

Nkrumah was to develop a distinct vision of Pan-Africanism that was both trans-Atlantic and trans-Saharan. In his most-cited work, *Consciencism: Philosophy and Ideology for Decolonization* (1964), he elaborated a detailed Pan-Africanist philosophy and political ideology, declaring Africa to be a "new society" with a multiple heritage of indigenous, Islamic, and Euro-Christian beliefs that must all be explored to create "harmonious growth and development."[42] In his 1961 book, *I Speak of Freedom: A Statement of African Ideology*, Nkrumah identified neoimperialism and neocolonialism to be the main threats to the newly independent African states.

Liberation now had to pass through economic development. Although politically independent, economic links to the West prevented economic development in Africa as the world market dictated the prices of the goods produced in African states. Pan-Africanism was also put forward as an intervention into the Cold War world order. Cold War politics meant that both former colonial powers and the US meddled in African politics. In Nkrumah's mind, the solution to these economic and political problems lay in the practical realization of Pan-African ideals. According to him, the United Nations Organisation could not offer a guarantee for the independence and territorial integrity of all states, as the organisation mirrored Cold War interests; it was "just as reliable an instrument for world order and peace as the Great Powers are prepared to allow it to be." Because political unification of African states was a claim to overcome fragmentation, Nkrumah's vision emphasized not a loose confederation designed for economic co-operation but a strong political union. "The economic development of the continent must be planned and pursued as a whole".[43]

In 1963, Nkrumah made a push for such a union at a Pan-African Conference in Addis Ababa, Ethiopia, which resulted in the formation of the Organization of African States. In preparation for the conference, Nkrumah circulated his views beforehand in a publication entitled *Africa Must Unite* (1963), which set out his radical Pan-Africanist vision. Ama Biney calls the book Nkrumah's "foremost documented intellectual appeal for continental unification."[44] In the book, Nkrumah sets out his idea that Africa could only escape neocolonial domination through a strong, united continental government that could co-ordinate economic programs, defence policy and foreign relations. He envisioned a bicameral African parliament, "with a lower house to discuss problems facing Africa and an upper house to assure equality of the associated states."[45] Nkrumah carried this vision forward into the Addis Ababa conference, which met in May 1963. At the meeting, Nkrumah argued that political union was necessary as soon as possible, because it was a prerequisite for social and economic development and freedom. He also made the case for a strongly centralized form of continental government. The outcome of the meeting, although it established the Organization of African Unity, disappointed Nkrumah because it represented a compromise between the radicals who supported him and the moderates who wanted a looser union and more regional and national autonomy. As a compromise, the Charter of the Organization of African Unity emphasized the sovereignty of African nation-states by upholding the principle of non-interference in domestic affairs. Despite this setback, Nkrumah continued to push for Pan-Africanism at the second Organization of African Unity summit in Egypt in 1964. At the summit, Nkrumah argued with Tanzanian President Julius Nyerere, who made a case for a gradualist approach to an African Union government that did not rule out regional groupings in the short and medium terms. By the end of the conference, most delegates agreed with Nyerere and they considered Nkrumah's idea for immediate centralization utopian and premature. Undaunted, Nkrumah returned home and spent £8 million to build a conference centre for the Organization's summit of 1965. Once again, however, he failed to build a consensus for his vision and the meeting disbanded without a commitment to developing the machinery for an African Union government.[46]

According to Toyin Falola, Nkrumah blamed capitalism for the imposition of colonial rule in Africa; for many of the excesses of European government, including the exploitation of Africans; and for the establishment of Western monopoly capitalism, which took more resources out of African than it brought and which had forced Africans to participate in world wars that had little to do with them. He believed that colonial exploitation had given rise to the possibility of creating a class-conscious group of workers who would be prepared to fight capitalism and colonialism and liberate themselves. Creating his own version of "African socialism," a vision that rested on the idea of developing an "African personality," Nkrumah envisaged an industrial society that would do away with dependence on the West. In particular, he called for African farmers to move away from production for export to production for internal consumption.[47] At the same time, the Convention People's Party would develop urban industries that would turn Ghana into an economic powerhouse for the whole continent. In creating this process of independence, development planning would play an important role. The next section will take up the question of the types of planning used by Nkrumah to in order to realize this dream of Ghana as the engine of African liberation.

DEVELOPMENT PLANNING IN GHANA, 1951–1966

Nkrumah's election in 1951 as Leader of Government Business gave him official control over national development, but the Gold Coast remained a colony between 1951 and 1957. During these years, the country's colonial status constrained Nkrumah's vision of development planning. One of the results was that he inherited a ten-year colonial development plan drafted by the colonial administration in 1950 to solicit funds under Britain's Colonial Welfare and Development Act of 1945. The colonial plan proclaimed an end goal of higher living standards[48] and focused on infrastructure and social welfare, but it was not very systematic. Instead, the ten-year consisted of a "shopping list" of priorities suggested by individual Departments (Agriculture, Health, Education, et cetera). In 1951, Nkrumah telescoped the colonial plan to five years but he did not alter its content.[49] The result was Nkrumah's "First Development Plan," which targeted infrastructure and social services rather than industrial development.[50]

Nkrumah's tone changed dramatically after he became prime minister of the fully independent Ghana in 1957. Most famously, he wrote in his *Autobiography* that "what other countries have taken three hundred years or more to achieve, a once dependent territory must try to accomplish this in a generation if it is to survive. Unless, as it were, 'jet-propelled,' it will lag behind and thus risk everything for which it had fought."[51] The goal of rapid industrialization would be achieved through two main initiatives: the Volta River Project—a hydroelectric dam on the Volta River first proposed during the colonial era—and the industrialization of the Ghanaian economy through a process of development planning.[52] Although Nkrumah called himself a Marxist socialist in his autobiography,[53] the scale of his development vision initially steered him toward Western support. In part this decision was pragmatic, especially with respect to the massive Volta River Project, which consisted of a hydroelectric dam, a bauxite-extraction facility and aluminium smelter, and a new harbour at a fishing village called Tema. The scale of the project demanded a level of foreign investment that could only come from the West.[54] Nkrumah thus went cap-in-hand to the American government, and after a series of discussions involving President Eisenhower and Edgar Kaiser (the head of the Kaiser Corporation, a major player in the aluminium industry), the project came together between October 1957 and August 1960. Kaiser created a consortium called the Volta Aluminum Company (Valco) to manage the project, and promises for financing came from the World Bank ($40 million), the United Kingdom ($14 million) and the US ($20 million).[55]

While Nkrumah negotiated the Volta River Project agreement, he also set out to create his Second Development Plan. For this he recruited W. Arthur Lewis, a West Indian economist trained at the London School of Economics, who came to Ghana in October 1957 on a two-year contract sponsored by the United Nations.[56] Lewis's economic vision was influenced heavily by Fabianism, a non-Marxist variant of socialism that influenced British colonial policies during the 1940s. Between 1943 and 1945, Lewis worked out his Fabian ideas as an economic advisor at the Colonial Office, and by the early 1950s he had worked out a model of African economic growth focusing on small-scale subsistence farming as a basis for industrialization. As peasants became more productive, rural labour would be released for industry. The development of industry then would feed back into farming through the growth of the market and the supply of cheap goods. Raising capital for this project would depend on a significant inflow of foreign investment from Britain and other Western countries. True to his Fabian sympathies, Lewis argued that this process could not be left to the forces of *laissez faire*, but rather had to be achieved through careful state planning.[57] In 1952, Nkrumah gave Lewis an opportunity to apply his ideas in practice by inviting him to the Gold Coast to produce a report on industrialization.[58]

Back in Ghana in 1957, Lewis worked with the Australian Robert Jackson[59] to draft Ghana's Second Five-Year Plan, which prioritized agricultural development as a basis for industrialization, financed by large inflows of foreign investment from Western sources. Nkrumah accepted the Plan, but he and the radical left-wing of the Convention People's Party increased its projected expenditures from £70–80 million (recommended by Lewis and Jackson) to £230 million.[60] Feeling that politics had trumped sound economic advice, Lewis resigned in frustration in December 1958.[61] However, the Second Development Plan unveiled by the Convention People's Party in 1959 retained his core ideas in prioritizing agricultural development, welcoming foreign capital from the West, and advocating a mixed economy with little direct state ownership. George Padmore liked its mixed approach, which he characterized as incorporating elements of both capitalism and communism.[62] For his part, Nkrumah hoped that the Second Plan would be practical, yet produce quick results to help jet-propel Ghana to economic self-sufficiency.[63]

This did not happen. In fact, between 1959 and 1961 the growth rate under the Second Development Plan failed to cover its expenditures and produced a £53 million deficit.[64] By this time, Lewis had left and Padmore had died (1959), which along with the Plan's failure allowed Convention People's Party radicals to increase their influence. Most importantly, Padmore's death opened up an opportunity for Tawia Adamafio—a former leader of the National Association of Socialist Students Organizations and self-proclaimed supporter of the Soviet Union—whom Nkrumah appointed general secretary of the Convention People's Party in June, 1959.[65] Under Adamafio's leadership, the Convention People's Party became more radically socialist and the Government began to tilt toward the Soviet Union.[66] During 1960, two Ghanaian delegations visited (May and August), securing promises of Soviet aid,[67] and Nkrumah met Khrushchev for the first time at the UN General Assembly in New York (September). In New York, Nkrumah promised the Soviet leader that "there was no other way for Africa than the way of socialism."[68] These meetings culminated in a Ghana-Soviet agreement signed in December.[69]

Nkrumah's leftward shift continued during 1961. Brezhnev, who at that point was chairman of the Presidium of the Supreme Soviet, visited in February and participated in the opening ceremony of Nkrumah's new ideological school, the Winneba Institute. In March, Nkrumah demoted the conservative Komla Gbedemah from the Ministry of Finance to the Ministry of Health. On April 8, possibly under Adamafio's influence, Nkrumah made a Dawn Broadcast in which he urged the population to accept socialism and promised to weed out corrupt Convention People's Party members—whom Nkrumah identified as pro-capitalist. After the Dawn Broadcast, Nkrumah once again purged the conservative wing of the Convention People's Party.[70] Between July and September 1961, Nkrumah visited the Soviet Union, the Eastern Bloc and China with a large delegation that included most of the Convention People's Party's left wing.[71] The trip convinced Nkrumah that Ghana needed scientific socialism adapted from the Soviet system if Ghana were to industrialize and become self-sufficient.[72] For instance, speaking in Moscow in 1961, Nkrumah declared that "your tremendous successes in industrial and agricultural development and the organization that helped to achieve them are now an important lesson not only for us, Ghanaians, but for the whole Africa."[73] In Crimea, Nkrumah told the Soviet Minister of Culture that he would create a new development plan based on "the experience of the USSR and other socialist countries."[74] Nkrumah also believed that the Soviet model of bringing together a large number of ethnic groups could be used as a model for a Pan-African Federation, and he spoke of "the 'Leninization' or the combination of anti-imperialism and the Soviet model of nationalism in his future vision of Africa."[75]

Nkrumah's "Soviet turn" signalled Ghana's formal break with the non-Marxist, reformist ideas of W. A. Lewis and the development models favoured by Fabian colonial officials during the 1940s. After returning to Ghana, Nkrumah recruited the Hungarian Marxist economist Joszef Bognar, formerly the Karl Marx Professor of Trade at Budapest University, for advice on development planning. He also created the National Planning Commission to oversee the drafting of a new development plan, controlled by leftists H. Miller-Craig, Joseph H. Mensah and T. T. Naer. Under Bognar's influence, the Commission worked on the new Plan during 1961 and early 1962.[76] The resulting planning process contrasted sharply with the Nigerian Government's decision to sponsor a fact-finding mission between 1965 and 1967 to produce adequate data as a foundation for planning.[77] In Ghana, the planners started at the end, identifying the growth rate deemed necessary to stimulate industrialization, then developed the Seven-Year Plan to meet this growth rate. Also, in contrast to Lewis's focus on agriculture—which also was the focus of Nigerian plan and the University of Michigan's computer simulation—Ghana's Seven-Year Plan envisioned a "big push" for rapid industrialization.[78]

As we have noted, Nkrumah's brand of socialism and Pan-Africanism prioritized economic growth as a means to achieving national sovereignty and African unity under a federation of socialist republics. The Seven-Year Plan therefore emerged primarily as a tool for Convention People's Party's preconceived vision of the future. The Party publicly announced this vision in 1962 when Nkrumah and Adamafio presented the Program for Work and Happiness. The Program outlined the foundation for the new Seven-Year Plan and its goal of laying the foundation for "a socialist state devoted to the welfare of the masses."[79] According to the Program, the Convention People's Party inherited a poor nation and faced the task of rapidly developing the productive power of the country. Because colonialism had deprived the country of "facilities for capital formation, it was clear from the very beginning that this prodigious task could only be accomplished successfully by the institution of socialism." As such, the Convention People's Party would prioritize "national economic planning based on the public ownership of the means of production and distribution." The planning process would be controlled by the government through the National Planning Commission, which would utilize scientific principles, including statistical information to guide planners. As a foundation, however, the success of planning pre-supposed the complete electrification of Ghana through the Volta River Project.[80]

Behind the scenes, Bognar and the National Planning Commission had worked to compile a draft Plan (written by J. H. Mensah) in 1962. The Plan took the idea of economic jet propulsion to new heights, starting with a growth target of 5.5% per year and declaring that it would require £840 million in expenditures.[81] Conscious of Nkrumah's drift to the left, the World Bank and the American government watched closely as the Planning Commission finalized the Seven-Year Plan in 1963. Worried that the National Planning Commission had produced an unrealistic Plan through a flawed process, World Bank consultants stationed in Ghana suggested that the Convention People's Party hold a conference of international development experts to examine and comment on the Plan, hoping that they would bring some sense into the discussion. The Convention People's Party agreed and the conference, held in March 1963, attracted a veritable who's who of development economists from the US, the Eastern Bloc and the emerging Third World. [82] The experts generally liked the Plan, although many believed that it was too optimistic in terms of its projected targets and expenditures.[83] Mensah agreed to revise the Plan for further review by Bognar, Lewis, Czelsow Borbrowski (Planning Commission of Poland) and K. N. Raj (Delhi School of Economics), but in doing so he increased the budget to £1 billion. In August, 1963, Mensah visited Geneva to present the new Plan to Raj, Lewis, Bobrowski, and Bognar. Raj and Lewis turned down the meeting, so that in the end the only advice came from the two Eastern Bloc economists.[84] The meeting had little impact, and ultimately the Convention People's Party's consultations with foreign development economists simply added an air of legitimacy to the Convention People's Party's preconceived political vision of rapid socialist industrialization. In Robert Tignor's opinion, this process reflected the Convention People's Party's view of "economists as mere technicians whose task it was to achieve the goals that the politicians set for the economy, no matter how unrealistic."[85] In other words, the process of planning and forecasting as embodied in the Seven-Year Plan exerted a performative function in relation to Nkrumah's future vision for Ghana and Africa.

Nkrumah launched the Seven-Year Plan in 1964, the same year that Ghana officially became a one-party state.[86] The end vision of the Plan was to create an industrialized, socialist country that would "(promote) the well-being of the masses" as well as lay the foundation for African economic unity and eventually the political unification of the continent.[87] Starting from this preordained vision of the future, Mensah and the planners worked to rationalize its necessity by turning to the statistics available on demographics, economics and education. Rather than developing a process of "planning without facts," as Wolfgang Stolper suggested for Nigeria during the 1960s (Baker, this volume), the Convention People's Party's experts appeared to anchor the Seven-Year Plan in concrete statistical information. However, Mensah warned that the available national statistics were far from complete. "The task of drawing up a comprehensive Plan," he wrote in the introduction to the Plan's final version, ". . . was made more difficult for the new planning organisation by the fact that we are as yet in the process of building up an adequate stock of statistical information about the country." As a result, the Planning Commission only had four sources of statistics at its disposal: "the 1960 census, the data on external trade, the consumption surveys, and the national income estimates." As a result, the Plan was far from perfect. "Much work remains to be done," continued Mensah, "in working out detailed programs of implementation and in planning individual projects. This Plan itself must be constantly reviewed as we gain more experience and collect more information." New statistical information would include "new data on agricultural and industrial production, on the balance of payments, on labour, wages and prices, an on the inter-relationships between different parts of the economy. There will also have grown up a cadre of Ghanaian planners who can use this information to draw up a technically more advanced type of Plan."[88]

But the statistical limitations of the Seven-Year Plan did not concern Mensah, for "in the final analysis, the real value of a plan is to give the nation a sense of direction and to institute a system of purposive discipline." For Mensah and Nkrumah, of course, the correct "direction" was toward socialist industrialization, and "discipline" referred to hard work and the population's willingness to make sacrifices in the name of national socialist development. The Plan, in other words, was primarily a political document, and the main goal of the Planning Commission to secure the public's support for the Party's preordained vision. "The task of holding the nation onto its chosen course and of drawing out of the people of this country their willing sacrifice and enthusiastic co-operation has been entrusted to a new Planning Commission."[89] It is not surprising, therefore, that Mensah presented a body of statistics demonstrating the need to abandon Lewis's agriculture-focused development vision in favour of the big push for socialist industrialization. The numbers (especially those contained in the 1960 census) indicated that Ghana's population was young and growing rapidly, and that the previous development plans had failed to stimulate enough economic growth to secure employment for the increasing number of youth looking for work.[90] By implication, Ghana faced a dark, Malthusian future unless the country radically changed its development policies to a big push for socialist industrialization. In this sense, Mensah used Ghana's available statistics to motivate the population to accept the predetermined vision of Nkrumah and the Convention People's Party.

In this sense, Nkrumah's Seven-Year Plan illustrates the optimism of the early 1960s that development planning and forecasting could be used as performative tools for securing visions of the future. In the case of the Convention People's Party, this optimism rested on a large dose of faith. Nkrumah believed that investing in the Volta dam would yield enough hydroelectric energy for the new state-owned industries, as well as revenues to plough into additional industrialization. The Convention People's Party planners believed that their growth targets would be achieved. But belief and faith do not necessarily translate into achievement, especially when development planners are guided by visions of the future rather than realism about political opposition in the present. In Nkrumah's case, after 1962 the American government began to oppose the Volta River Project as Nkrumah drifted toward a more radical socialist agenda. Up to 1962, Presidents Eisenhower and Kennedy supported Convention People's Party despite Nkrumah's socialist rhetoric and his tour of the Eastern Bloc. Relations cooled substantially, however, over the next two years as Nkrumah tightened his grip on power[91] and became more openly critical of the US.[92] In return, the American government balked at providing further assistance. Although the Volta River Project went ahead, the Valco Consortium refused to build a smelter in Ghana and imposed onerous profit-sharing conditions on the contract. In the end, the Volta River Project actually resembled a neocolonial deal.[93]

As for the Seven-Year Plan, not only did the opposition resist it on economic grounds, but there was increasing political opposition to Nkrumah's centralization of power and the authoritarian tendencies within the Convention People's Party. By 1960, Parliament had become the sole instrument of the Party and the party reorganized local government boundaries to take over regional and village politics.[94] In April, the Party held a plebiscite that transformed Ghana into a republic with Nkrumah as president under a highly centralized constitution. In 1962, as we already know, Nkrumah took over the judiciary after the assassination attempt. During the same year Nkrumah declared formally that village-level organizations, including those led by chiefs, would be absorbed into the Party.[95] In January 1964, the Party held another referendum that declared Ghana a single-party state.[96] Nkrumah's tightening grip on power between 1960 and 1964 provoked opposition, and resistance to the Convention People's Party increased after 1964 as the Seven-Year Plan failed to deliver on its promises. State-run industries failed to turn a profit and indeed made substantial losses. "Instead of creating wealth to further his other schemes and ambitions," writes Rooney, "they became a bottomless pit eating up public expenditure on an ever increasing scale."[97] State farms and fisheries similarly failed, absorbing capital for machinery and infrastructure but failing to produce products or on a scale sufficient to cover their budgets. In the meantime, Convention People's Party officials and supporters benefited handsomely from the state-run schemes, skimming funds, building and buying houses, and generally enjoying the highest levels of conspicuous consumption. Coupled with increasing authoritarianism, these development failures and growing corruption fed into resistance against the Party that reached a breaking point in early 1966. On February 23–24, while Nkrumah was out of the country on a peace mission in Vietnam, a military coup toppled his government and removed the Convention People's Party from power.[98] The Seven-Year Plan died with the coup.

CONCLUSION

Between 1951 and 1966, Kwame Nkrumah developed a vision for the future of Ghana (and Africa) on the basis of Pan-Africanism and socialism. Nkrumah developed this vision within the context of colonialism and the Cold War, which conjured up the spectre of a probable future marked by neocolonialism and superpower domination. Nkrumah had attended the 1955 Bandung Conference that envisioned a third way between the USA and USSR for newly independent colonial states including Ghana. Against the dark prospect of the Cold War, Pan-Africanism and socialism offered an alternative future of continental independence and development. Following his key motto to "seek ye first the political kingdom and all other things will be added to it,"[99] Nkrumah envisioned a non-aligned future of Ghana and Africa. This was to be achieved through his Pan-Africanist program that called for a united Africa with a continental government. This particular vision can be traced to Nkrumah's anticolonial activities during the 1940s and his engagement with the transnational network of socialist and radical Pan-Africanist intellectuals in the US and UK. Nkrumah, together with his influential advisor George Padmore, were in the forefront when the hitherto African American dominated Pan-African movement turned its focus to the African continent and became the main vehicle for the call for the independence of the African colonies. Nkrumah's ultimate goal was a strong, centralized union of African states outside the orbit of both the capitalist West and communist East. Defining neocolonialism as the biggest threat to African independence, he argued that only a unified continent was to achieve full economic independence. Part of his political vision was a future strong African union government to be charged with the economic planning of the future of the continent. Branding his concept for economic and social development as African socialism, he wanted to highlight the African roots of a more humane socialism, namely one that claim to rest on African communitarian traditions. Nkrumah strongly believed that securing a politically free and economically strong United States of Africa depended on the industrial development of Ghana. Rejecting a capitalist future for the country because of its colonial association, Nkrumah developed a socialist blueprint for industrialization. Nkrumah had engaged with socialist ideas since his student days in the US, but the vision came to the forefront after the visits of Convention People's Party members, and then Nkrumah himself, to the Eastern Bloc and China between 1959 and 1961. Surrounded by committed socialist economic planners, including the Hungarian Bognar, Nkrumah embarked on his bold vision with the 1964 Seven-Year Plan.

In contrast to the Nigerian experiment with modelling alternative futures, Nkrumah's Seven-Year Plan began with a specific vision of the future that he had developed before arriving back in Ghana in 1947. In this schema, Nkrumah invoked development planning as a tool for achieving that future vision: a modern, industrialized Ghana at the head of the United States of Africa. Nkrumah's desire for this future was deeply rooted in resistance to colonialism and neocolonialism. In this sense, Nkrumah's program represented a global trend after World War Two in which future visions looked toward securing one of a number of possible futures, in the interests of opening up political space and increasing social power. In the early 1960s, these visions were marked by substantial optimism over the potential for shaping the future in a desired direction. Caught up in this optimism, Nkrumah believed that development planning and forecasting would ensure the success of rapid industrialization under the Seven-Year Plan. By 1966, however, Nkrumah's hopes for an industrial Ghana and a united Africa lay in ruins, and he found himself in a sad exile trying to determine what went wrong.

NOTES

1. Wolfgang Frederick Stolper, *Planning Without Facts: Lessons in Resource Allocation from Nigeria's Development: With an Input-Output Analysis of the Nigerian Economy, 1959-60* (Cambridge, MA: Harvard University Press, 1966).
2. Kwame Nkrumah, *Ghana: The Autobiography of Kwame Nkrumah* (New York: Nelson, 1957), 14.
3. Kwame Anthony Appiah, "Pan-Africanism," in Kwame Anthony Appiah and Henry Louis Gates Jr., eds., *Africana: The Encyclopedia of the African and African American Experience* (New York: Basic Civitas Books 1999), 1484.
4. Toyin Falola, *Nationalism and African Intellectuals* (Rochester, NY: University of Rochester Press 2001), 144.
5. Henry Sylvester Williams (1869–1911) was the son of immigrant parents from Barbados settled in Trinidad. He studied law at Dalhousie University in Halifax, Nova Scotia before moving to London, where he earned his living as an official lecturer for the Temperance Society. In 1898, he founded the African Association, whose membership was restricted to those of African descent. In 1900, Williams and his association organized the first Pan-African Congress in London, which was attended by representatives from Africa, the Caribbean, the USA and Black people residing in Britain. He was called to the bar in 1902 and worked as a lawyer in South Africa from 1903 to 1905. After his return to London in 1905, he joined the National Liberal Club and the Fabian Society. In 1908, he returned to Trinidad.
6. William Edward Burghardt DuBois (1868–1963) was an African American activist, writer, social scientist, critic and public intellectual. He was the co-founder of the National Association for the Advancement of Colored People (NAACP) and the Pan-African Congress as well as the editor of the NAACP magazine *Crisis*. He was one of the most influential African Americans before the Civil Rights Movement of the 1960s. Du Bois organized the Pan-African Congresses in 1919, 1921, 1923 and 1927.
7. Hakim Adi and Marika Sherwood, *Pan-African History: Political Figures from Africa and the Diaspora Since 1787* (London: Routledge, 2003), viii.

8. P. Olisanwuche Esedebe, *Pan-Africanism: The Idea and the Movement, 1776–1991* (Washington, DC: Howard University Press, 1994), 55–58.
9. Adi and Sherwood, *Pan-African History*, viii.
10. Falola, *Nationalism*, 153–154.
11. George Padmore, *The Life and Struggles of the Negro Toiler* (London: The RILU Magazine for the International Trade Union Committee of Negro Workers, 1931), 126.
12. See further Holger Weiss, *Framing a Radical African Atlantic: African American Agency, West African Intellectuals and the International Trade Union Committee of Negro Workers* (Leiden: Brill 2014), 4.
13. Kwadwo Afari-Gyan, "Kwame Nkrumah, George Padmore and W.E.B. Du Bois," *Research Review*, New Series 7, no. 1/2 (1991), 1–10.
14. Weiss, *Framing a Radical African Atlantic*, 706. On Padmore and Kouyaté, see further Brent Hayes Edwards, *The Practice of Diaspora: Literature, Translation, and the Rise of Black Internationalism* (Cambridge, MA: Harvard University Press, 2003).
15. Jonathan Derrick, *African's Agitators: Militant Anti-Colonialism in Africa and the West, 1918–1939* (London: Hurst and Company, 2008), 390.
16. R. L. Okonkwo, "The Garvey Movement in British West Africa," *Journal of African History* 21 (1980), 109.
17. Esedebe, *Pan-Africanism*, 55–58.
18. Ama Biney, *The Political and Social Thought of Kwame Nkrumah* (London: Palgrave Macmillan, 2011), 26.
19. Francis Nwai-Kofi Nkrumah, "Education and Nationalism in Africa," *Educational Outlook* 18, no. 1 (1943), 40.
20. Ibid., 2, 38.
21. Marika Sherwood, *Kwame Nkrumah: The Years Abroad 1935–1947* (Legon, Ghana: Freedom Publications, 1996), 78–79; Biney, *Political and Social Thought*, 25–26.
22. Sherwood, *Kwame Nkrumah*, 64–95.
23. Biney, *Political and Social Thought*, 29.
24. Afari-Gyan, "Kwame Nkrumah," 1, 3.
25. Esedebe, *Pan-Africanism*, 142–144; Sherwood, *Kwame Nkrumah*, 114–117; Biney, *Political and Social Thought*, 31.
26. Although he did not join the British Communist Party, Nkrumah certainly had close friendships with British communists, including Emile Burns. We also know that the Secretariat's office had Moscow-published materials on its shelves. Biney, *Political and Social Thought*, 32–33.
27. Bankole Awoonor-Renner, *West African Soviet Union* (London: WANS Press 1946), 15–20.
28. Basil Davidson, *Black Star: A View of the Life and Times of Kwame Nkrumah* (New York: Praeger, 1973), 50; Sherwood, *Kwame Nkrumah*, 122–127.
29. Nkrumah, "The Circle" (reproduced in Nkrumah, *Ghana*, Appendix B, 303–304)
30. Biney, *Political and Social Though*, 35; Sherwood, *Kwame Nkrumah*, 189.
31. Sean Hanretta, "'Kaffir' Renner's Conversion: Being Muslim in Public in Colonial Ghana," *Past & Present* 210 (2011), 187–220.
32. The new name Ghana came from the empire in the West African Savannah that existed between the fourth and thirteenth centuries.
33. Afari-Gyan, "Kwame Nkrumah," 4–5.
34. Ibid.
35. Kofi Buenor Hadjor, *Nkrumah and Ghana: The Dilemma of Post-Colonial Power* (London: Kegan Paul International 1988), 92.
36. Biney, *Political and Social Thought*, 136–139.

37. George Padmore, *The Gold Coast Revolution: The Struggle of an African People from Slavery to Freedom* (London: D. Dobson, 1953).
38. George Shepperson, "Pan-Africanism and 'Pan-Africanism': Some Historical Notes," *Phylon* 23, no. 4 (1962), 346–358.
39. Kwame Nkrumah, *Revolutionary Path* (New York: International Publishers, 1973), 43.
40. See for example Nkrumah, *Ghana*, 43–44.
41. Nkrumah, *Autobiography*, 290.
42. Falola, *Nationalism*, 156.
43. Kwame Nkrumah, *I Speak of Freedom: A Statement of African Ideology* (New York: Praeger, 1961), ix–xii
44. Biney, *Political and Social Thought*, 124–125.
45. Ibid., 124.
46. Ibid., 146–150.
47. Falola, *Nationalism*, 128.
48. To fend off the labour problem and nationalist agitation (see Frederick Cooper, *Decolonization and African Society: The Labor Question in French and British Africa* Cambridge, UK: Cambridge University Press, 1996.).
49. D. K. Greenstreet, "Public Administration: A Comparative Analysis of the 1950 Colonial Ten-Year Development Plan and the 1951 'Nkrumah' Development Plan of the Gold Coast (Ghana)," *Economic Bulletin of Ghana*, 2nd Series 3, no. 4 (1973), 31–32, 52.
50. In fact, only nine percent of the budget actually was spent on productive activities. Paul Nugent, *Africa Since Independence*, Second Edition (Houndmills, UK: Palgrave Macmillan, 2012), 172; Douglas Rimmer, *Staying Poor: Ghana's Political Economy, 1950–1990* (Oxford: Pergamon Press, 1992), 62.
51. In fact, Nkrumah penned this phrase before independence, as it appeared in the book's preface written in October 1956. Nkrumah, Ghana, x.
52. Kwame Nkrumah, *Ghana*, 155, 12.
53. Kwame Nkrumah, *Ghana*, 12.
54. Stephanie Decker, "Corporate Political Activity in Less Developed Countries: The Volta River Project in Ghana, 1958–66," *Business History* 53, no. 7 (2011), 1002.
55. Thomas J. Noer, "The New Frontier and African Neutralism: Kennedy, Nkrumah, and the Volta River Project," *Diplomatic History* 8, no. 1 (1984): 63.
56. Robert L. Tignor, *W. Arthur Lewis and the Birth of Development Economics* (Princeton: Princeton University Press, 2006), 144, 153.
57. Yoichi Mine, "The Political Element in the Works of W. Arthur Lewis: The 1954 Lewis Model and African Development," *Developing Economies* 64, no. 3 (2006), 335–344.
58. Lewis met Nkrumah in 1946 at a conference organized by the Fabian Colonial Bureau, and Nkrumah invited him to Ghana shortly after assuming office as leader of government business. The result of Lewis's visit was the influential *Report on Industrialization in the Gold Coast* (Accra: Gold Coast Government, 1953). Tignor, *W. Arthur Lewis*, 144, 153.
59. Nkrumah recruited Jackson in 1953 to advise on the Volta River Project, and he also chaired the Ghana's development commission between 1956 and 1961. Jackson previously had served as an officer of the British Treasury Board and as the assistant secretary-general of the United Nations. W. Scott Thompson, *Ghana's Foreign Policy, 1957–1966: Diplomacy, Ideology and the New State* (Princeton: Princeton University Press, 1969), 21.
60. To appease Lewis and the CPP conservatives, an agreement was reached—brokered partly by Lewis—to divide the plan into Phase 1 (£126 million of high priority items) and Phase 2 (£104 million in "wish list" items). Tignor, *W. Arthur Lewis*, 161–165, 180–181.

61. Lewis left Ghana after serving only fourteen months of his UN contract. Tignor, *W. Arthur Lewis*, 169.
62. Afari-Gyan, "Kwame Nkrumah," 5, 7.
63. Tignor, *W. Arthur Lewis*, 161.
64. Bob Fitch and Mary Oppenheimer, *Ghana: End of an Illusion* (New York: Monthly Review Press 1966), 90–91. Nkrumah stated to radicalize the CPP and the national trade unions as early as 1958 with the appointments of Kwasi Amoaka-Atta as secretary-general of the CPP, John Tettegah as the secretary-general of the Trades Union Congress and Appiah Danquah and secretary-general of the United Ghana Farmers Council. These men in turn began to appoint leftists to the Bank of Ghana and the Commercial Bank of Ghana. Tignor, *W. Arthur Lewis*, 162–165, 180–181.
65. David Rooney, *Kwame Nkrumah: Vision and Tragedy* (Accra: Sub-Saharan Publishers, 2010), 238, 243.
66. Emmanuel Hansen, "Background to the Revolution," *Transition* 35 (1968), 26.
67. Rooney, *Kwame Nkrumah*, 238, 243.
68. Khrushchev's phrase, quoted in Sergey Mazov, *A Distant Front in the Cold War: The USSR in West Africa and the Congo, 1956–1964* (Stanford: Stanford University Press, 2010), 145.
69. Rooney, *Kwame Nkrumah*, 243.
70. Rooney, *Kwame Nkrumah*, 238, 243–245; Tignor, *W. Arthur Lewis*, 182–183.
71. Thompson, *Ghana's Foreign Policy*, 173; Rooney, *Kwame Nkrumah*, 246–247.
72. Hansen, "Background," 26; Rooney, *Kwame Nkrumah*, 186; K. Awooner, "Kwame Nkrumah: Symbol of Emergent Africa," *Africa Report* 17, no. 6 (1972): 3–4.
73. Quoted in Mazov, *Distant Front*, 202.
74. Quoted in ibid., 204.
75. Ibid.
76. Tignor, *W. Arthur Lewis*, 184–185.
77. See Baker's chapter in this volume.
78. Nugent, *Africa*, 173.
79. Nkrumah, *Revolutionary Path*, 145, 397; Rimmer, *Staying Poor*, 71.
80. Colin Legum, "Socialism in Ghana: A Political Interpretation," in William H. Friedland and Carl G. Rosenberg Jr., eds., *African Socialism* (Stanford: Stanford University Press, 1964), 142; Convention People's Party, *Programme for Work and Happiness* (Accra: Government Printer, 1962), 4–17.
81. Tignor, *W. Arthur Lewis*, 185–187.
82. The attendees included influential economists from the Eastern Bloc (e.g., Bognar from Hungary and Czelsow Bobrowski from Poland), as well as delegates who worked in the US (Baranski, Albert Hirschman), Great Britain (W. A. Lewis, Nicholas Kaldor, Dudley Seers), and India (K. N. Raj). Tignor, *W. Arthur Lewis*, 187.
83. Tony Killick, *Development Economics in Action: A Study of Economic Policies in Ghana* (New York: Routledge 2010), 58–60.
84. Tignor, *W. Arthur Lewis*, 186–189.
85. Ibid., 192.
86. After a referendum in January 1964. Rimmer, *Staying Poor*, 71.
87. Government of Ghana, *Seven-Year Development Plan* (Accra: Office of the Planning Commission, 1964), vi, x.
88. Ibid., vii–viii.
89. Ibid., viii.
90. See, for example, Government of Ghana, *Seven-Year Development Plan*, 5–12, 53–55.

91. For example, in 1962 Nkrumah sacked the chief justice and abolished Ghana's independent judiciary after the courts acquitted several CPP members—including Adamafio—after an attempt on Nkrumah's life. Davidson, *Black Star*, 182.
92. Thompson, *Ghana's Foreign Policy*, 266–267, 300.
93. Davidson, *Black Star*, 197.
94. Rimmer, *Staying Poor*, 69–70.
95. K. Nkrumah, "Address at the First Seminar at the Winneba Ideological Institute, 3 February 1962," in Nkrumah, *Revolutionary Path*, 178.
96. Rooney, *Kwame Nkrumah*, 323.
97. Ibid., 258.
98. Ibid., 258–268, 339–340; Fitch and Oppenheimer, *Ghana*, 3–5.
99. Nkrumah coined this phrase in the Convention People's Party's 1951 manifesto entitled *Towards the Goal*. See Dennis Austin, *Politics in Ghana, 1946–60* (London: Oxford University Press, 1964), 131. He reiterated it in *Africa Must Unite* (New York: International Publishers, 1970), 50.

REFERENCES

Adi, Hakim and Marika Sherwood. *Pan-African History: Political Figures from Africa and the Diaspora Since 1787*. London: Routledge, 2003.

Afari-Gyan, Kwadwo. "Kwame Nkrumah, George Padmore and W.E.B. Du Bois." *Research Review*, New Series 7 no. 1/2 (1991): 1–10.

Appiah, Kwame Anthony. "Pan-Africanism." In *Africana: The Encyclopedia of the African and African American Experience*, edited by Kwame Anthony Appiah and Henry Louis Gates Jr., 325–328. New York: Basic Civitas Books, 1999.

Austin, Dennis, *Politics in Ghana, 1946–60*. London: Oxford University Press, 1964.

Awooner, K. "Kwame Nkrumah: Symbol of Emergent Africa." *Africa Report* 17, no. 6 (1972): 22–25.

Awoonor-Renner, Bankole. *West African Soviet Union*. London: WANS Press, 1946.

Biney, Ama. *The Political and Social Thought of Kwame Nkrumah*. London: Palgrave Macmillan, 2011.

Convention People's Party. *Programme for Work and Happiness*. Accra: Government Printer, 1962.

Cooper, Frederick, *Decolonization and African Society: The Labor Question in French and British Africa*. Cambridge, UK: Cambridge University Press 1996.

Davidson, Basil. *Black Star: A View of the Life and Times of Kwame Nkrumah*. New York: Praeger, 1973.

Decker, Stephanie. "Corporate political activity in less developed countries: The Volta River Project in Ghana, 1958–66." *Business History* 53, no. 7 (2011): 993–1017.

Derrick, Jonathan. *African's Agitators: Militant Anti-Colonialism in Africa and the West, 1918–1939*. London: Hurst and Company, 2008.

Esedebe, P. Olisanwuche. *Pan-Africanism: The Idea and the Movement, 1776–1991*. Washington, DC: Howard University Press, 1994.

Falola, Toyin. *Nationalism and African Intellectuals*. Rochester, NY: University of Rochester Press, 2001.

Fitch, Bob and Mary Oppenheimer. *Ghana: End of an Illusion*. New York: Monthly Review Press, 1966.

Government of Ghana. *Seven-Year Development Plan*. Accra: Office of the Planning Commission, 1964.

Greenstreet, D. K. "Public Administration: A Comparative Analysis of the 1950 Colonial Ten-Year Development Plan and the 1951 'Nkrumah' Development Plan of the Gold Coast (Ghana)." *Economic Bulletin of Ghana*, 2nd Series 3, no. 4 (1973): 31–55.

Hadjor, Kofi Buenor. *Nkrumah and Ghana: The Dilemma of Post-Colonial Power*. London: Kegan Paul International, 1988.

Hanretta, Sean. " 'Kaffir' Renner's Conversion: Being Muslim in Public in Colonial Ghana." *Past & Present* 210 (2011): 187–220.

Hansen, Emmanuel. "Background to the Revolution." *Transition* 35 (1968): 24–28.

Hayes Edwards, Brent. *The Practice of Diaspora: Literature, Translation, and the Rise of Black Internationalism*. Cambridge, MA: Harvard University Press, 2003.

Killick, Tony. *Development Economics in Action: A Study of Economic Policies in Ghana*. New York: Routledge, 2010.

Legum, Colin. "Socialism in Ghana: A Political Interpretation." In *African Socialism*, edited by William H. Friedland and Carl G. Rosberg Jr., 131–159. Stanford: Stanford University Press, 1964.

Lewis, W. Arthur. *Report on Industrialization in the Gold Coast*. Accra: Gold Coast Government Printer, 1953.

Mazov, Sergey. *A Distant Front in the Cold War: The USSR in West Africa and the Congo, 1956–1964*. Stanford: Stanford University Press, 2010.

Mine, Yoichi. "The Political Element in the Works of W. Arthur Lewis: The 1954 Lewis Model and African Development." *Developing Economies*, 64, no. 3 (2006): 329–355.

Nkrumah, Kwame. "Address at the First Seminar at the Winneba Ideological Institute, 3 February 1962." In *Revolutionary Path*, 170–180. New York: International Publishers, 1973.

Nkrumah, Kwame. *Ghana: The Autobiography of Kwame Nkrumah*. New York: Nelson, 1957.

Nkrumah, Kwame. *I Speak of Freedom: A Statement of African Ideology*. New York: Praeger, 1961.

Nkrumah, Kwame. *Africa Must Unite*. New York: International Publishers, 1970.

Nkrumah, Kwame. *Revolutionary Path*. New York: International Publishers, 1973.

Nkrumah, Francis Nwai-Kofi. "Education and Nationalism in Africa." *Educational Outlook* 18, no. 1 (1943): 32–40.

Noer, Thomas J. "The New Frontier and African Neutralism: Kennedy, Nkrumah, and the Volta River Project." *Diplomatic History* 8, no. 1 (1984): 61–79.

Nugent, Paul. *Africa Since Independence*, Second Edition. Houndmills, UK: Palgrave Macmillan, 2012.

Okonkwo, R. L. "The Garvey Movement in British West Africa." *Journal of African History* 21 (1980): 105–117.

Padmore, George. *The Life and Struggles of the Negro Toiler*. London: The RILU Magazine for the International Trade Union Committee of Negro Workers, 1931.

Padmore, George. *The Gold Coast Revolution: The Struggle of an African People from Slavery to Freedom*. London: D. Dobson, 1953.

Rimmer, Douglas. *Staying Poor: Ghana's Political Economy, 1950–1990*. Oxford: Pergamon Press, 1992.

Rooney, David. *Kwame Nkrumah: Vision and Tragedy*. Accra: Sub-Saharan Publishers, 2010.

Shepperson, George. "Pan-Africanism and 'Pan-Africanism': Some Historical Notes." *Phylon* 23, no. 4 (1962): 346–358.

Sherwood, Marika. *Kwame Nkrumah: The Years Abroad 1935–1947*. Legon, Ghana: Freedom Publications, 1996.

Stolper, Wolfgang Frederick. *Planning Without Facts: Lessons in Resource Allocation from Nigeria's Development: With an Input-Output Analysis of the Nigerian Economy, 1959-60*. Cambridge, MA: Harvard University Press, 1966.
Thompson, W. Scott. *Ghana's Foreign Policy, 1957–1966: Diplomacy, Ideology and the New State*. Princeton: Princeton University Press, 1969.
Tignor, Robert L. *W. Arthur Lewis and the Birth of Development Economics*. Princeton: Princeton University Press, 2006.
Weiss, Holger. *Framing a Radical African Atlantic: African American Agency, West African Intellectuals and the International Trade Union Committee of Negro Workers*. Leiden, The Netherlands: Brill, 2014.

Contributors

Jenny Andersson is a senior research fellow of the French research council CNRS and an affiliated researcher with the Centre d'études européennes at Sciences Po in Paris. She is the Principal Investigator of the Futurepol project (A political history of the future: Knowledge production and future governance in the post-war period) funded by the European Research Council. Andersson has published several books on European social democracy, among them *The Library and the Workshop: Social Democracy and Capitalism in an Age of Knowledge* (Stanford: Stanford University Press, 2010). She is the author more recently of "The Great Future Debate and the Struggle for the World," *American Historical Review* 177, no. 5 (2012): 1411–1430.

Stefan Cihan Aykut is a political scientist and holds a PhD in history of science. He currently works at LATTS, Université Paris-Est Marne-la-Vallée. His research interests include the construction of climate change as a global public problem, international climate negotiations, and energy policy, where he focusses on the making of national energy futures in economic expertise and governmental commissions. His most recent publication is *Gouverner le climat? 20 ans de négociations internationales* (Paris: Presses de Sciences Po, 2015, with Amy Dahan).

Kevin T. Baker is a PhD candidate in the History and Science in Human Culture programs at Northwestern University. His work focuses on the role that the systems and computer sciences have played in shaping modern political and economic thought. His dissertation will examine the history of sustainable development policies in the 1970s and 1980s and the global debates surrounding the Club of Rome's *Limits to Growth* report.

Ana-Maria Cătănuş is a historian and scientific researcher with the National Institute for the Study of Totalitarianism of the Romanian Academy. She has published on the history of the Romanian communist regime, and dissidence in the Soviet Union and Romania. Her recent work is *Vocaţia libertăţii. Forme de disidenţă în România anilor 1970–1980/Vocation for* *freedom: Forms of dissidence in Romania in the 1970s and the 1980s* (Bucharest: National Institute for the Study of Totalitarianism Publishing House, 2014).

Jeff Grischow is associate professor of history and the associate director of the Tshepo Institute for the Study of Contemporary Africa at Wilfrid Laurier University in Waterloo, Ontario, Canada. Originally interested in colonial development, Grischow published *Shaping Tradition: Civil Society, Community and Development in Colonial Northern Ghana, 1899–1957* (Leiden: Brill, 2006). His most recent work focuses on the history of disability in Africa, including "Kwame Nkrumah, Disability and Rehabilitation In Ghana, 1957–66," *Journal of African History* 52, no. 2 (2011): 179–199.

Eglė Rindzevičiūtė holds a PhD in culture studies and is chargée de recherché at the Centre d'études européennes at Sciences Po. She has published extensively on the history of Soviet and post-Soviet governance, including *Constructing Soviet Cultural Policy: Cybernetics and Governance in Lithuania after World War II* (Linköping: Linköping University Press, 2008). Her current work involves a study of transnational relations among East and West experts during the Cold War with a focus on the International Institute of Applied Systems Analysis, Austria.

Vítězslav Sommer is junior researcher at the Centre d'études européennes at Sciences Po. As a member of the ERC project Futurepol, he is currently working on the institutional and social background of the Czechoslovak "future studies." His main research topics are state socialism and post-socialism in East-Central Europe. His most recent publication is *The "Club of Politically Engaged Conformists"? The Communist Party of Czechoslovakia, Popular Opinion and the Crisis of Communism, 1956* (CWIHP Working Paper 66, March 2013, with Kevin McDermott).

Sverker Sörlin is professor of environmental history at the Royal Institute of Technology, Stockholm, Sweden. He assumed the first chair of Environmental History in Scandinavia (1993–2004) and is a member of the Institute for Advanced Study, Princeton (2013–2014). Sörlin's core area of research is in the roles and functions of knowledge in environmentally informed modern societies. His recent work includes an edited book *The Future of Nature: Documents of Global Change* (New Haven, CT: Yale University Press, 2013, with Libby Robin and Paul Warde).

Elodie Vieille Blanchard was trained as a mathematician and a historian of science. She holds a PhD in history of science. In her dissertation she examined the debate about the limits to growth in 1945–1990, with a focus on mathematical modelling. She currently teaches mathematics to handicapped pupils in the Paris area and on a voluntary basis works toward the development of vegetarianism in France.

Paul Warde is lecturer in environmental history at the University of Cambridge, having previously been professor of environmental history at the University of East Anglia. He has written extensively on environmental, economic and energy history in the early modern and modern periods. Recent work includes *Power to the People: Energy in Europe over the Last Five Centuries* (Princeton: Princeton University Press, 2013, with Astrid Kander and Paolo Malanima); and *The Future of Nature: Documents of Global Change* (New Haven, CT: Yale University Press, 2013, with Libby Robin and Sverker Sörlin).

Holger Weiss is professor of general history at Åbo Akademi University in Finland. He has published on global and Atlantic history, African history, environmental history, slavery and the trans-Atlantic slave trade, and Islam and Islamic economy in West Africa. His most recent book is *Framing a Radical African Atlantic: African American Agency, West African Intellectuals and the International Trade Union Committee of Negro Workers* (Leiden: Brill, 2014).

Index

Note: page numbers with *f* indicate figures.

For Product Safety Concerns and Information please contact our EU
representative GPSR@taylorandfrancis.com
Taylor & Francis Verlag GmbH, Kaufingerstraße 24, 80331 München, Germany

www.ingramcontent.com/pod-product-compliance
Ingram Content Group UK Ltd.
Pitfield, Milton Keynes, MK11 3LW, UK
UKHW022002180926
475701UK00003B/146
* 9 7 8 0 3 6 7 2 6 3 8 2 9 *